AF781594

Management for Professionals

More information about this series at https://link.springer.com/bookseries/10101

Hui Liu · Nikolaos Nikitas · Yanfei Li ·
Rui Yang

Big Data in Energy Economics

Science Press
Beijing

Springer

Hui Liu
School of Traffic and Transportation Engineering
Central South University
Changsha, Hunan, China

Nikolaos Nikitas
Department of Structural Engineering, Data Sciences and Wind Energy
University of Leeds
Leeds, UK

Yanfei Li
College of Mechanical and Electrical Engineering
Hunan Agricultural University
Changsha, Hunan, China

Rui Yang
School of Traffic and Transportation Engineering
Central South University
Changsha, Hunan, China

ISSN 2192-8096 ISSN 2192-810X (electronic)
Management for Professionals
ISBN 978-981-16-8964-2 ISBN 978-981-16-8965-9 (eBook)
https://doi.org/10.1007/978-981-16-8965-9

Jointly published with Science Press
The print edition is not for sale in China (Mainland). Customers from China (Mainland) please order the print book from: Science Press.

© Science Press 2022
This work is subject to copyright. All rights are solely and exclusively licensed by the Publisher, whether the whole or part of the material is concerned, specifically the rights of translation, reprinting, reuse of illustrations, recitation, broadcasting, reproduction on microfilms or in any other physical way, and transmission or information storage and retrieval, electronic adaptation, computer software, or by similar or dissimilar methodology now known or hereafter developed.
The use of general descriptive names, registered names, trademarks, service marks, etc. in this publication does not imply, even in the absence of a specific statement, that such names are exempt from the relevant protective laws and regulations and therefore free for general use.
The publishers, the authors, and the editors are safe to assume that the advice and information in this book are believed to be true and accurate at the date of publication. Neither the publishers nor the authors or the editors give a warranty, express or implied, with respect to the material contained herein or for any errors or omissions that may have been made. The publishers remain neutral with regard to jurisdictional claims in published maps and institutional affiliations.

This Springer imprint is published by the registered company Springer Nature Singapore Pte Ltd.
The registered company address is: 152 Beach Road, #21-01/04 Gateway East, Singapore 189721, Singapore

Preface

The world is now experiencing the third energy revolution from traditional fossil energy to new energy sources. The energy revolution drove the industrial revolution and also promoted the development of social productivity and human civilization. However, the contradictions between economic growth, resource consumption, and ecological balance pose new challenges to energy utilization.

Building a global energy Internet and vigorously developing big data technology are effective ways to solve the problem of irrational energy exploitation. The precise matching and optimal allocation of resources under the new energy supply and demand relationship are the necessary conditions for achieving a win-win situation for energy, economy, and environment. There are various economic phenomena such as markets, prices, supply, and demand in the process of energy development and use. In turn, economic markets are inextricably linked to energy markets in terms of policy making, development planning, commodity trading, and economic activity. The cross-study of "Energy + Economy + Big Data" is particularly important, which provides new ideas for energy economics research. Based on the massive multivariate data generated in the energy and economic fields, this book innovatively proposes the theory of big data in energy economics to promote the application of big data analysis in energy conversion and management.

The issue of big data in energy economics from the book analyses the economic impact of energy production and consumption patterns, supply and demand price markets from a data science perspective. It provides an important reference for the development of multiple disciplines like energy, economics, artificial intelligence, data science, data engineering, engineering informatics, computational intelligence, etc. The book systematically explains the key technologies of the global energy Internet and big data in energy economics, including signal processing, machine learning, deep learning, optimization models, ensemble learning, feature engineering, correlation analysis, intelligent clustering, pattern recognition, etc. The issue of big data proposed in the book focuses on the aspects of data decomposition, data forecasting, data boosting, data integration, data clustering, data identification, etc. In addition, a large number of experimental case studies are provided in the book, with application scenarios including crude oil market, coal market, wind power market, photovoltaic market, electricity market, etc.

The studies in the book are supported by the National Natural Science Foundation of China, the National Key R&D Program of China, the Innovation Drive of Central South University of China. The publication of the book is funded by the graduate text book project of Central South University of China. In the process of writing the book, Ye Li, Chao Chen, Chengming Yu, Zijie Cao, Yucheng Yin, Xirui Chen, and other team members have done a lot of experimental verification and other work, the authors would like to express heartfelt appreciations.

Changsha, China
October 2021

Prof. Dr. -Ing. habil. Hui Liu

Contents

1 Introduction 1
1.1 Overview of Research Progress in Energy Economics 1
1.1.1 History of Energy Economics 1
1.1.2 Framework for Big Data in Energy Economics 2
1.1.3 Strategies and Measures for the Development of Big Data in China's Energy Economics 6
1.1.4 Strategies and Measures for the Development of Big Data in World's Energy Economics 9
1.2 Key Technologies of Energy Internet in Energy Economics 14
1.2.1 Concept of Energy Internet 14
1.2.2 Reasons for Building a Global Energy Internet 16
1.2.3 Key Technologies of Energy Internet 17
1.3 Big Data Demand Analysis for Energy Economics 20
1.3.1 Summary of Key Technical Tools 21
1.3.2 Application Scenarios of Big Data Technology 29
1.4 Scope of This Book 31
References 34

2 Big Data Analysis of Energy Economics in Oil Market 43
2.1 Introduction 43
2.2 Influencing Factors Analysis of Oil Prices 45
2.2.1 Data Description of Crude Oil Prices Influencing Factors 46
2.2.2 Correlation Analysis of the Factors Affecting Crude Oil Prices 46
2.3 Big Data Forecasting of Oil Prices 50
2.3.1 Base Forecasting Models 50
2.3.2 Crude Oil Futures and Spot Prices Time Series Forecasting Model 53
2.3.3 Performance Metrics 54
2.3.4 Results and Discussions 55
2.4 Econometric Analysis of Oil Prices 57

2.4.1 Energy Economic Analysis of Crude Oil Market 57
2.4.2 Big Data Prediction Technology 61
2.4.3 Policies and Recommendations 62
2.5 Conclusions 62
References 64

3 Big Data Analysis of Energy Economics in Coal Market 67
3.1 Introduction 67
3.2 Influencing Factors Analysis of Coal Prices 68
3.2.1 Data Description of Coal Prices Influencing Factors 69
3.2.2 Correlation Analysis of the Factors Affecting Coal Prices 72
3.3 Big Data Forecasting of Coal Prices 73
3.3.1 The Components of the Proposed Model 73
3.3.2 Multi-factor Coal Price Hybrid Forecasting Model 80
3.3.3 Performance Metrics 81
3.3.4 Results and Discussions 81
3.4 Econometric Analysis of Coal Prices 86
3.4.1 Energy Economic Analysis of the Coal Market 86
3.4.2 Big Data Prediction Technology 87
3.4.3 Policies and Recommendations 88
3.5 Conclusions 91
References 92

4 Big Data Analysis of Energy Economics in Wind Power Market 95
4.1 Introduction 95
4.2 Multi-temporal and Spatial Scale Wind Power Big Data Forecasting 96
4.2.1 Description of Original Wind Dataset 96
4.2.2 Framework of Wind Power Forecasting Models 97
4.2.3 Analysis of Wind Power Forecasting Models 98
4.3 Conversion Efficiency of Wind Power Energy 104
4.4 Market Economy Analysis of Wind Power Application 107
4.4.1 Market Economy Analysis of Wind Power Application in China 107
4.4.2 Market Economy Analysis of Wind Power Application in America 111
4.4.3 Market Economy Analysis of Wind Power Application in Europe 112
4.5 Conclusions 113
References 114

5 Big Data Analysis of Energy Economics in Photovoltaic Power Generation Market 117
5.1 Introduction 117
5.2 Big Data Forecasting of Photovoltaic Power Generation 117

5.2.1 Big Data Processing Engines 118
5.2.2 Forecasting Strategy and Methods 119
5.2.3 Forecasting Models 120
5.3 Photovoltaic Power Consumption by Small and Medium-Sized Users 123
5.3.1 Dataset Description 123
5.3.2 Experiments 123
5.4 Photovoltaic Power Consumption in Urban Public Areas 126
5.4.1 Dataset Description 127
5.4.2 Experiments 128
5.5 Market Economy Analysis of Photovoltaic Systems 129
5.5.1 Dispatch of Photovoltaic Power Integration 129
5.5.2 Optimization Model of Photovoltaic Power Integration 130
5.5.3 Single- and Multi-objective Optimization Algorithms 133
5.6 Conclusions 135
References 136

6 Big Data Analysis of Power Market Energy Economics 137
6.1 Introduction 137
6.2 Big Data Forecasting of Urban Electricity Price 138
6.2.1 Electricity Price Forecasting Method Based on Empirical Mode Decomposition and Extreme Learning Machine 139
6.2.2 Electricity Price Forecasting Method Based on Wavelet Packet Decomposition and Deep Belief Network 143
6.2.3 Big Data Processing of Electricity Price Based on Empirical Wavelet Transform and Long Short-Term Memory Network 150
6.3 Correlation Analysis of Urban Energy Consumption and Economic Growth 155
6.3.1 Grey Correlation Model in the Energy Economy 155
6.3.2 Grey Correlation Analysis of Economic Growth and Energy Consumption Varieties 156
6.3.3 Grey Correlation Analysis of Economic Growth and Energy Consumption Industrial Structure 157
6.4 Metering Charge Adjustment Analysis of City Electricity Prices 159
6.4.1 Background of the K-means Algorithm for Characteristic Analysis of Electricity Price 159
6.4.2 Analysis of User Electricity Price Consumption Characteristics Based on the K-means Algorithm 161
6.4.3 Optimization Design of Residential Stepped Electricity Price 162

6.5 Conclusions 165
References 166

7 Big Data Management of Smart City Energy Conservation and Emission Reduction 169
7.1 Introduction 169
7.1.1 Background and Introduction 169
7.1.2 Dataset Description 170
7.2 Non-intrusive Load Identification of Electrical Equipment 171
7.2.1 Nonintrusive Load Identification Based on Signal Decomposition 172
7.2.2 Non-intrusive Load Identification Based on Electrical Switching Event Classification 177
7.2.3 Non-intrusive Load Identification Based on Multi-label Classification 184
7.3 Guide to Smart City Electricity Behavior 187
7.3.1 Smart Grid Planning of a City 187
7.3.2 Urban Public Electricity Behavior Research 191
7.4 Analysis of Energy Conservation and Emission Reduction of Smart Cities 192
7.5 Conclusions 193
References 194

8 Optimization Analysis of Clean Energy Transformation 197
8.1 Introduction 197
8.1.1 Global Status of Clean Energy Development 197
8.1.2 International Experience in the Transformation of Clean Energy Industry 200
8.2 Efficiency Analysis of Energy Utilization Under Diversified Development 202
8.2.1 Evaluation Indexes and Methods of Energy Efficiency 202
8.2.2 Analysis of Influencing Factors and Mechanism of Energy Efficiency 203
8.2.3 International Comparative Analysis of Energy Efficiency 207
8.3 Analysis of Reasonable Energy Consumption Patterns 211
8.3.1 Challenges Facing Energy Consumption 211
8.3.2 Analysis of Key Factors Affecting Clean Energy Consumption 212
8.3.3 Reform Strategy of Clean Energy Consumption Patterns 216
8.4 Economic Analysis of Clean Energy Transformation 219
8.4.1 Mechanisms for Developing Clean Energy to Affect Economic Growth 220
8.4.2 Ways to Promote a Low-Carbon Economy Based on Clean Energy 227

8.5 Conclusions 228
References 229

9 Global Energy Internet Green and Low-Carbon Energy Economic Innovation 233
9.1 Introduction 233
9.2 Reform and Innovation of the New Energy System Under the Energy Internet 236
9.2.1 Comparison of Conventional Energy System and New Energy System 236
9.2.2 Production in New Energy System 237
9.2.3 Supply and Marketing in New Energy System 240
9.3 Energy Saving and Emission Reduction Under the Energy Internet 243
9.3.1 Energy Saving and Emission Reduction in Production Process 244
9.3.2 Energy Saving and Emission Reduction in Supply and Marketing Process 246
9.4 Healthy Construction of the Ecological Environment Under the Energy Internet 249
9.4.1 Land Ecology and Photovoltaic Power 249
9.4.2 Hydropower and Ecology 251
9.4.3 Biological Energy and Ecology 252
9.5 Conclusions 253
References 255

Nomenclature

AC	Alternating Current
AdaBoost	Adaptive Boosting
AE	Autoencoder
AI	Artificial Intelligence
ANNs	Artificial Neural Networks
AR	Autoregressive
ARIMA	Autoregressive Integrated Moving Average
ASEAN	Association of Southeast Asian Nations
AVC	Automatic Voltage Control
BDE	Binary Differential Evolution
BGWO	Binary Grey Wolf Optimization
BHHO	Binary Harris Hawk Optimization
BiLSTM	Bi-directional Long Short-Term Memory
BOA	Bayesian Optimization Algorithm
BPNN	Back-Propagation Neural Network
BPSO	Binary Particle Swarm Optimization
BRICS	Brazil, Russia, India, China, and South Africa
CCHP	Combined Cooling Heating and Power
CCUS	Carbon Capture, Utilization, and Storage
CEI	City Energy Internet
CITL	Community-Independent Transaction Log
CNN	Convolutional Neural Network
COVID-19	Coronavirus Disease 2019
DAG	Directed Acyclic Graph
DBN	Deep Belief Network
DC	Direct Current
DEA	Data Envelopment Analysis
DWT	Dynamic Time Warping
ECE	Economic Commission of Europe
EEMD	Ensemble Empirical Mode Decomposition
EI	Energy Internet
EIA	Energy Information Administration

ELM	Extreme Learning Machine
EM	Expectation Maximum
EMD	Empirical Mode Decomposition
EMEA	Europe, the Middle East, and Africa
EU	European Union
EUETS	European Union Emission Trading Scheme
EV	Electric Vehicles
EWT	Empirical Wavelet Transform
FF-ANN	Feedforward Artificial Neural Networks
FOB	Free on Board
GA	Genetic Algorithm
GARCH	Generalized Autoregressive Conditional Heteroskedasticity
GDP	Gross Domestic Product
GEI	Global Energy Internet
GEIDCO	Global Energy Interconnection Development and Cooperation Organization
GHG	Global Greenhouse Gas
GIS	Geographic Information System
GRA	Grey Relational Analysis
GRU	Gated Recurrent Unit
GST	Grey System Theory
GWEC	Global Wind Energy Council
GWO	Grey Wolf Optimization
HBET	Hydropower Biological Evaluation Toolset
HHT	Hilbert-Huang Transform
HMM	Hidden Markov Model
HSIC	Hilbert-Schmidt Independence Criterion
ICEEMDAN	Improved Complete Ensemble Empirical Mode Decomposition with Adaptive Noise
IEA	International Energy Agency
IF	Interval Forecasting
IMF	Intrinsic Mode Function
IRENA	International Renewable Energy Agency
KNN	K-Nearest Neighbour
LME	London Metal Exchange
LR	Linear Regression
LSTM	Long Short-Term Memory
MA	Mergers and Acquisitions
MAE	Mean Absolute Error
MAPE	Mean Absolute Percentage Error
MAR	Missing at Random
MF	Multi-Factor
MI	Mutual Information
MLP	Multi-Layer Perceptron
ML-RBF	Multi-Label Radial Basis Function

MOBOA	Multi-Objective Bayesian Optimization Algorithm
MODWPT	Maximal Overlap Discrete Wavelet Packet Transform
MODWT	Maximal Overlap Discrete Wavelet Transform
MOGWO	Multi-Objective Grey Wolf Optimizer
MOO	Multi-Objective Optimization
MOPSO	Multi-Objective Particle Swarm Optimization
mRMR	Minimum Redundancy Maximum Relevance
MT	Machine Translation
NEI	National Energy Internet
NIE	Newly Industrialized Economies
NILM	Non-intrusive Load Monitoring
NLP	Natural Language Processing
NSGA-II	Non-dominated Sorting Genetic Algorithm II
NYMEX	New York Mercantile Exchange
OECD	Organization for Economic Cooperation and Development
OPEC	Organization of the Petroleum Exporting Countries
P2G	Power-To-Gas
PCC	Pearson Correlation Coefficient
PCC	Point of Common Coupling
PF	Point Forecasting
PPP	Purchasing Power Parity
PSO	Particle Swarm Optimization
PV	Photovoltaic
RBF	Radial Basis Function
RBM	Restricted Boltzmann Machine
RD	Research and Development
RELM	Rank Extreme Learning Machine
RF	Random Forests
RMSE	Root Mean Square Error
RNN	Recurrent Neural Network
RW	Random Walk
SLFN	Single-Hidden Layer Feedforward Neural Network
SOM	Self-Organizing Map
SOO	Single-Objective Optimization
SR	Speech Recognition
SSA	Singular Spectral Analysis
SSE	Sum of Squared Error
SVMs	Support Vector Machines
TES	Total Energy Supply
TGC	Tradable Green Certificates
TOPSIS	Technique for Order Preference by Similarity to an Ideal Solution
UHV	Ultra-High Voltage
UNSD	United Nations Statistics Division
VAR	Vector Autoregressive
VPP	Virtual Power Plant

WB	World Bank
WD	Wavelet Decomposition
WDI	World Development Indicators
WFLO	Wind Farm Layout Optimization
WHO	World Health Organization
WOA	Whale Optimization Algorithm
WPD	Wavelet Packet Decomposition
WT	Wavelet Transform
WTI	West Texas Intermediate

List of Figures

Fig. 1.1 Understanding of big data in energy economics in this book 5
Fig. 1.2 Framework of big data in energy economics in this book 6
Fig. 1.3 Conceptualization of energy big data center 8
Fig. 1.4 The framework of the Energy Internet 15
Fig. 1.5 The impact of big data analysis technology on energy economics 20
Fig. 1.6 The summary classification of data preprocessing technology 21
Fig. 1.7 The classification summary of forecasting strategies 24
Fig. 1.8 The summary classification of non-intrusive load identification technology 27
Fig. 2.1 The historical prices data from February 8, 1991 to July 23, 2021 48
Fig. 2.2 Correlation analysis of crude oil futures and spot prices 49
Fig. 2.3 Model structure of BiLSTM neural network 51
Fig. 2.4 Model structure of DBN neural network 52
Fig. 2.5 The deep learning framework for forecasting crude oil futures and spot prices 53
Fig. 2.6 The 10-step forecasting results of crude oil futures and spot prices data 57
Fig. 2.7 The stacking diagrams of 1-step to 10-step error indices under different prices time series data 58
Fig. 3.1 The historical prices data from June 1991 to June 2021 71
Fig. 3.2 Correlation analysis of the factors affecting coal price 72
Fig. 3.3 The multi-factor hybrid model framework for coal prices forecasting 80
Fig. 3.4 The 3-step forecasting results of Coal (Australia) time series data under different models 84
Fig. 3.5 The stacking diagrams of 1-step to 10-step error indices under different decomposition methods 85
Fig. 4.1 Location of wind turbines in the dataset 97

Fig. 4.2 Wind power data of four target sites 98
Fig. 4.3 Framework of spatial scale wind power forecasting models 99
Fig. 4.4 Fitness curves of four optimization algorithms for spatial feature selection 100
Fig. 4.5 Spatial feature selection results for target site #1 101
Fig. 4.6 Spatial feature selection results for target site #2 102
Fig. 4.7 Spatial feature selection results for target site #3 103
Fig. 4.8 Spatial feature selection results for target site #4 104
Fig. 4.9 The 1-step wind power forecasting results of different spatial selection methods for target site #1 105
Fig. 4.10 The 1-step wind power forecasting results of different spatial selection methods for target site #2 106
Fig. 4.11 The 1-step wind power forecasting results of different spatial selection methods for target site #3 107
Fig. 4.12 The 1-step wind power forecasting results of different spatial selection methods for target site #4 108
Fig. 5.1 The cumulative and annual PV installations (2011–2019) 118
Fig. 5.2 Diagram of PV power system 118
Fig. 5.3 Machine learning pipeline in Spark 119
Fig. 5.4 Forecasting framework of PV power generation 121
Fig. 5.5 Structure of GRU cell 122
Fig. 5.6 Structure of an MLP 122
Fig. 5.7 PV power output within one day 124
Fig. 5.8 Dataset visualization and splits 124
Fig. 5.9 Loss curve of the training 125
Fig. 5.10 Forecasting of the models 125
Fig. 5.11 Forecasting of the models for the first day in the test set 126
Fig. 5.12 Dataset visualization and splits 127
Fig. 5.13 Loss curve of the training 128
Fig. 5.14 Forecasting of the models 129
Fig. 5.15 Flowchart of NSGA-II 134
Fig. 5.16 Pareto front with two objective functions 135
Fig. 6.1 The competitive mode of the electricity market 138
Fig. 6.2 The example of EMD feature extraction 142
Fig. 6.3 Basic structure of single-hidden layer feedforward neural network 142
Fig. 6.4 Electricity price forecasting model framework based on the EMD and ELM 144
Fig. 6.5 Example of the EMD-ELM electricity price prediction model 144
Fig. 6.6 Decomposition results of three layers WPD 145
Fig. 6.7 Example of the EMD decomposition of electricity price data 147
Fig. 6.8 Electricity price forecasting framework based on the WPD and DBN 149

Fig. 6.9 Example of the WPD-DBN electricity price forecasting model ... 149
Fig. 6.10 Example of EWT decomposition of electricity price data ... 152
Fig. 6.11 LSTM memory unit structure diagram ... 152
Fig. 6.12 Electricity price forecasting model framework based on the EWT and LSTM ... 154
Fig. 6.13 Example of the EWT-LSTM electricity price forecasting model ... 154
Fig. 6.14 Grey correlation between different energy types and GDP growth rate ... 157
Fig. 6.15 Grey correlation between energy consumption and GDP in different industries ... 158
Fig. 6.16 The ideal clustering result of users' electricity load ... 161
Fig. 7.1 The power supply information and area information of the appliances ... 171
Fig. 7.2 The current amplitudes ... 176
Fig. 7.3 The recall value of the two algorithm of Current #1 ... 177
Fig. 7.4 The recall value of the two algorithm of Current #2 ... 177
Fig. 7.5 Structure of CNN ... 179
Fig. 7.6 Confusion chart of the results predicted by CNN ... 183
Fig. 7.7 Confusion chart of the results predicted by LSTM ... 183
Fig. 7.8 The framework of city smart grid planning ... 188
Fig. 7.9 Analysis of the effect of energy conservation and emission reduction of Jiangxi Province in China ... 193
Fig. 8.1 Annual average change in global energy production, 1971–2019 ... 198
Fig. 8.2 Total primary energy supply shares, 1971–2019 ... 199
Fig. 8.3 World power generation structure by fuel, 1971–2019 ... 199
Fig. 8.4 Energy efficiency indexes ... 202
Fig. 8.5 Influencing factors and mechanism of energy efficiency ... 203
Fig. 8.6 Energy use per $1,000 GDP ... 208
Fig. 8.7 Physical energy efficiency, 2000–2018 ... 210
Fig. 8.8 Wind power and load curve in a province on a typical day ... 213
Fig. 8.9 Wind and solar power curtailment rate by region, 2015 ... 214
Fig. 8.10 Reform strategy of clean energy consumption patterns ... 216
Fig. 8.11 Industries impacted by structural demand and forms of impacts ... 223
Fig. 8.12 Mechanism of clean energy technology output effect ... 225
Fig. 9.1 Primary energy consumption ... 234
Fig. 9.2 The model of EI information interaction ... 235
Fig. 9.3 Model of energy flows interexchange ... 236
Fig. 9.4 The comparisons of the conventional energy system and new energy system ... 238
Fig. 9.5 The basic structure of distributed energy ... 239
Fig. 9.6 The idea graph of the VPP ... 241

Fig. 9.7 The basic structure of microgrids 242
Fig. 9.8 The concept graph of the prosumer 243
Fig. 9.9 World CO_2 emissions from 1990 to 2018 244
Fig. 9.10 The process of utilization for biomass 254

List of Tables

Table 1.1	Some applications of big data technology in energy economics	7
Table 2.1	Data description of crude oil influencing factors	47
Table 2.2	The statistical parameters of the crude oil dataset	48
Table 2.3	The parameters of the models	54
Table 2.4	Multi-step forecasting results of crude oil futures and spot prices	56
Table 3.1	Data description of coal price influencing factors	70
Table 3.2	The statistical parameters of the coal price and other related factors dataset	71
Table 3.3	The algorithm parameters of different methods	82
Table 3.4	Multi-step forecasting results of Coal (Australia) time series data	83
Table 4.1	Statistics of wind power data in target sites	97
Table 4.2	Evaluation of different wind power forecasting models for target site #1	109
Table 4.3	Evaluation of different wind power forecasting models for target site #2	109
Table 4.4	Evaluation of different wind power forecasting models for target site #3	110
Table 4.5	Evaluation of different wind power forecasting models for target site #4	110
Table 5.1	Statistics of the dataset	124
Table 5.2	Error metrics of the forecasting	126
Table 5.3	Statistics of the dataset	127
Table 5.4	Error metrics of the forecasting	129
Table 5.5	Description of items in the optimization model	131
Table 5.6	Description of items in Eq. (5.4)	131
Table 6.1	The rank of gray correlation between energy consumption type and GDP growth rate	157
Table 6.2	Rank of gray correlation between different industries and GDP	160

Table 6.3 Tiered level and consumer distribution under the current scheme 164
Table 6.4 Tiered electricity pricing and electricity bills distribution under the adjusted scheme 165
Table 7.1 The information of selected appliances types in the WHITED dataset 172
Table 7.2 Appliances in the REDD dataset 173
Table 7.3 Information of the two current signals in comparison experiments 176
Table 7.4 Information of the samples used in experiments 182
Table 7.5 Experiments of the introduced models 186
Table 7.6 Confusion matrix of classification indicators 187
Table 8.1 Energy use (kg of oil equivalent) per $1,000 GDP (constant 2017 PPP) by region, 1990–2015 208
Table 8.2 Comparison of physical energy efficiency 209
Table 8.3 Clean energy generation curtailment by region, 2015 213
Table 8.4 Clean energy installed capacity and penetration rate by region, 2015 215
Table 9.1 Total production of world crude steel Units/KT 245
Table 9.2 Sales of new energy vehicles in China, 2021. 1–7 248
Table 9.3 Electricity generation from biofuels and waste by source, World 1990–2018 253

Introduction

1

1.1 Overview of Research Progress in Energy Economics

1.1.1 History of Energy Economics

Throughout the history of human evolution, energy and development have been inextricably linked. Energy has been influencing economic growth and social welfare in different forms since the beginning of its involvement in production and life. Energy is also an essential driving force for economic and social development. As an important strategic resource and the basic industry, energy plays a very important role in promoting and guaranteeing the sustainable, rapid, and healthy development of society and the improvement of people's life. Energy economics is a young science developed under this background. People's perception of the development of energy economics can be roughly divided into three stages (Banks, 2012).

(a) Phase I: The supply of energy production is relatively abundant, and human beings use and consume natural resources in large quantities to achieve high development.

During the first industrial revolution in the mid-eighteenth century, mankind's rapid economic growth was premised on the massive consumption of natural resources, mainly fossil energy. Western economics began to flourish from this moment, and the change from classical to neoclassical laid the foundation for the emergence of energy economics and the birth of fundamental ideas. From the beginning of the first industrial revolution to the 1830s and then to the second industrial revolution in the early twentieth century, the world's population soared and the demand for resources increased. The opening of a new era of human electrification has led to the rapid development of global productivity. The main research task of energy economics in this phase is to achieve efficient production,

© Science Press 2022

H. Liu et al., *Big Data in Energy Economics*, Management for Professionals,
https://doi.org/10.1007/978-981-16-8965-9_1

conversion, and delivery of energy products to meet the material needs of a period of rapid economic development. Prior to the 1970s, there was a more optimistic attitude toward energy supply. Energy seemed to be an inexhaustible resource.

(b) Phase II: The energy crisis broke out and problems such as energy shortage and resource waste were exposed.

From the 1970s to the early 1980s, the waste of resources caused by one-sided pursuit of economic development aroused people's attention. Traditional economic views and economic development models can no longer solve the growing energy paradox. Equitable distribution and effective allocation of energy resources, energy efficiency improvement, and coordinated development of energy economy become more and more critical. The focus of research in energy economics is beginning to shift to the development of alternative and complementary relationships between different energy sources and the interrelationship between systematically integrated energy sources and the economy. In 1975, *H. A. Merklein's* "Energy Economics" was published, which marked the formal emergence of energy economics as a new branch of economics (Merklein, 1977).

(c) Phase III: The harmonization of economic growth, resource consumption, ecological imbalance, environmental degradation, and other external impact issues associated with energy development and utilization has raised concerns.

Since the mid-1980s, energy sustainability has become a central issue of research. More and more scholars are devoting their efforts to the issue of external diseconomy, such as energy and environment. The study of coordinated and sustainable development between energy, environment, and economic growth has become a current hot topic in energy economics. Econometric models and applications on energy demand and supply, substitution between energy factors and other factors of production, and the relationship between energy and economic growth are being used more frequently. Some large-scale macroeconomic models on energy, environment, and economic growth have been established and widely adopted. Universities in many countries and regions are offering energy economics disciplines and energy economics courses. Energy economics is entering a booming phase.

In general, the three phases of energy economics show that people's understanding has gone through a process from shallow to deep, from local to global, from simple to complex, and from concrete to abstract. Energy economics is also developed and matured in the process of deepening people's understanding.

1.1.2 Framework for Big Data in Energy Economics

Since the beginning of the twenty-first century, sustainable green use of energy has been widely promoted. Issues such as climate change and environmental pollution caused by the unreasonable development of energy have become the focus of the international community's concerns. To achieve the organic coordination of energy development, economic growth, ecological balance, and environmental protection, people's research focus has turned to the transformation of emerging technologies.

To fundamentally solve the problems of coordination and sustainable development between energy, environment, and economic growth, it is necessary to transform the way of energy and economic development. The development of science and technology has injected new and strong impetus into the progress of energy economics. The rapid development of Artificial Intelligence (AI) big data technology provides powerful scientific and technological means and new ideas for the study of energy economics. More and more energy companies are beginning to apply big data technology to strategic decision-making, technology research and development, production and operation, safety, and environmental protection. They are trying to extract more value from big data resources. Technological reforms will promote the economic development of the energy industry in the direction of sustainable development, which will in turn affect the healthy growth of the global economy. The combination of computational intelligence and energy conversion and management makes energy use smarter. It is imperative to develop big data technology.

Big data is a comprehensive technology that takes the whole data collection as the research object. It is a combination of sensing technology, information and communication technology, computer technology, data analysis technology, and specialized field technology. Big Data is also the inheritance and development of the traditional data mining and data analysis technology (Sagiroglu & Sinanc, 2013). Combining big data technology with energy and economy results in energy big data and economic big data.

Energy big data is the technology and idea to collect, process, analyze and apply data from energy fields such as electricity, oil and gas, and other fields such as population, geography, and meteorology in an integrated way. Energy big data is not only the in-depth application of big data technology in the energy field, but also the deep integration of energy production, transmission, storage, consumption, transaction, and related technology revolution with the concept of big data. Energy big data will accelerate the development of energy industry and business model innovation (Zhou et al., 2016a).

Economic big data is a technology and idea to grasp economic monitoring, forecasting, and early warning analysis from different levels such as theory, method, model, and data analysis, which can achieve a more objective and scientific use of methodological models and data analysis conclusions to provide support for macro policy decisions. The combination of big data and economy provides the technical means to calculate macroeconomic indicator series in national economic accounting, population, people's life, investment, energy, price, finance, domestic trade, foreign economy, agriculture, industry, transportation, telecommunication, education, health, etc. The impact of big data on traditional economics research methods will revolutionize the entire economics, sociology, and public administration (Einav & Levin, 2014).

It is worth noting that energy and the economy are not independent but inseparable and interdependent. There are various economic phenomena in the exploration, extraction, and use of energy, such as markets, prices, supply, and demand. Energy economics is an emerging discipline that is understood and expressed differently by different scholars (Wei & Liao, 2016). Chinese scholar *Sun Hongzheng* believes that energy economics is a science that uses economic methods to study energy balance, analyze energy use effects, conduct a feasibility analysis of energy development, processing, and conservation, and formulate energy policies. Chinese economist *Li Hong* believes that the research of energy economics mainly focuses on the relationship between energy and economic growth, the relationship between energy and environmental pollution, the optimal allocation of energy, and energy conservation and circular economy. British energy economic experts *Michael G. Webb and Martin J. Ricketts* think that energy economics focuses on the allocation of energy resources and the energy conversion issues in the process of social production and consumption (Webb & Ricketts, 1980). British scholars *Richard J Eden, Michael V Posner*, and others believe that energy economics is concerned with the availability of energy resources and their relevance to economic activity, the relationship between energy demand and economic growth, and the impact of potential or real energy shortages on economic activity (Eden et al., 1981). In addition, some American scholars think that the main task of energy economics is to formulate energy policies.

The development of energy economics is undergoing a process of exploration and improvement. In the face of the booming opportunities of energy big data and economic big data, the research on the intersection of "Energy + Economy + Big Data" has become significant. Under the wave of data science in the new era, the laws of energy and economy are being decrypted by people in a digital way, and the big data of energy economics is born. The understanding of big data in energy economics is provided as shown in Fig. 1.1.

This book explains what big data in energy economics is. Big data in energy economics is a technology and idea that highly integrates, interconnects, transparently opens, and mutually shares the massive information data of economy, society, politics, ecology, environment, climate, security, engineering, science and technology, price, market, trade, policy, and health, etc., which are closely related to energy to varying degrees. Under the system of big data in energy economics, the production, transmission, demand, pricing, consumption, and organizational structure of energy market, as well as the policy making, development planning, commodity trading, and economic activities of the economic market will undergo a radical change. Precise matching and resource allocation optimization under the new energy supply and demand relationship are indispensable to achieve a win–win situation for energy, economy, and environment.

"Big Data in Energy Economics" analyzes the supply and demand price markets for renewable and non-renewable energy sources and the macro-energy economy from a data science perspective, and elaborates the impact of energy production and consumption on the economy. Energy economics professionals can use the

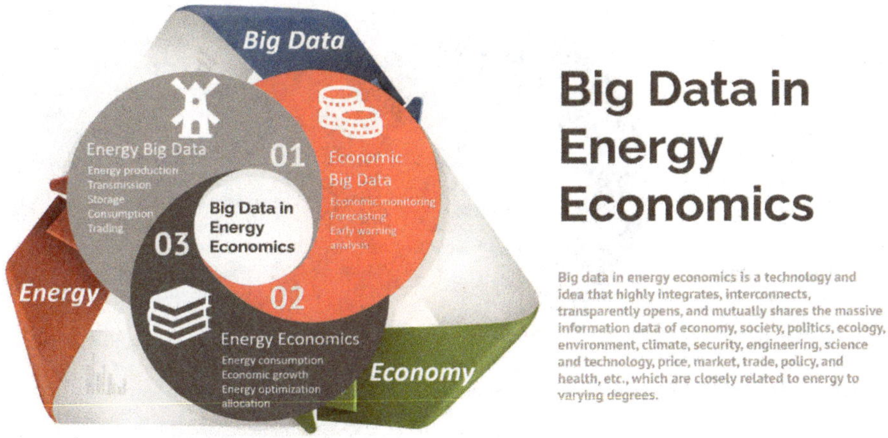

Fig. 1.1 Understanding of big data in energy economics in this book

results of big data modeling and analysis to formulate a reasonable and feasible electricity price and billing mechanism to improve the absorption capacity of the grid. Scientific power equipment scheduling and cost–benefit analysis can avoid losses caused by accidental damage to equipment. Energy users can adjust their electricity consumption behavior through the modeling results, and achieve the effect of energy saving and emission reduction while reasonably reducing electricity expenses. To achieve new energy innovation, promote energy structure transformation, help economic restructuring, accelerate the sound carbon trading mechanism and energy saving and emission reduction regulations, and achieve a win–win situation for the energy economy, it is essential to develop the theory of big data modeling for smart energy economy.

The contradiction between social development and energy consumption has prompted urgent changes in energy production, consumption, organizational structure, investment, and conversion efficiency. The large-scale application of technologies such as Energy Internet (EI), cloud computing, and big data is a necessary means to achieve accurate matching and configuration optimization of energy resources under the new supply and demand situation. The framework diagram of big data in energy economics is provided as shown in Fig. 1.2.

Some scholars have found that the most widely used artificial intelligence technology in the energy field is energy price forecasting and energy consumption and demand forecasting (Ghoddusi et al., 2019). Complex features such as nonlinearity, strong non-stationarity, hysteresis dependence, and volatility clustering make it very difficult to predict energy commodity prices and supply and demand conditions. Machine learning technology has played a huge role in it. Its strong adaptability overcomes the shortcomings of traditional statistical models and econometric models. Some applications of big data technology in the field of energy economy are summarized as shown in Table 1.1.

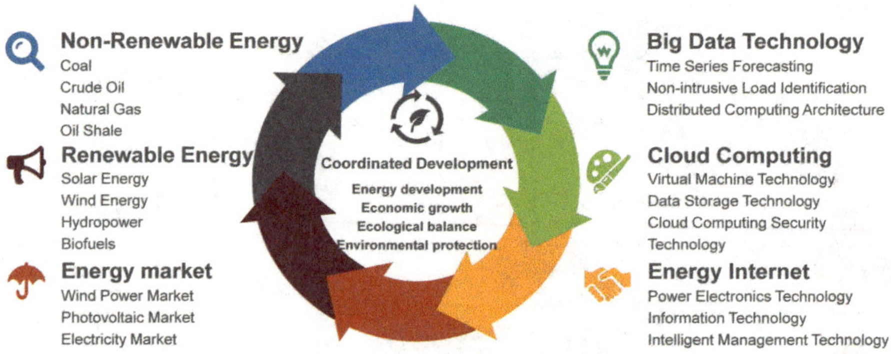

Fig. 1.2 Framework of big data in energy economics in this book

With the gradual integration of internet technology and internet thinking with the energy economy system, people are beginning to realize the huge application potential of big data in the energy economy across the whole spectrum of the energy industry. The new era is of great significance in promoting the development of renewable energy, and stimulating the vitality of cross-border integration and innovative development momentum in the energy industry (Zhou et al., 2016a).

1.1.3 Strategies and Measures for the Development of Big Data in China's Energy Economics

With the development trend of the global urgent need to realize the energy transition, the "Internet+" smart energy economy has become a hot spot in the energy sector. The deep integration of the energy grid and the internet is an important transformative force for solving current energy problems and reshaping the global energy landscape. Big data in energy economics integrates massive energy data, economic data, and big data technology. It is an important means of building an "Internet+" smart energy economy. In China, to promote the development of energy economics, the government, enterprises, universities, and scientific research institutions are all contributing their strength.

1.1.3.1 Government and Enterprises Join Forces to Vigorously Build Energy Big Data Centers

With the explosive growth of energy data and the continuous improvement of data processing capabilities, energy big data has become a new element of social production. The construction of energy big data centers has become an important means for the digital transformation of the energy and power industry to achieve carbon neutrality (Zhu et al., 2021). Under the new power system and dual-carbon

Table 1.1 Some applications of big data technology in energy economics

Application field		References
Energy price prediction applications	Crude oil price prediction	(Busari & Lim, 2021; Wang et al., 2020b; Chai et al., 2019)
	Coal price prediction	(Alameer et al., 2020; Ding et al., 2021; Matyjaszek et al., 2019)
	Natural gas price prediction	(Wang et al., 2021a, 2020a)
	Electricity price prediction	(Chang et al., 2019; Li & Becker, 2021; Memarzadeh & Keynia, 2021)
Energy market consumption and demand prediction applications	Fossil fuel market demand prediction	(Yanguas Parra et al., 2021; Al-Fattah, 2021; Hafezi et al., 2021)
	Wind power market demand prediction	(Yang et al., 2021b; Arasteh & Riahy, 2019)
	Photovoltaic power generation market demand prediction	(Liu et al., 2021a; Yin et al., 2022)
	Electricity market consumption prediction	(Liu et al., 2020)
	Smart grid load prediction	(Hafeez et al., 2021; Jiang et al., 2021b; England & Alouani, 2020)
	Energy supply and transportation market demand prediction	(Roozbeh Nia et al., 2021)
Other applications	Market policy analysis	(Bjerregaard & Møller, 2019; Cieplinski et al., 2021; Marmolejo-Duarte & Onecha-Pérez, 2021)
	Market trading strategy analysis	(Wang et al., 2021b; Yao et al., 2021; Xu et al., 2021)
	Energy system structure analysis	(Zhang et al., 2018b)
	Model calibration	(Fumo et al., 2021; Guo et al., 2021; Hou et al., 2021)
	Data preprocessing	(Haces-Fernandez et al., 2021; Xiao et al., 2021; Liu et al., 2021b)

goals (peak carbon dioxide emissions, carbon neutral), the amplification, superposition, and multiplication effects of energy data elements have become increasingly prominent. Energy big data centers have blossomed all over China.

In China, the government is cooperating with enterprises to vigorously build energy big data centers across the country. More than 15 provinces have completed the inauguration and establishment of energy big data centers, and have received strong support from the State Grid and China Southern Power Grid. These provinces include Shanxi, Guizhou, Gansu, Henan, Anhui, Zhejiang, Sichuan, Fujian, Shaanxi, Chongqing, Hunan, Shandong, Liaoning, Tianjin, Qinghai, etc.

The energy big data center is an important carrier to promote the deep integration of the energy revolution and the digital revolution (Jiang et al., 2016). It is also an energy data convergence center, a value creation center, and an openly shared service center. There are four main benefits for accelerating the construction of energy big data centers (Li et al., 2020). (a) It helps to promote the integration and fusion of big data in energy fields such as water, electricity, gas, and oil. (b) It is important to break the "information barriers" in the industry, rationalize the allocation and utilization of energy resources, and establish a widely interconnected, integrated, and open energy internet ecology. (c) It has an important supporting role for regional energy planning, economic operation, energy conservation, and emission reduction, and the overall mining of potential data values. (d) It is important for sharing data resources, mining data value, assisting the professionals in scientific decision-making, promoting management quality and efficiency, promoting the improvement of energy efficiency of the whole society, and the construction of a new type of smart city. Figure 1.3 shows a conceptual view of the energy big data center.

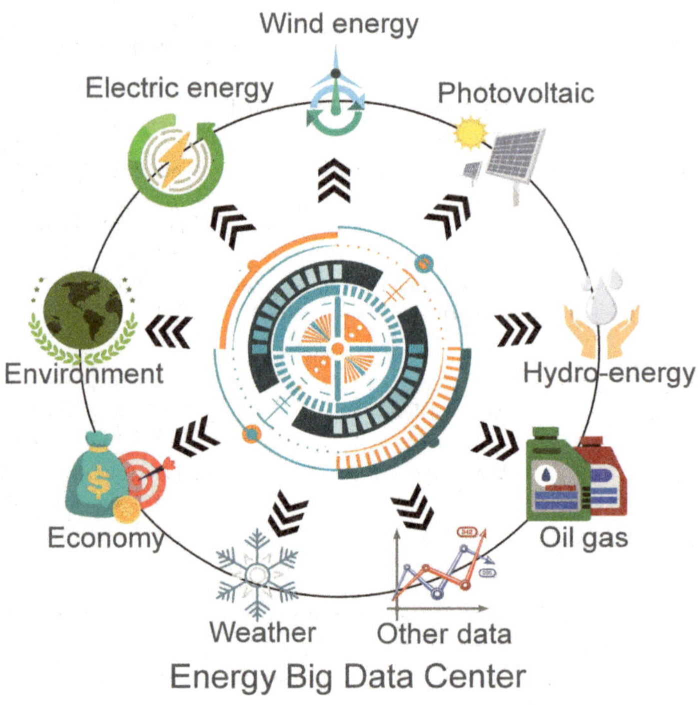

Fig. 1.3 Conceptualization of energy big data center

1.1.3.2 Colleges and Universities Actively Explore New Development Models of Energy Economics

Faced with the two major energy problems, the imbalance between energy supply and demand and the environmental pollution caused by the burning of fossil fuels, Chinese universities and institutions are also actively contributing to the reform and development of energy economics (Wei et al., 2010). Many higher education institutions in Changsha, Xiamen, Beijing, Shanghai, Shenzhen, and other places have established energy research institutes specializing in energy economics issues.

Facing the strategic needs of national medium and long-term energy economy development, universities have launched many researches on sustainable energy development, optimal allocation of energy resources, energy strategy and policy formulation, low-carbon economy and new energy exploration, energy digitization, distributed energy research and development, smart power grid construction, shale oil and gas exploration, oil and gas research, clean and efficient utilization of coal, geothermal resources development, and solar cell development.

Solving the energy problem is a matter of urgency in China. The purpose of these institutes is to create a high quality, reliable, economical, and sustainable energy system for the future of society. The energy research institutes cover a wide range of research areas including power generation, energy storage and transmission, energy efficiency, and energy policy. The establishment of the energy research institutes will promote the development of energy-related disciplines, build interdisciplinary platforms, and facilitate cooperation and exchange between China and the world (Liu et al., 2019a).

1.1.4 Strategies and Measures for the Development of Big Data in World's Energy Economics

In the face of new changes in the pattern of energy supply and demand, and new trends in international energy development, countries around the world are contributing to the development of energy technology (Cheng et al., 2017). Accelerating energy transformation and developing and utilizing clean energy such as solar energy, wind energy, and water energy have become the general direction of energy development in the world. Under the active initiative of China, the proposal of "seizing the new round of energy restructuring and energy technology change trend, building Global Energy Internet (GEI) and realizing green low-carbon development" has been highly praised and widely responded by the international community.

1.1.4.1 Promoting the "Global Energy Internet China Initiative" to Become an International Consensus

Since its establishment in Beijing, China, the Global Energy Interconnection Development and Cooperation Organization (GEIDCO) has done a lot of work in concept dissemination, planning and research, international cooperation, and project promotion, which has contributed to the progress of the GEI.

GEIDCO vigorously spread the GEI China Initiative to the world. Its leaders held talks with the UN Secretary-General and dozens of heads of state and government through the European Union, the African Union, the Arab League, the Association of Southeast Asian Nations (ASEAN), and other international organizations, government departments, enterprises, institutions, and universities of relevant countries. GEIDCO has held hundreds of international conferences around the world to in-depth publicize the concept and development results of the GEI, which has been widely recognized and supported by all parties. The GEI has been incorporated into the working framework of implementing the United Nations' 2030 Agenda for Sustainable Development, promoting the implementation of the Paris Agreement, promoting global environmental governance, solving the problem of health without electricity, poverty, and the construction of the "The Belt and Road". GEI has also included the outcome documents of the 9th Clean Energy Ministerial Conference, the 54th Summit of the Economic Community of West African States, and the 8th Ministerial Conference of the China-Arab Cooperation Forum. The China Initiative put forward by GEIDCO has produced tremendous influence and driving force (Feng et al., 2017).

UN Secretary-General Guterres emphasized that GEI is the core of human sustainable development and the key to global inclusive growth. It is essential to achieve the goals of the United Nations' 2030 Agenda for Sustainable Development and the Paris Agreement. China's GEI Initiative reflects China's leadership and responsibility in promoting energy transition and addressing climate change, and they will do their utmost to promote its implementation.

1.1.4.2 Global Energy Internet is an Effective Carrier for Building a Community with a Shared Future for Mankind

Building a community of human destiny provides a fundamental way to promote global governance and sustainable human development. At present, the Coronavirus Disease 2019 (COVID-19) pandemic is spreading globally, which highlights the importance and urgency of accelerating the construction of a community with a shared future for mankind. Building a community with a shared future for mankind requires coordinated efforts and innovative breakthroughs in energy, economy, society, environment, and other fields. The essence of GEI is a global energy community with a shared future, which will effectively promote the global energy transition, respond to climate change, protect the ecological environment, and promote economic prosperity and social progress. GEI is important content and carrier for building a community with a shared future for mankind (Wang et al., 2020d).

GEIDCO Chairman *Liu Zhenya* said in a keynote speech entitled "Global Energy Internet Changes the World" that the cooperative organization has carried out in-depth research on the economy and society, energy resources, power supply and demand in more than 100 countries around the world. At this meeting, GEIDCO proposed a global energy internet backbone network architecture planning, which is generally divided into three phases, and will eventually form the "Nine Horizontals and Nine Verticals" energy arteries covering five continents.

According to the plan, 18 major synchronized joint power grids will be gradually formed in the world. "Nine Horizontals" are 9 major channels, including Arctic Energy Interconnection, Asia-Europe North Horizontal, Asia Europe South Horizontal, Asia-Africa North Horizontal, Asia-Africa South Horizontal, North America North Horizontal, North America South Horizontal, South America North Horizontal, and South America South Horizontal. "Nine Verticals" are 9 major channels, including Europe-Africa West Vertical, Europe-Africa Middle Vertical, Europe-Africa East Vertical, Asia West Vertical, Asia Middle Vertical, Asia East Vertical (Asia–Pacific Channel), America West Vertical, America Middle Vertical, and America East vertical.

The cooperative organization plans to build a "Five Horizontals and Five Verticals" interconnection channel in 2035, and Asia-Europe-Africa will take the lead in realizing intercontinental interconnection. Investment estimates show that the total investment from 2018 to 2050 is US$38 trillion, of which power investment is US$27 trillion and power grid investment is US$11 trillion. By 2050, 177,000 km of new transmission lines will be added to the backbone grid, with a cumulative investment of US$390 billion. According to projections, clean energy generation will account for more than 81% of global electricity generation by 2050. This will boost global economic growth by an average of 0.2% points per year and reduce the global average cost of electricity by 2.8 cents/kWh compared to today. By that time, more than 100 million jobs will have been created globally. The global energy internet backbone network architecture planning is provided on the official page of (GEIDCO).

Building a GEI will narrow the regional gap and achieve coordinated development of regional economies. Relying on GEI, it can promote South-South and North–South cooperation, and transform the advantages of wind and solar energy resources in backward regions such as Africa, Asia, and South America into economic advantages. This not only provides sufficient power supply for the local area, thereby solving the problem of energy shortage. It can also enjoy the spillover effects of the development of developed economies through the income generated by the output of electricity. The construction of GEI has narrowed the regional differences in world economic development. By strengthening interconnection, GEI will promote the establishment of a new global division of labor system for energy production, exchange, distribution, and consumption (Yu et al., 2010). Utilizing the advantages of different regions can speed up the process of regional economic integration and promote the peaceful and harmonious development of all countries (Li et al., 2021).

- GEI is the development direction of the world's energy reforms and adjustments (Inshakov et al., 2019).

  (a) Promote a revolution in energy production with clean energy alternatives. Promote the large-scale development of solar, wind, and water energy to accelerate the replacement of fossil energy generation. This approach is to achieve a shift in energy production from fossil-energy-led to clean-energy-led, so that coal, oil, and gas return to their basic properties as industrial raw materials and create greater value.

(b) Promote open sharing of energy with interconnection. Play a key role in the optimal allocation of large grid resources, promote the construction of a widely interconnected global energy sharing platform, achieve large-scale allocation and efficient use of clean energy, and promote the green low-carbon development of countries.

(c) Promote energy consumption revolution with electric energy substitution. Promote the replacement of coal, oil, gas, and firewood with electricity, significantly increase the electrification level of the whole society, and convert water and carbon dioxide into hydrogen, methane, and other fuels and raw materials through electricity to meet the needs of sustainable human development.

- GEI is a powerful engine that promotes technological industry innovation and high-quality economic development (Liu & Chen, 2018).

(a) Promote technological innovation. GEI integrates energy and electricity, artificial intelligence, big data, cloud computing, blockchain, and other advanced technologies, which will drive the transformation of traditional industrial economy to new digital economy, sharing economy, and platform economy, accelerate the cultivation of new business models, and achieve high-quality economic development.

(b) Promote industrial upgrading. GEI has a wide range of industrial driving forces and will promote the development of new energy, new materials, high-end equipment, intelligent manufacturing, electric vehicles, and other emerging industries, thereby creating new growth points for China and the global economy.

(c) Pull green investment. The total investment in the construction of GEI is expected to exceed 35 trillion U.S. dollars. It will not only promote the quality and upgrade of power infrastructure, but also promote the integrated development of energy, information, transportation, and form a global interconnection, co-construction, sharing, green, and intelligent infrastructure network.

(d) Contribute to economic globalization. GEI will promote electricity trade to become the main form of world energy trade, which will create a new pattern of global green energy trade and inject strong new momentum into world economic development.

- GEI is an important carrier for promoting international cooperation and building a harmonious world (Yu, 2019).

(a) Promote world peace and development. Relying on the interconnection of power grids to promote the development and sharing of clean energy, resource exporting countries, countries passing through, and receiving countries can all benefit, enhance mutual trust between countries, and reduce conflicts caused by fossil energy competition.

(b) Contribute to the construction of the "The Belt and Road". GEI can promote cross-border and transcontinental power interconnection and the integration of energy policies of various countries, expand power transactions and green investment, enhance technical exchanges and talent training of various countries, and promote deepening and solidification of the "The Belt and Road" construction.

(c) Promote South-South cooperation and North–South cooperation. GEI gives full play to the complementary advantages of developed and developing countries in technology, market, and resources, which will promote the transformation of the advantages of clean energy resources in developing countries and regions into economic advantages. Create a new pattern of global governance that is peaceful, inclusive, and win–win, and ultimately achieve global inclusive growth and common prosperity.

(d) Help solve the problems of population without electricity, poverty, and health. Reduce the cost of electricity consumption by US$2 trillion each year, and allow everyone to enjoy clean, economical, and efficient sustainable energy, so that the poor can obtain education and employment opportunities, and get rid of poverty as soon as possible. Adequate green energy will reduce environmental pollution and improve medical and health conditions, and enhance human well-being.

- GEI is an innovative solution to tackle climate change and build a beautiful world (Wu et al., 2021; Zhihong et al., 2018).

(a) Implement carbon emission reduction targets. The Paris Agreement is an important document for the global response to climate change. GEI provides a technologically advanced, economical, operable, reproducible, and statistical system solution for the implementation of the Paris Agreement. By accelerating actions such as interconnection, clean substitution, electric power substitution, and energy efficiency improvement, global carbon emissions can reach their peak by around 2025. By 2035, clean energy will surpass fossil energy to become the main energy supply, and global carbon emissions will drop by about 50% from the peak. By 2050, clean energy will account for more than 80% of primary energy, which can achieve the 2 °C or even 1.5 °C temperature rise control target of the Paris Agreement.

(b) Improve the economics of emission reduction. GEI can coordinate the global resource difference, season difference, time difference, and price difference, promote the large-scale development of clean energy and the large-scale efficient allocation, greatly reduce the cost of clean energy development, and significantly improve the investment efficiency, so that the competitiveness of clean energy exceeds that of fossil energy. Economics will be an endogenous driver to accelerate carbon reduction.

(c) Comprehensively improve the ecological environment. By 2050, global emissions of sulfur dioxide, nitrogen oxides, and fine particulate matter will drop by 70%. The air quality will be significantly improved, solid waste emissions and marine pollution caused by the extraction, transportation, and use of fossil energy will be greatly reduced, and problems such as land desertification, forest degradation, and deterioration of biodiversity will also be effectively controlled.
(d) Innovative mechanisms and models. GEI will promote the coordination of national plans and project cooperation, promote the construction of a global electricity-carbon market, realize the coordination of electricity trading and carbon trading, and promote energy transformation, environmental governance, and carbon emission reduction in various countries through mechanism innovation.

1.2 Key Technologies of Energy Internet in Energy Economics

Accelerating the transformation of energy is the most urgent task for mankind. After nearly half a century of technological development and the renewal of people's concept on energy utilization, energy economics is moving in the direction of green, efficient, and sustainable development. To fundamentally solve the three major problems of increasingly tight resources, increasing environmental pollution, and global warming, it is necessary to transform economic development methods. Cultivating a low-carbon economy with low energy consumption, low pollution, and low emissions is a realistic and feasible solution. The construction of the GEI has accelerated the implementation of this program (Huang et al., 2017; Zhou et al., 2016b). The framework of the energy internet in this book is provided as shown in Fig. 1.4.

1.2.1 Concept of Energy Internet

As Rifkin described, people believe that the EI is a new energy system with the smart grid as the core, renewable energy as the foundation, and the internet as the link, through the high integration of energy and information, to achieve efficient and clean use of energy (Rifkin, 2011).

EI is a new model that combines distributed power generation, energy storage systems, and loads into numerous micro-energy networks, thus forming a new method of production and sales. EI adopts advanced information technology, intelligent terminals, and platforms to enable two-way flow and peer-to-peer exchange of energy and information. EI is an intelligent platform that upgrades the traditional power grid to have a powerful energy resource optimization configuration function. It can widely integrate the traditional power grid and intelligent technology

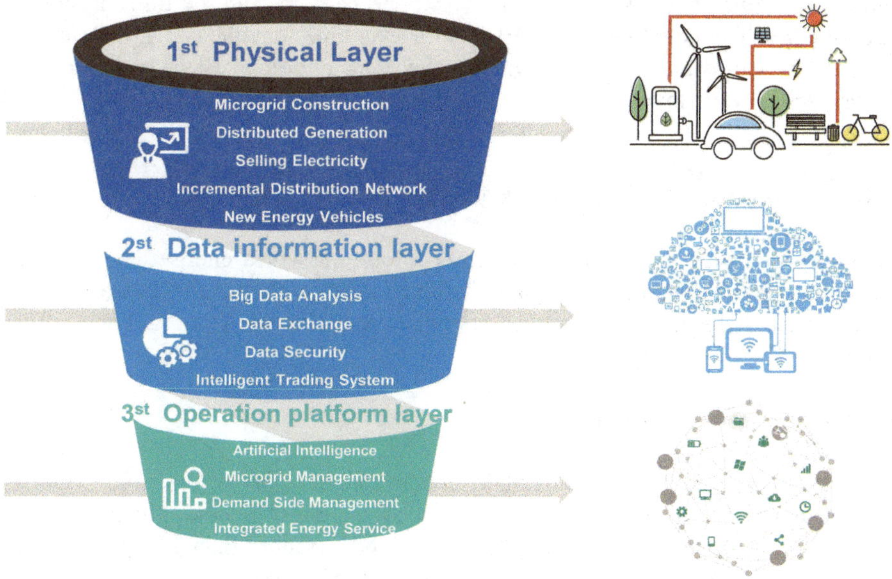

Fig. 1.4 The framework of the Energy Internet

to make use of advanced power transmission technology. Like the traditional internet, EI has three basic characteristics of intelligence, platformization, and sharing. The specific explanation is provided as follows (Cheng et al., 2017):

- Intelligentization: EI collects energy big data intelligently and opens the closed loop of information.
- Platformization: Break down industry barriers and accept various market players.
- Sharing: Assisting in sharing industry resources and realizing industry complementarity.

Research institutions all over the world are investing manpower and material resources in the research of EI. For example, the FREEDM project in the United States has established an energy network architecture with smart functions to absorb many distributed energy sources based on mature electronic technology (Huang et al., 2010). This verifies the possibility that EI can form an independent operating system. In addition, the German eTelligence project uses internet technology to build an energy regulation and real-time power trading system. The system utilizes load adjustment to balance and suppress the intermittent and volatility of new energy output, thereby improving the system's ability to absorb new energy. This verifies the possibility that people can adjust energy allocation through the real-time power trading system. In China, the government proposed the concept of GEI. It is a strong and globally interconnected smart grid with

Ultra-High Voltage (UHV) grid as the backbone network and the transmission of clean energy. The development of EI is an irreversible inevitable trend. Countries all over the world are actively exploring the contribution of EI to mankind, and history will always favor changers.

1.2.2 Reasons for Building a Global Energy Internet

GEIDCO identifies four major challenges to current world energy development, including resource scarcity challenges, climate change challenges, environmental pollution challenges, and non-electric population challenges (Cao & Yang, 2013; Ma et al., 2016).

- Resource scarcity challenges.

If countries around the world maintain the current intensity of resource exploitation, the reserves of natural resources will face a huge test. Shockingly, the world's proven coal reserves are only enough to be mined for more than 110 years, while oil and natural gas can only be mined for more than 50 years.

- Climate change challenges.

Fossil energy generates large amounts of greenhouse gases, which is also the main cause of climate change. Taking 2018 data as an example, carbon dioxide emissions from the use of fossil energy are approximately 33.1 billion tons, an increase of 1.7% year-on-year, which is a record high. If greenhouse gas emissions are not effectively controlled, global temperature rise is expected to exceed 3 degrees Celsius by the end of this century, which will seriously threaten the survival of human beings.

- Environmental pollution challenges.

After many industrial revolutions, the human economy has developed rapidly. Unfortunately, this has also led to serious environmental pollution, including land pollution, marine pollution, air pollution, water pollution, noise pollution, and so on. The deterioration of the ecological environment has severely affected the growth and reproduction of organisms.

- Non-electric population challenges.

The global energy progress report jointly compiled by the International Energy Agency (IEA), the International Renewable Energy Agency (IRENA), the United Nations Statistics Division (UNSD), the World Bank (WB), and the World Health Organization (WHO) show that as of 2019, there are still 840 million people around the world still do not have access to electricity. There are also billions of people who can only rely on fuelwood, coal, animal dung, etc. for heating and cooking. It is difficult to guarantee their basic needs such as education, medical care, production, and living.

Building a GEI will narrow the regional gap and achieve coordinated development of regional economies. The development of GEI will drive strategic emerging industries such as new energy, new materials, smart manufacturing, and electric vehicles. Promoting the development of GEI can absorb a large amount of excess capacity, including steel, cement, wind power, photovoltaics, etc. Building GEI can promote the development of infrastructure, including power supply, power grid, equipment, scientific research, information, and other fields. The GEI is a major innovation in the energy strategy. It will promote the transformation of the world's energy pattern from fossil energy as the leading to clean energy as the leading. GEI has opened a new path for the world's energy green and low-carbon development, which will surely bring great benefits to all mankind.

1.2.3 Key Technologies of Energy Internet

Under the situation of increasingly severe energy crisis, renewable clean energy sources such as solar energy, wind energy, and water energy have become the focus of people's attention in the new era (Zha et al., 2014). However, renewable energy generally has the characteristics of uneven distribution, discontinuous production, and strong randomness, which makes it quite difficult to use and manage. With the development of internet technology, the use of network technology can well connect these scattered energy sources to achieve peer-to-peer exchange and shared networks. Simply put, EI is a kind of network engineering about the connection and interaction of energy points. The development characteristics of EI are related to the characteristics of renewable energy and the internet. It has a high penetration rate of renewable energy, nonlinear random characteristics, multiple big data characteristics, and multi-scale dynamic characteristics. The characteristics of EI determine the application scope and application status of energy internet technology. Relying on the development of EI can realize the scientific management and effective control of energy, thereby improving the utilization rate of energy and promoting social development.

After entering the energy 4.0 era, technologies such as the Internet of Things, blockchain, digital twins, and edge computing have emerged and have been applied in energy production, transportation, distribution, and retail (Zha et al., 2014). The energy industry has entered an era of specialization, intelligence, and networking. This section will summarize the key technologies of EI, and the specific contents are provided as follows (Qingyang et al., 2020; Wang et al., 2020e).

1.2.3.1 New Energy Power Generation Technology

Common new energy power generation technologies include solar power generation, wind power generation, ocean power generation, biomass power generation, and geothermal power generation. Based on traditional renewable energy such as wind, solar, and biomass energy, new energy also includes resources such as shale gas and small reactor nuclear power. New energy power generation technologies also include various high-efficiency power generation technologies, operation control technologies, and energy conversion technologies.

1.2.3.2 Large-Capacity Long-Distance Power Transmission Technology

Large-capacity long-distance power transmission is the fundamental core of the global energy revolution, and it is also a supporting means to solve large-scale energy bases for the delivery of renewable energy power generation. Taking China as an example, the UHV backbone network is the foundation of the power transmission system. The use of high-voltage direct current to interconnect renewable energy bases can build a nationwide Alternating Current (AC) and Direct Current (DC) hybrid super grid, thereby improving the flexibility, complementarity, safety, and reliability of the power supply (Xiao et al., 2017).

Large-capacity long-distance power transmission technology includes flexible and controllable multi-terminal DC transmission technology, flexible DC transmission technology, DC grid technology, submarine cable technology, operation control technology, etc.

1.2.3.3 Advanced Energy Storage Technology

Different from the user-side nodes of traditional power grids, user-side nodes in EI (such as homes or communities) generally have power generation capabilities, and therefore need to be equipped with a certain scale of distributed energy storage system. In addition, due to the high penetration of renewable energy, a larger scale centralized energy storage system is necessary to maintain the stable operation of the grid-side or generation-side systems of EI.

Advanced energy storage technologies include compressed air energy storage, flywheel energy storage, battery energy storage, superconducting energy storage, supercapacitor energy storage, ice heat storage, hydrogen storage, Power-To-Gas (P2G), and other energy storage technologies. The widespread application of energy storage technology and new energy-saving materials in the power system will have a fundamental impact on the traditional power system in all aspects of power generation, transmission, distribution, and utilization. It is the key direction of electrical technology research and development.

1.2.3.4 Demand Response Technology

Demand response means that users respond to electricity prices and other incentives to change the way they use electricity. By implementing demand response, it is possible to reduce the load demand in a short period of time and adjust the load in a certain period in the future to achieve peak shifting and valley filling.

The technology also needs to formulate corresponding electricity price policies and market mechanisms. A demand response system needs to be established, including a master station system, a communication network, and an intelligent terminal. According to the open interconnection protocol, the two-way interaction of tariff incentive signals, user selection, and execution information is realized to achieve the purpose of independent and controllable user load. In the energy internet, the optimal scheduling of multiple customer-side demand response resources will improve the efficiency of comprehensive energy utilization.

1.2.3.5 Advanced Information Technology

Advanced information technology includes intelligent perception technology, cloud computing technology, big data analysis technology, etc., which represent the development direction of information technology in the energy field (Fu et al., 2017).

Intelligent sensing technology uses smart sensors to obtain the operating status data of the power transmission and distribution network, electrified transportation network, information communication network, and natural gas network in the EI. At the same time, it can also obtain the operating status parameters of various networked energy-consuming equipment, distributed power sources, and microgrids on the user side. After processing, aggregation, and analysis, these sensor data can provide advanced control strategies for EI's electric energy trading, new energy quota trading, distributed energy and electric vehicle charging facility detection, operation and maintenance, and energy-saving services.

Cloud computing technology uses the powerful interconnection capabilities of cloud computing to support generators (including distributed power and microgrid operators), network operators, users, wholesale or retail power sales companies, and other market entities at any time and place to conduct trading activities.

Big data analysis technology is introduced for in-depth analysis of massive new data such as pipeline network security monitoring, market economy operation, energy trading and user electric energy metering, gas metering, distributed power generation, and electric vehicles in the energy internet. Reliable data analysis can greatly help environmental protection, energy conservation, and economic growth. A detailed description of big data analysis techniques in energy economics is provided in Sect. 1.3.

1.2.3.6 Other Key Technologies

Other key technologies of EI include micro-energy grid technology, advanced power electronics technology, standardization technology, etc. (Sun et al., 2017). The main technologies of the micro-energy grid include multi-energy coordinated planning, multi-energy conversion, optimized coordinated control and management, and distributed power generation forecasting. Advanced power electronic technology includes high-voltage, large-capacity or small-capacity, low-loss power electronic device technology, control technology, and new equipment technology. The importance of standardized technologies stems from the fact that the EI involves multiple energy conversions, transactions, services, and multiple market entities in multiple segments. Following up with the corresponding technical standards and regulations for energy trade can ensure the proper operation of the energy internet.

1.3 Big Data Demand Analysis for Energy Economics

Big data analysis, AI machine learning, and forecasting are important technical system support for the EI to realize the characteristics of life forms. Energy internet integrates operating data, weather data, meteorological data, grid data, and electricity market data for big data analysis, load prediction, power generation prediction, and machine learning modeling. Reasonable and effective big data analysis can open up and optimize the operational efficiency of energy production and consumption, so as to promote the real-time dynamic adjustment of the demand side and the supply side. Figure 1.5 shows the impact of big data analysis technology on energy economics.

Based on GEI, the development of big data analysis technology is the research focus of energy internet and energy economics measurement analysis. The focus of this book also revolves around big data technology. Section 1.3 summarizes and analyzes the key technical tools and application scenarios of the big data method.

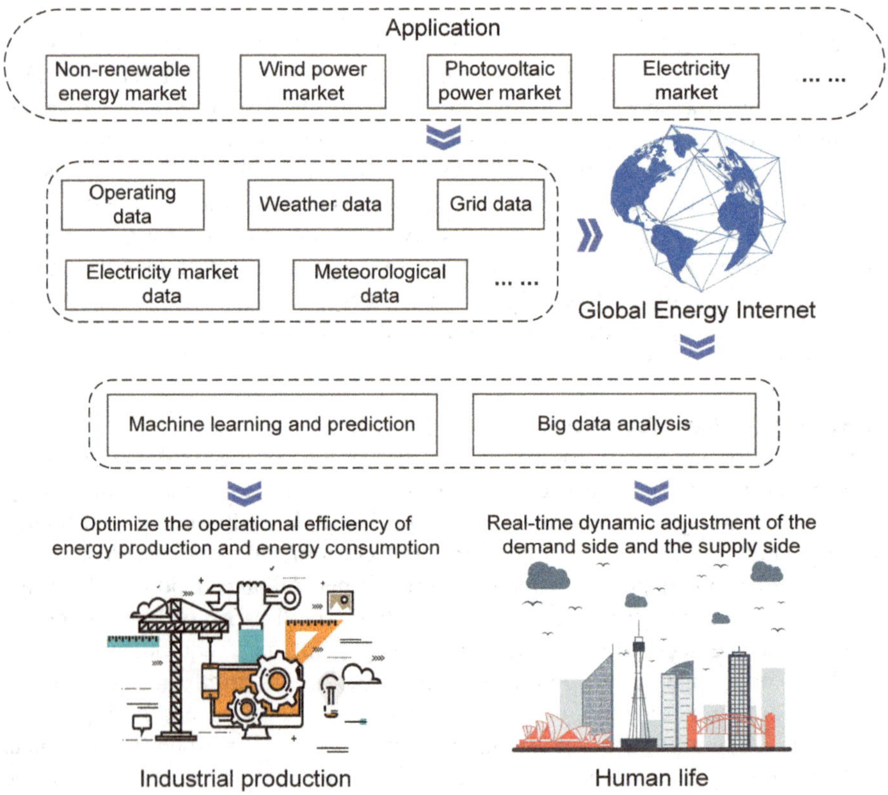

Fig. 1.5 The impact of big data analysis technology on energy economics

1.3.1 Summary of Key Technical Tools

1.3.1.1 Data Preprocessing Technology

Since errors and noise are inevitably generated during the acquisition and transmission of energy data, the collected data often have missing values and outliers, and they undermine the information quality of energy data. To solve this dilemma, missing data imputation methods and outlier detection methods are widely used.

As a sensitive variable, the price of energy commodities is affected by many factors. To carry out in-depth predictive modeling of price time series, these influencing factors need to be taken into consideration. Different influencing factors have different degrees of impact on energy prices. The correlation analysis methods are developed to screen high correlation factors. The summary classification of data preprocessing technology is presented in Fig. 1.6.

Missing Data Imputation Methods

According to the number of input variables in the model, missing data imputation methods are divided into two categories, namely univariate data imputation method and multivariate data imputation method.

(a) For a certain energy commodity price or production capacity raw data, the univariate data imputation method can be used for data preprocessing. Common methods include interpolation, Expectation Maximum (EM), artificial

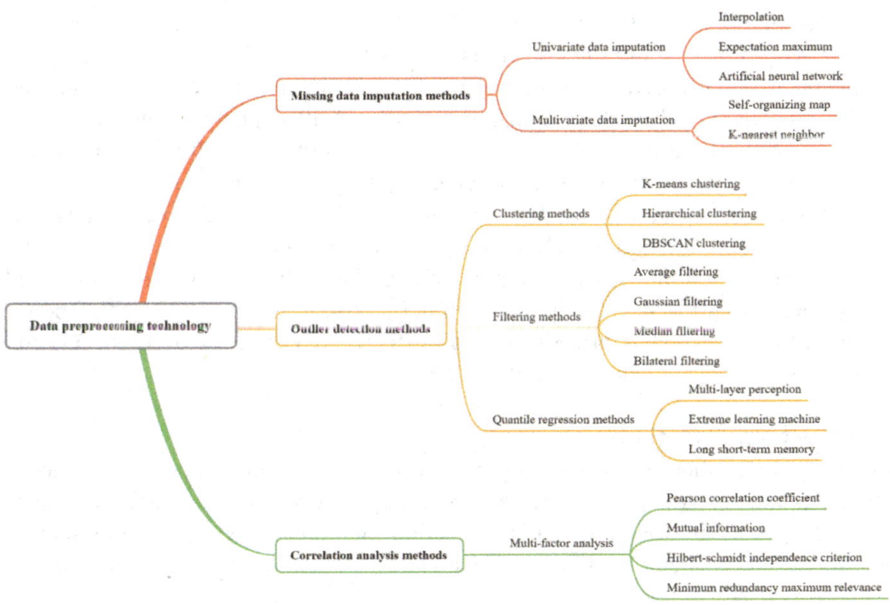

Fig. 1.6 The summary classification of data preprocessing technology

neural network. The interpolation method has become the simplest and most practical missing value imputation method because of its good missing value completion performance under small volume data. Generally, according to the different interpolation functions required in different situations, the interpolation method can be divided into many kinds, including linear interpolation, spline interpolation, Lagrangian polynomial interpolation, Newton interpolation, Hermite interpolation, Cubic spline interpolation, and so on. For large volumes of data that the interpolation method is not good at, the EM method can be used to maximize the likelihood function to estimate the missing data. The EM method can adapt to the Missing at Random (MAR) mode and effectively complete unbiased estimation of parameters. The prerequisite for the successful use of the EM method is that the method assumes that the entire energy data set obeys a certain conditional distribution, to generate the parameters of the distribution to optimize the likelihood function. Furthermore, the excellent learning ability of artificial neural networks allows them to process complex nonlinear data (Borges et al., 2020). Missing data filling methods based on neural networks include Multi-Layer Perceptron (MLP) (Fallah et al., 2020), AutoEncoder (AE), etc. (García-Laencina et al., 2010; Nelwamondo et al., 2013).

(b) For energy commodity price data under the comprehensive influence of multiple factors, a multivariate data imputation method can be used for data preprocessing. The factors that affect the price of energy commodities are often very diverse and complex. When the data of multiple variables is lost, a more reasonable method of filling the missing data of multiple factors is required. The Self-Organizing Map (SOM) method can estimate missing values by minimizing the Euclidean distance between valid data and the template obtained by completive learning. The model structure of SOM is simple, and the time overhead of the training process is small. But the downside is that SOM ignores the correlation between data attributes, which makes it unable to deal with the extreme situation where all the data in the multivariate vector is missing at the same time (Folguera et al., 2015). It is worth mentioning that the K-Nearest Neighbor (KNN) method can obtain a set of neighbors in the cluster to which incomplete samples belong. The filling result can be obtained by calculating the mean value of each neighboring sample on the corresponding attribute. KNN is proven to be an effective method for multiple missing data completion (Oehmcke et al., 2016).

Outlier Detection Methods

Outlier detection methods are usually used to correct for outliers present in energy data, such as abnormal power fluctuations during power generation and anomalies during data transmission. The outlier detection methods introduced in this section include clustering methods, filtering methods, and quantile regression methods.

(a) The clustering method detects outliers by calculating the difference between the behavior of outlier data and normal data, which is a highly effective method

(Duan et al., 2009). The original energy data is divided into multiple clusters with different characteristics after clustering calculation, and the data that is different from the normal cluster is regarded as an outlier (Ben-Gal, 2005).

(b) The filtering method identifies outliers by analyzing local time characteristics (Yang et al., 2021a). Commonly, for example, the Kalman filter uses the calculated distance between the actual data and the Kalman filter estimated data to distinguish values with larger distances as outliers (Ting et al., 2007).

(c) Quantile regression method estimates the upper and lower bounds of the data by calculating the autocorrelation function (Pradeepkumar & Ravi, 2017; Xu et al., 2015). When the value exceeds the specified bounds, the data is defined as an outlier. It is worth mentioning that neural networks are widely used in quantile regression calculations, so that the outlier calculation results obtained are more reliable (Cho et al., 2008; Hao & Naiman, 2007; Niu et al., 2021).

Correlation Analysis Methods

The correlation analysis method calculates the correlation between two variables by examining the degree of linear correlation between them. Under normal circumstances, the calculated correlation value is normalized to $[-1, 1]$, which reflects the direction and degree of mutual change between the two variables. Common correlation analysis methods include Pearson correlation coefficient (Sedgwick, 2012), Mutual Information (MI) (Gierlichs et al., 2008; Kraskov et al., 2004), Hilbert–Schmidt Independence Criterion (HSIC) (Gretton et al., 2005), Minimum Redundancy Maximum Relevance (mRMR) (Radovic et al., 2017; Ramírez-Gallego et al., 2017) and so on.

1.3.1.2 Forecasting Technology

As the core technology of energy economic big data analysis, forecasting has played an important role. The development, conversion, and use of energy inevitably need to use forecasting technology to assist in formulating corresponding methods and policies. The market supervision of energy commodities, the development of international trade, and monetary policy all require the help of predictive technology. This section analyzes the application of temporal horizon, prediction scale, and forecasting mechanism in the field of energy economy.

Temporal Horizon

According to the different perspectives of energy financial time series data, forecasting methods can be divided into two categories, namely short-term forecasts, and long-term forecasts. The temporal horizon for short-term forecasts usually ranges from a few days to a few months. Long-term forecasts can often cover one year or even many years. As the forecast range increases, its forecasting effect will become more and more vague. For short-term forecasting, the fitting effect of the forecast curve and the actual curve is ideal. This can help relevant personnel to formulate sophisticated energy financial plans to deal with short-term emergencies. For long-term forecasts, it can represent the trend of energy financial changes in the future for a long period of time. This can help the government and relevant

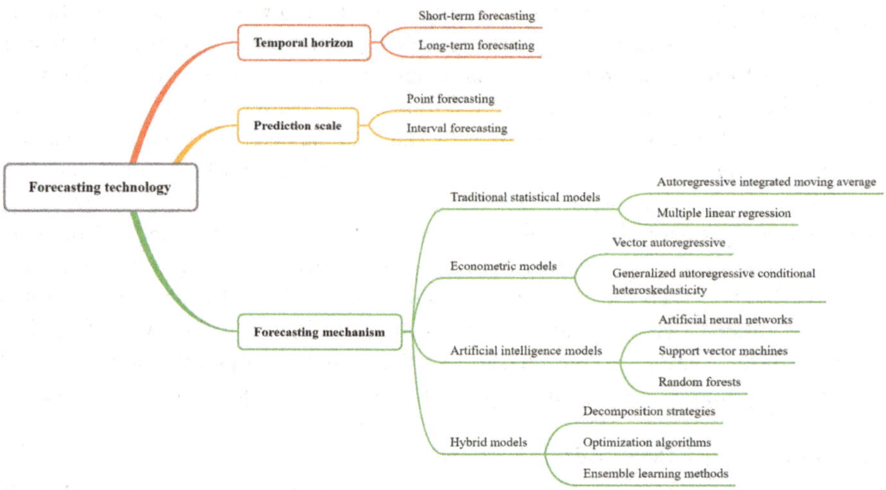

Fig. 1.7 The classification summary of forecasting strategies

personnel formulate reasonable macro policies to combat the possible economic depression.

Prediction Scale

The forecast of energy economic data includes deterministic forecasts and probabilistic forecasts. Most studies usually emphasize the deterministic prediction of energy financial series, that is, Point Forecasting (PF). PF requires significant autocorrelation of the components of the energy financial series data, so that deterministic predictive modeling can be carried out. Uncertain components are unpredictable. The advantage of PF is that a relatively accurate prediction value can be obtained, which is convenient for simple and intuitive viewing of the prediction result. To fit the uncertain components and obtain more reliable prediction results, probabilistic prediction modeling is required, that is, Interval Forecasting (IF). IF will generate a prediction interval, and it will try to keep the actual value within the interval with greater confidence. The relevant practitioners of energy finance can obtain a more reasonable and reliable guidance from the distribution of the prediction interval.

Forecasting Mechanism

According to the prediction mechanism of the model, the prediction methods in the field of energy economics can be divided into traditional statistical models, econometric models, and artificial intelligence models. Figure 1.7 shows the classification summary of forecasting strategies.

For a long time in the past, some traditional statistical models and econometric models have been used as mainstream energy financial analysis methods, including Autoregressive Integrated Moving Average (ARIMA) model (Choi et al.,

2011), Vector AutoRegressive (VAR) model (Mirmirani & Li, 2004), Generalized AutoRegressive Conditional Heteroskedasticity (GARCH) model (Zhang et al., 2015), and so on. Although these models are widely used in energy financial time series forecasting, they require a premise, that is, restrictive distribution assumptions (Alameer et al., 2020; Eckert et al., 2019; Sun et al., 2018). This limitation will make it difficult to fully mine the non-linear features hidden in the data, which will lead to the poor prediction effect of the model. After years of technological innovation, AI models have become a current research hotspot. The AI model with self-learning mechanism has powerful nonlinear feature mapping capabilities. Common machine learning models include Artificial Neural Networks (ANNs) (Singhal & Swarup, 2011), and Support Vector Machines (SVMs) (Zhao et al., 2015), Random Forests (RF) (Sadorsky, 2021), and so on.

As the demand for prediction accuracy continues to increase, some improved methods have been developed and merged into hybrid models, including decomposition strategies, optimization algorithms, ensemble learning methods, and so on.

(a) The decomposition strategy is often used to reduce the volatility of the original energy financial series, to facilitate the prediction model to better learn the changing law of the series. After a series with high instability is decomposed into multiple subseries, its volatility is significantly reduced. Build multiple sub-predictors to predict each decomposition component separately, and then superimpose these prediction results to get a smaller prediction error. Common decomposition methods include Empirical Wavelet Transform (EWT) (Wang and Wang, 2020), Empirical Mode Decomposition (EMD) (He et al., 2015), Ensemble Empirical Mode Decomposition (EEMD) (Qin et al., 2019), Singular Spectral Analysis (SSA) (Zhang et al., 2021), and so on (Jiang et al., 2021a).

(b) The parameter setting of the prediction algorithm is a complicated and difficult process. Parameter setting requires high prior knowledge of algorithm users. The complex structure of neural network poses huge challenges to parameter settings, including learning rate, normalization parameters, number of neural network layer, number of neurons, number of learning rounds, batch data size, output neuron encoding method, selection of cost function, initialization of weights, types of neuron activation functions, scale of model training data, etc.

To obtain the global optimal parameter setting results, meta-heuristic algorithms are developed for parameter optimization of the model, including single-objective optimization algorithms and multi-objective optimization algorithms. Using a single objective optimization function can make the single evaluation index predicted by the model reach a satisfactory degree. But the downside is that other evaluation indicators may not perform as expected. Common single-objective optimization algorithms include Whale Optimization Algorithm (WOA) (Mirjalili & Lewis, 2016), Particle Swarm Optimization (PSO) algorithm (Wang et al.,

2018), Bayesian Optimization Algorithm (BOA) (Pelikan et al., 1999), Grey Wolf Optimization (GWO) algorithm (Teng et al., 2019). The use of multi-objective optimization algorithms can balance the performance of the model in multiple aspects at the same time (Deb, 2014). Common multi-objective optimization algorithms include Multi-Objective Bayesian Optimization Algorithm (MOBOA) (Khan et al., 2002), Non-dominated Sorting Genetic Algorithm II (NSGA-II) (Deb et al., 2000), Multi-Objective Grey Wolf Optimizer (MOGWO) (Mirjalili et al., 2016), Multi-Objective Particle Swarm Optimization (MOPSO) (Zhang et al., 2004).

(c) The factors affecting the changes of energy economic indicators are very diverse and complex, and traditional single predictors cannot meet the needs of big data analysis work. To better predict the fluctuation trend of energy futures prices, more robust methods need to be explored. Ensemble learning organically integrates several weak predictors through specific methods to obtain a predictor with strong comprehensive ability. The typical algorithms of integrated learning include bagging method, boosting method, and stacking method (Ribeiro & dos Santos Coelho, 2020). Based on the bootstrap method, bagging first performs permutation sampling on energy financial data and converts it into multiple data sets. Then, bagging builds multiple learning models to learn each data set. Finally, these prediction results are integrated to get a final average output (Breiman, 1996a). Different from the average mechanism of bagging, boosting is a combination of the stronger ones in the learner (Schapire, 2003). The stacking method is to train a main model to integrate other auxiliary models. Before stacking, several different benchmark models are trained, and the output of these training models is stacked as input to the advanced model. The final result is a stacked output of the main model (Breiman, 1996b).

1.3.1.3 Non-Intrusive Load Identification Technology

A key part of the energy economy is the accurate measurement of electric energy. Reasonable and efficient electricity consumption can help alleviate the energy crisis. The electric power load monitoring is the top priority for energy-saving work to be carried out. Electric energy measurement can help people understand the switching status and energy consumption of electricity load, so as to formulate a reasonable energy saving plan, reduce energy consumption and reduce expenditure. Non-intrusive Load Monitoring (NILM) technology is the core technology to realize electricity consumption monitoring. Research on NILM can promote the process of urban smart electricity consumption. The implementation of NILM usually includes four parts, namely data acquisition, event detection, feature extraction, and load identification (Zoha et al., 2012). Figure 1.8 shows the summary classification of non-intrusive load identification technology.

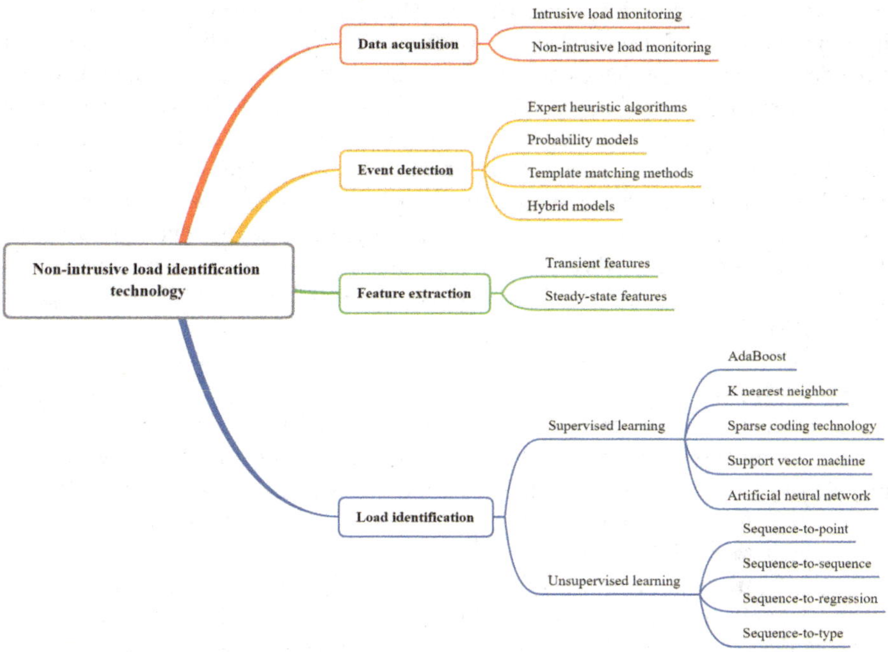

Fig. 1.8 The summary classification of non-intrusive load identification technology

Data Acquisition

Traditional intrusive load monitoring implements real-time monitoring of electricity consumption characteristics by installing load sensors for each electric device. The operability of this method is very poor, and the implementation and maintenance costs are extremely high. Different from the traditional intrusive load monitoring, NILM can complete the load identification of the entire area based on the total electricity meter data (Aladesanmi & Folly, 2015). It infers information such as the operating gear of the user's load and electrical parameters by accurately identifying the data value series obtained by the smart meter. This data acquisition method avoids the high cost of installing many sensors. The installation cost is very low, and the maintenance of the equipment is also very convenient (Klemenjak & Goldsborough, 2016).

Event Detection

To analyze the obtained bus power data, it is necessary to perform event detection on power consumption behavior, including event-based methods and non-event-based methods. Event-based detection methods can be roughly divided into expert heuristic algorithms, probability models, template matching methods, and hybrid models (Anderson et al., 2012). The expert heuristic algorithm judges whether the data state is steady or dynamic according to the difference between the standard deviation and the preset threshold. The probability model uses the probability density function of the Gaussian distribution to determine the probability of an event in the detection window. To a certain extent, this method solves the dilemma that a single real event is indistinguishable from noise (Yang et al., 2014). The template matching method establishes a typical template library to match the current event. Calculating the event distance has become a key step in the template matching method. Commonly used calculation methods include Euclidean distance (Shaw et al., 2008), dynamic time warping method (Basu et al., 2015). The hybrid model obtained by building the hybrid framework is more robust than the traditional event detection model, and can better filter impulse noise (Afzalan et al., 2019; De Baets et al., 2017).

Feature Extraction

There are many types of electrical loads in the non-intrusive load monitoring and decomposition system of the smart grid. Load feature selection and feature extraction usually directly affect the load recognition results. Therefore, the results of feature extraction and selection are very important for the subsequent classification and recognition modules.

Feature extraction methods in NILM system are mainly divided into two categories, namely transient features, and steady-state features. For the extraction of transient features, there are common features such as transient current and voltage now of load switching. The most commonly used steady-state characteristics include active power (Powers et al., 1991), reactive power (Hart, 1992), odd and even harmonics (Bouhouras et al., 2019; Dong et al., 2013), V-I binary curve (De Baets et al., 2018; Liu et al., 2019b), and time–frequency analysis (Gillis & Morsi, 2017; Tabatabaei et al., 2017).

Load Identification

Load identification is an important step in the NILM method, and its essence is to let the NILM system learn the load operation mode of each electrical appliance, so as to achieve the goal of identifying the load. Load identification methods can be divided into two types: supervised and unsupervised. The more mainstream supervised recognition algorithms include AdaBoost algorithm (Zhiren et al., 2019), K nearest neighbor algorithm (Khan et al., 2019), sparse coding technology (Singh & Majumdar, 2017), SVMs (Hernandez et al., 2021), ANNs method (Zhang et al., 2018a) and so on. The load identification algorithm based on unsupervised learning does not require manual assistance to obtain electrical appliance characteristics, which increases the practicability of the algorithm. After the load identification step is completed, the classification result of the load status can be obtained, which is helpful for the relevant personnel in the industry to formulate a reasonable electricity consumption policy.

1.3.2 Application Scenarios of Big Data Technology

The future development direction of the energy economy is bound to serve smart cities, and more scientific and reasonable energy utilization is a new trend in social and economic development. Promoting the optimization of clean energy transition is a key link in the development of a green and low-carbon economy. When big data meets energy economics, the implementation of operational load forecasting in the energy industry, new energy demand forecasting, fault forecasting in health management, energy efficiency behavior analysis, and energy optimization scheduling will all be promoted.

Relying on the internet, the internet of things, big data, cloud computing, and other new technologies, the real-time monitoring, data analysis, and optimization of energy production, storage, and use can be carried out. Combining digital, networked, and intelligent means can realize safe, efficient, green, and smart applications of energy. The application of big data analysis technology will deeply influence the development of smart energy. This book focuses on the introduction of big data technology in application scenarios such as traditional non-renewable energy and new clean energy. Common application scenarios include crude oil market, coal market, natural gas market, wind power market, photovoltaic power market, electricity market.

1.3.2.1 Non-Renewable Energy

As representatives of non-renewable fossil fuels, crude oil, coal, and natural gas are currently the main sources of human energy. However, as the amount of extraction and use is rapidly soaring, their explorable remaining reserves are beginning to be in short supply. Reasonable mining and efficient utilization are beginning to become particularly important. The rise in energy prices will have an impact on the global economic system, including the increase in the production cost of downstream products, the shift of the country's Phillips curve, and inflationary pressures. This influence is quite complex, as reflected in the varying degrees of influence in the long and short term. The causes of energy prices are also very diverse and complex, including supply and demand, monetary policy, international emergencies, new energy developments, etc.

To have a more reliable macro control of the market changes in crude oil prices, coal prices, and natural gas prices, it is necessary to introduce big data analysis techniques (Zhao & Lifeng, 2020). The technical analysis system can combine a variety of influencing factors to predict the complex nonlinear fluctuations of energy prices, thereby providing scientific guidance for reducing economic losses. The impact of energy big data on non-renewable energy has a major role in promoting the energy transition, and stimulating the innovation and development of the energy industry.

1.3.2.2 Renewable Energy

Wind Power Market

Wind power has a wide global distribution and huge reserves. At present, more than 100 countries around the world have built wind power projects, mainly in Asia, Europe, and the Americas. As an important part of wind power, the offshore wind power is also developing rapidly with the support of technology and policies, which vigorously accelerates the global wind power development process. With abundant and stable offshore wind resources, the development of wind farms from land to sea has become a new trend worldwide. The efficient use of wind power resources requires the power of big data, among which forecasting technology is the top priority (Li et al., 2015).

Accurate big data wind power forecasting can promote integrated wind power dispatch planning (Guirguis et al., 2017). It can successfully distribute the power demand of the system to multiple wind turbines at a relatively low cost to achieve power balance. This ensures the stability of the system while reducing maintenance costs. In addition, with the help of Wind Farm Layout Optimization (WFLO) technology, the location layout of wind turbines can be globally optimized, which greatly increases the utilization rate of wind energy (Guirguis et al., 2016). Reasonable resource optimization can increase the total capacity of wind power in the whole life cycle, and the efficient conversion of wind energy increases corporate profits.

Photovoltaic Power Market

For the photovoltaic industry, photovoltaic big data is like a gold mine that is gradually being tapped (Mayer & Gróf, 2021). Through the acquisition, management, analysis, and application of photovoltaic big data, its potential value is gradually injecting new impetus into the development of the photovoltaic industry. Data technology affects the planning and design, construction and acceptance of photovoltaic projects, monitoring and control, operation and maintenance and management, asset evaluation and trading (Bazilian et al., 2013).

Photovoltaic big data can provide useful data for scientific research on photovoltaic systems, and its functions can be divided into four points. First, photovoltaic big data can improve the operation level of photovoltaic power plants, thereby reducing the operation and maintenance costs of photovoltaic enterprises, and improving the power generation efficiency of photovoltaic power plants. Second, photovoltaic big data can optimize the business process of photovoltaic enterprises, so that the resources and activities of the production and operation process can be rationally arranged, thereby improving the economic efficiency and market competitiveness of the enterprise. Third, through real-time monitoring and forecasting of photovoltaic power plants, photovoltaic power generation data can be involved in the real-time power dispatch of the entire grid, thereby providing a reference for energy consumption and energy production distribution. Fourth, by cooperating with demand-side response management, photovoltaic power generation can realize dynamic deployment of production, transmission, and consumption on demand, so as to improve efficiency, save energy and reduce emissions (Antonanzas et al., 2016).

Electricity Market

The electricity market connects all links of energy production, transmission, and consumption (Ventosa et al., 2005). It also covers many fields such as market transactions, grid operation, electricity production, energy use, hydrometeorology, and financial services. The electricity market not only has a wide range of data sources, but the scale of data is also extremely large. Based on the massive data information of the electricity market, data mining and knowledge discovery are key technologies to solve the problems of large-scale energy demand matching, individual service demand perception, and market risk identification and monitoring. Big data technology is also an effective way to support the transfer and exchange of data value based on the service concept of "Internet+Electricity trading". Among them, electricity price forecasting and load forecasting are an important part of energy management in the electricity market (Yang et al., 2020). Data-driven big data analysis methods include clustering, regression, factor analysis, time series forecasting, multi-dimensional scale analysis, and relational models (Wang et al., 2020c).

The massive power market big data is facing great difficulties in processing. Building a distributed computing platform and developing memory computing and streaming computing technologies are key methods to solve this problem. In the context of the EI, real-time data analysis and visualization of the electricity market can promote user-side energy efficiency management and demand-side module response. Sophisticated data-based power market supervision can greatly save power waste.

1.4 Scope of This Book

This book is suitable for the general direction of the smart energy economy. It systematically introduces the key technologies of the global energy internet and big data in energy economics, as well as many urgently needed application scenarios such as the crude oil market, coal market, wind power market, photovoltaic power market, and electricity market. This book utilizes a combination of theoretical analysis and case studies to explore the feasibility and effectiveness of AI methods in the field of energy economics.

This book introduces the smart energy economic system from the unique perspective of big data. The latest theoretical models are systematically discussed to advance scientific research and practical applications. This book explores the conversion and utilization of non-renewable and new clean energy sources through big data modeling and analysis, and provides insights in building the energy internet and green low-carbon energy economic transformation. The contents of the 9 chapters are integrated together to form the energy economy big data system under the background of the smart city energy internet. More specifically, the organization of each chapter of this book is provided as follows:

***Chapter* 1: Introduction**

This chapter first introduces the development of energy economics, and describes big data in energy economics. Then the concept, significance, and key technologies of energy internet are analyzed. The core tools and typical application scenarios of big data analysis in energy economics are explained in a systematic way. Finally, the general framework and scope of the book are presented.

***Chapter* 2: Big Data Analysis of Energy Economics in Oil Market**

This chapter uses West Texas Intermediate spot prices, European Brent spot prices, and New York Mercantile Exchange crude oil future prices as experimental cases. The factors affecting crude oil price fluctuations and the impact of crude oil price changes on the global economy are analyzed. The mainstream mean absolute error, mean absolute percentage error, and root mean square error evaluation indicators are used to compare the prediction performance of bi-directional long short-term memory and deep belief network. Finally, an econometric analysis of crude oil prices is given.

***Chapter* 3: Big Data Analysis of Energy Economics in Coal Market**

This chapter focuses on the forecast analysis of coal prices and its influencing factors, including crude oil, iron ore, gold, silver, copper, natural gas, AUD&EUR exchange rates, AUD&CNY exchange rates, AUD&USD exchange rates. Pearson correlation coefficient is used to evaluate the correlation between these factors and coal prices. The decomposition strategy and the AdaBoost.MRT method are utilized to improve the prediction accuracy of the deep belief network. Finally, an econometric analysis of crude oil prices is presented.

***Chapter* 4: Big Data Analysis of Energy Economics in Wind Power Market**

This chapter first proposes a spatial scale wind energy prediction framework. Then, four different binary optimization algorithms are embedded into it. A set of actual datasets collected from wind farms are utilized to test the forecasting performance of each model. The conversion efficiency of wind energy is briefly introduced. Finally, the market economy of wind power applications is analyzed for China, America, and Europe.

***Chapter* 5: Big Data Analysis of Energy Economics in Photovoltaic Power Generation Market**

This chapter first introduces the photovoltaic power generation prediction technology, and gives examples in two application scenarios. Then, an economic dispatch model of a power generation system with photovoltaic power integration is demonstrated. Finally, the intelligent solving methods are presented and explained.

***Chapter* 6: Big Data Analysis of Energy Economics in Power Market**

This chapter analyzes and studies the energy economic big data from the following three aspects. First, the electricity price is predicted based on artificial intelligence big data technology, including extreme learning machine, deep belief network, and long short-term memory. Subsequently, the correlation between energy consumption and economic growth is studied. Finally, the chapter analyzes the adjustment of metering fees of urban electricity prices based on the K-means algorithm.

***Chapter* 7: Big Data Management of Energy Conservation and Emission Reduction in Smart Cities**

Smart city is the inevitable direction of urban development. As an important part of the healthy and green development of cities, energy saving and emission reduction are closely related to the sustainable development of the economy. Load identification technology is one of the core technical difficulties to be solved in realizing the overall layout planning of smart power grid. Besides, smart grid construction is a necessary means of urban power consumption behavior management and energy conservation and emission reduction implementation. This chapter introduces the load identification methods and gives feasible suggestions for urban power consumption behavior planning and energy saving and emission reduction efficiency.

***Chapter* 8: Optimization Analysis of Clean Energy Transformation**

This chapter summarizes the effective practices of developed countries to promote the clean energy transition. Combining with the existing problems in China, the strategic path of clean energy industry transformation to promote low-carbon economy is put forward. From the perspective of economy and technology, energy utilization efficiency at China and other countries under diversified development is compared. The mechanism and qualitative analysis are made on the root of the difficulty of clean energy consumption in typical regions of China. Finally, the mechanisms of clean energy consumption affecting economic growth are studied.

***Chapter* 9: Global Energy Internet: Green and Low-Carbon Energy Economic Innovation**

This chapter first introduces the concept of energy internet. Then the differences of new energy system and conventional energy system are revealed and the characteristics of new energy internet are analyzed. After that, different innovations of energy saving and carbon emission reduction are summarized. At last, the integration of energy system and ecology system is discussed.

References

Afzalan, M., Jazizadeh, F., & Wang, J. (2019). Self-configuring event detection in electricity monitoring for human-building interaction. *Energy and Buildings, 187*, 95–109. https://doi.org/10.1016/j.enbuild.2019.01.036

Al-Fattah, S. M. (2021). Application of the artificial intelligence GANNATS model in forecasting crude oil demand for Saudi Arabia and China. *Journal of Petroleum Science and Engineering, 200*, 108368. https://doi.org/10.1016/j.petrol.2021.108368

Aladesanmi, E. J., & Folly, K. A. (2015). Overview of non-intrusive load monitoring and identification techniques. *IFAC-PapersOnLine, 48*(30), 415–420.

Alameer, Z., Fathalla, A., Li, K., Ye, H., & Jianhua, Z. (2020). Multistep-ahead forecasting of coal prices using a hybrid deep learning model. *Resources Policy, 65*, 101588.

Anderson, K. D., Berges, M. E., Ocneanu, A., Benitez, D., & Moura, J. M. F. (2012). Event detection for non intrusive load monitoring. In *IECON 2012–38th Annual Conference on IEEE Industrial Electronics Society* (pp. 3312–3317). IEEE.

Antonanzas, J., Osorio, N., Escobar, R., Urraca, R., Martinez-de-Pison, F. J., & Antonanzas-Torres, F. (2016). Review of photovoltaic power forecasting. *Solar Energy, 136*, 78–111.

Arasteh, F., & Riahy, G. H. (2019). MPC-based approach for online demand side and storage system management in market based wind integrated power systems. *International Journal of Electrical Power and Energy Systems, 106*, 124–137. https://doi.org/10.1016/j.ijepes.2018.09.041

Banks, F. E. (2012). *Energy economics: A modern introduction.* Springer Science & Business Media.

Basu, K., Debusschere, V., Douzal-Chouakria, A., & Bacha, S. (2015). Time series distance-based methods for non-intrusive load monitoring in residential buildings. *Energy and Buildings, 96*, 109–117. https://doi.org/10.1016/j.enbuild.2015.03.021

Bazilian, M., Onyeji, I., Liebreich, M., MacGill, I., Chase, J., Shah, J., Gielen, D., Arent, D., Landfear, D., & Zhengrong, S. (2013). Re-considering the economics of photovoltaic power. *Renewable Energy, 53*, 329–338.

Ben-Gal, I. (2005). Outlier detection. In *Data mining and knowledge discovery handbook* (pp. 131–146). Springer.

Bjerregaard, C., & Møller, N. F. (2019). The impact of EU's energy labeling policy: An econometric analysis of increased transparency in the market for cold appliances in Denmark. *Energy Policy, 128*, 891–899. https://doi.org/10.1016/j.enpol.2019.01.057

Borges, C. E., Kamara-Esteban, O., Castillo-Calzadilla, T., Andonegui, C. M., & Alonso-Vicario, A. (2020). Enhancing the missing data imputation of primary substation load demand records. *Sustainable Energy, Grids and Networks, 23*, 100369. https://doi.org/10.1016/j.segan.2020.100369

Bouhouras, A. S., Gkaidatzis, P. A., Panagiotou, E., Poulakis, N., & Christoforidis, G. C. (2019). A NILM algorithm with enhanced disaggregation scheme under harmonic current vectors. *Energy and Buildings, 183*, 392–407. https://doi.org/10.1016/j.enbuild.2018.11.013

Breiman, L. (1996). Bagging predictors. *Machine Learning, 24*(2), 123–140.

Breiman, L. (1996). Stacked regressions. *Machine Learning, 24*(1), 49–64.

Busari, G. A., & Lim, D. H. (2021). Crude oil price prediction: A comparison between AdaBoost-LSTM and AdaBoost-GRU for improving forecasting performance. *Computers and Chemical Engineering, 155*, 107513. https://doi.org/10.1016/j.compchemeng.2021.107513

Cao, J., Yang, M. (2013). Energy internet-towards smart grid 2.0. In *2013 Fourth International Conference on Networking and Distributed Computing* (pp. 105–110). IEEE.

Chai, J., Wang, Y., Wang, S., & Wang, Y. (2019). A decomposition–integration model with dynamic fuzzy reconstruction for crude oil price prediction and the implications for sustainable development. *Journal of Cleaner Production, 229*, 775–786. https://doi.org/10.1016/j.jclepro.2019.04.393

Chang, Z., Zhang, Y., & Chen, W. (2019). Electricity price prediction based on hybrid model of adam optimized LSTM neural network and wavelet transform. *Energy, 187*, 115804. https://doi.org/10.1016/j.energy.2019.07.134

Cheng, L., Qi, N., Zhang, F., Kong, H., & Huang, X. (2017). Energy internet: Concept and practice exploration. In *2017 IEEE Conference on Energy Internet and Energy System Integration (EI2)* (pp. 1–5). IEEE.

Cho, H. J., Kim, Y., Jung, H. J., Lee, S.-W., & Lee, J. W. (2008). OutlierD: An R package for outlier detection using quantile regression on mass spectrometry data. *Bioinformatics, 24*(6), 882–884.

Choi, T.-M., Yu, Y., & Au, K.-F. (2011). A hybrid SARIMA wavelet transform method for sales forecasting. *Decision Support Systems, 51*(1), 130–140.

Cieplinski, A., D'Alessandro, S., & Marghella, F. (2021). Assessing the renewable energy policy paradox: A scenario analysis for the Italian electricity market. *Renewable and Sustainable Energy Reviews, 142*, 110838. https://doi.org/10.1016/j.rser.2021.110838

De Baets, L., Ruyssinck, J., Develder, C., Dhaene, T., & Deschrijver, D. (2017). On the Bayesian optimization and robustness of event detection methods in NILM. *Energy and Buildings, 145*, 57–66. https://doi.org/10.1016/j.enbuild.2017.03.061

De Baets, L., Ruyssinck, J., Develder, C., Dhaene, T., & Deschrijver, D. (2018). Appliance classification using VI trajectories and convolutional neural networks. *Energy and Buildings, 158*, 32–36.

Deb, K. (2014). Multi-objective optimization. In *Search methodologies* (pp. 403–449). Springer.

Deb, K., Agrawal, S., Pratap, A., & Meyarivan, T. (2000). A fast elitist non-dominated sorting genetic algorithm for multi-objective optimization: NSGA-II. *International Conference on Parallel Problem Solving from Nature* (pp. 849–858). Springer.

Ding, L., Zhao, Z., & Han, M. (2021). Probability density forecasts for steam coal prices in China: The role of high-frequency factors. *Energy, 220*, 119758. https://doi.org/10.1016/j.energy.2021.119758

Dong, M., Meira, Xu, W., & Chung, C. Y. (2013). Non-intrusive signature extraction for major residential loads. *IEEE Transactions on Smart Grid, 4*(3), 1421–1430.https://doi.org/10.1109/TSG.2013.2245926

Duan, L., Xu, L., Liu, Y., & Lee, J. (2009). Cluster-based outlier detection. *Annals of Operations Research, 168*(1), 151–168.

Eckert, F., Hyndman, R. J., & Panagiotelis, A. (2019). Forecasting Swiss exports using Bayesian forecast reconciliation. KOF Working Papers.

Eden, R. J., Posner, M. V., Bending, R., Crouch, E., & Stanislaw, J. (1981). *Energy economics: Growth, resources and policies*. CUP Archive.

Einav, L., & Levin, J. (2014). Economics in the age of big data. *Science, 346*(6210).

England, B. S., & Alouani, A. T. (2020). Real time voltage stability prediction of smart grid areas using smart meters data and improved Thevenin estimates. *International Journal of Electrical Power and Energy Systems, 122*, 106189. https://doi.org/10.1016/j.ijepes.2020.106189

Fallah, B., Ng, K. T. W., Vu, H. L., & Torabi, F. (2020). Application of a multi-stage neural network approach for time-series landfill gas modeling with missing data imputation. *Waste Management, 116*, 66–78. https://doi.org/10.1016/j.wasman.2020.07.034

Feng, L., Xu, X., Wang, W., Wang, L., Zhang, H., Li, W., & Zhang, Y. (2017). The design and implementation of global energy interconnection digital research platform. In *2017 IEEE Conference on Energy Internet and Energy System Integration (EI2)* (pp. 1–5). IEEE.

Folguera, L., Zupan, J., Cicerone, D., & Magallanes, J. F. (2015). Self-organizing maps for imputation of missing data in incomplete data matrices. *Chemometrics Intelligent Laboratory Systems, 143*, 146–151.

Fu, R., Gao, F., Zeng, R., Hu, J., Luo, Y., & Qu, L. (2017). Big data and cloud computing platform for energy Internet. In *2017 China International Electrical and Energy Conference (CIEEC)* (pp. 681–686). IEEE.

Fumo, N., Torres, M. J., & Broomfield, K. (2021). A multiple regression approach for calibration of residential building energy models. *Journal of Building Engineering, 43*, 102874. https://doi.org/10.1016/j.jobe.2021.102874

García-Laencina, P. J., Sancho-Gómez, J.-L., & Figueiras-Vidal, A. R. (2010). Pattern classification with missing data: A review. *Neural Computing Applications, 19*(2), 263–282.

GEIDCO Global Energy Interconnection Development and Cooperation Organization. (2021). Global energy internet backbone network architecture planning. https://en.geidco.org.cn/research/plan/. Accessed 14 October 2021.

Ghoddusi, H., Creamer, G. G., & Rafizadeh, N. (2019). Machine learning in energy economics and finance: A review. *Energy Economics, 81*, 709–727. https://doi.org/10.1016/j.eneco.2019.05.006

Gierlichs, B., Batina, L., Tuyls, P., & Preneel, B. (2008). Mutual information analysis. *International Workshop on Cryptographic Hardware and Embedded Systems* (pp. 426–442). Springer.

Gillis, J. M., & Morsi, W. G. (2017). Non-intrusive load monitoring using semi-supervised machine learning and wavelet design. *IEEE Transactions on Smart Grid, 8*(6), 2648–2655. https://doi.org/10.1109/TSG.2016.2532885

Gretton, A., Bousquet, O., Smola, A., & Schölkopf, B. (2005). Measuring statistical dependence with Hilbert-Schmidt norms. *International Conference on Algorithmic Learning Theory* (pp. 63–77). Springer.

Guirguis, D., Romero, D. A., & Amon, C. H. (2016). Toward efficient optimization of wind farm layouts: Utilizing exact gradient information. *Applied Energy, 179*, 110–123. https://doi.org/10.1016/j.apenergy.2016.06.101

Guirguis, D., Romero, D. A., & Amon, C. H. (2017). Gradient-based multidisciplinary design of wind farms with continuous-variable formulations. *Applied Energy, 197*, 279–291. https://doi.org/10.1016/j.apenergy.2017.04.030

Guo, J., Liu, R., Xia, T., & Pouramini, S. (2021). Energy model calibration in an office building by an optimization-based method. *Energy Reports, 7*, 4397–4411. https://doi.org/10.1016/j.egyr.2021.07.031

Haces-Fernandez, F., Li, H., & Ramirez, D. (2021). A layout optimization method based on wave wake preprocessing concept for wave-wind hybrid energy farms. *Energy Conversion and Management, 244*, 114469. https://doi.org/10.1016/j.enconman.2021.114469

Hafeez, G., Khan, I., Jan, S., Shah, I. A., Khan, F. A., & Derhab, A. (2021). A novel hybrid load forecasting framework with intelligent feature engineering and optimization algorithm in smart grid. *Applied Energy, 299*, 117178. https://doi.org/10.1016/j.apenergy.2021.117178

Hafezi, R., Akhavan, A. N., Pakseresht, S., & David, A. W. (2021). Global natural gas demand to 2025: A learning scenario development model. *Energy, 224*, 120167. https://doi.org/10.1016/j.energy.2021.120167

Hao, L., & Naiman, D. Q. (2007). *Quantile regression*, vol. 149. Sage.

Hart, G. W. (1992). Nonintrusive appliance load monitoring. *Proceedings of the IEEE, 80*(12), 1870–1891.

He, K., Yu, L., & Tang, L. (2015). Electricity price forecasting with a BED (Bivariate EMD Denoising) methodology. *Energy, 91*, 601–609.

Hernandez, A. S., Ballado, A. H., & Heredia, A. P. D. (2021). Development of a non-intrusive load monitoring (NILM) with unknown loads using support vector machine. In *2021 IEEE International Conference on Automatic Control and Intelligent Systems (I2CACIS)* (pp. 203–207). IEEE.

Hou, D., Hassan, I. G., & Wang, L. (2021). Review on building energy model calibration by Bayesian inference. *Renewable and Sustainable Energy Reviews, 143*, 110930. https://doi.org/10.1016/j.rser.2021.110930

Huang, A. Q., Crow, M. L., Heydt, G. T., Zheng, J. P., & Dale, S. J. (2010). The future renewable electric energy delivery and management (FREEDM) system: The energy internet. *Proceedings of the IEEE, 99*(1), 133–148.

Huang, C., Chen, S., & Yan, Z. (2017). Electricity trading in global energy internet. In *2017 IEEE Conference on Energy Internet and Energy System Integration (EI2)* (pp. 1–5). IEEE.

Inshakov, O. V., Bogachkova, L. Y., & Popkova, E. G. (2019). The transformation of the global energy markets and the problem of ensuring the sustainability of their development. In *Energy sector: A systemic analysis of economy, foreign trade and legal regulations* (pp. 135–148). Springer.

Jiang, H., Wang, K., Wang, Y., Gao, M., & Zhang, Y. (2016). Energy big data: A survey. *IEEE Access, 4*, 3844–3861.

Jiang, P., Liu, Z., Wang, J., & Zhang, L. (2021). Decomposition-selection-ensemble forecasting system for energy futures price forecasting based on multi-objective version of chaos game optimization algorithm. *Resources Policy, 73*, 102234. https://doi.org/10.1016/j.resourpol.2021.102234

Jiang, Y., Liu, M., Peng, H., & Bhuiyan, Md. Z. A. (2021). A reliable deep learning-based algorithm design for IoT load identification in smart grid. *Ad Hoc Networks, 123*, 102643. https://doi.org/10.1016/j.adhoc.2021.102643

Khan, M. M. R., Siddique, Md. A. B., & Sakib, S. (2019). Non-intrusive electrical appliances monitoring and classification using K-nearest neighbors. In *2019 2nd International Conference on Innovation in Engineering and Technology (ICIET)* (pp. 1–5). IEEE.

Khan, N., Goldberg, D. E., & Pelikan, M. (2002). Multi-objective Bayesian optimization algorithm. In *Proceedings of the 4th Annual Conference on Genetic and Evolutionary Computation* (pp. 684–684). Citeseer.

Klemenjak, C., & Goldsborough, P. (2016). Non-intrusive load monitoring: A review and outlook. arXiv preprint arXiv:01191

Kraskov, A., Stögbauer, H., & Grassberger, P. (2004). Estimating mutual information. *Physical Review E, 69*(6), 066138.

Li, D., Zou, Q., Liu, X., Kong, W., & Meng, J. (2021). Review on the functional form and development trend of energy internet. In *2021 International Conference on Artificial Intelligence, Big Data and Algorithms (CAIBDA)* (pp. 128–132). IEEE.

Li, W., & Becker, D. M. (2021). Day-ahead electricity price prediction applying hybrid models of LSTM-based deep learning methods and feature selection algorithms under consideration of market coupling. *Energy, 237*, 121543. https://doi.org/10.1016/j.energy.2021.121543

Li, W., Bai, H., Hua, Y., Wang, Y., Bu, F., Han, D., Li, M., Yang, J., & Wu, R. (2020). The operation mode of provincial energy big data center. In *2020 10th International Conference on Power and Energy Systems (ICPES)* (pp. 590–595). IEEE.

Li, Z., Wu, W., Zhang, B., & Wang, B. (2015). Adjustable robust real-time power dispatch with large-scale wind power integration. *IEEE Transactions on Sustainable Energy, 6*(2), 357–368. https://doi.org/10.1109/tste.2014.2377752

Liu, C., Wu, W.-Z., Xie, W., & Zhang, J. (2020). Application of a novel fractional grey prediction model with time power term to predict the electricity consumption of India and China. *Chaos, Solitons and Fractals, 141*, 110429. https://doi.org/10.1016/j.chaos.2020.110429

Liu, G., Qu, L., Zeng, R., & Gao, F. (2019a). Energy internet in China. In *The energy internet* (pp. 265–282). Elsevier.

Liu, H., Gao, Q., & Ma, P. (2021). Photovoltaic generation power prediction research based on high quality context ontology and gated recurrent neural network. *Sustainable Energy Technologies and Assessments, 45*, 101191. https://doi.org/10.1016/j.seta.2021.101191

Liu, H., Yang, R., Wang, T., & Zhang, L. (2021). A hybrid neural network model for short-term wind speed forecasting based on decomposition, multi-learner ensemble, and adaptive multiple error corrections. *Renewable Energy, 165*, 573–594. https://doi.org/10.1016/j.renene.2020.11.002

Liu, L., & Chen, S. (2018). The application of artificial intelligence technology in energy internet. In *2018 2nd IEEE Conference on Energy Internet and Energy System Integration (EI2)* (pp. 1–5). IEEE.

Liu, Y., Wang, X., & You, W. (2019). Non-intrusive load monitoring by voltage-current trajectory enabled transfer learning. *IEEE Transactions on Smart Grid, 10*(5), 5609–5619. https://doi.org/10.1109/TSG.2018.2888581

Ma, Y., Wang, X., Zhou, X., Gao, Z., Wu, Y., Yin, J., & Xu, X. (2016). An overview of energy internet. In *2016 Chinese Control and Decision Conference (CCDC)* (pp. 6212–6215). IEEE.

Marmolejo-Duarte, C., & Onecha-Pérez, B. (2021). The spatial uneven difussion of energy-efficiency transparency policy. An analysis of the multi-family market in Barcelona. *Energy Policy, 156*, 112462. https://doi.org/10.1016/j.enpol.2021.112462

Matyjaszek, M., Riesgo, F. P., Krzemień, A., Wodarski, K., & Fidalgo, V. G. (2019). Forecasting coking coal prices by means of ARIMA models and neural networks, considering the transgenic time series theory. *Resources Policy, 61*, 283–292. https://doi.org/10.1016/j.resourpol.2019.02.017

Mayer, M. J., & Gróf, G. (2021). Extensive comparison of physical models for photovoltaic power forecasting. *Applied Energy, 283*, 116239.

Memarzadeh, G., & Keynia, F. (2021). Short-term electricity load and price forecasting by a new optimal LSTM-NN based prediction algorithm. *Electric Power Systems Research, 192*, 106995. https://doi.org/10.1016/j.epsr.2020.106995

Merklein, H. A. (1977). *Energy economics*. UNKNO.

Mirjalili, S., & Lewis, A. (2016). The whale optimization algorithm. *Advances in Engineering Software, 95*, 51–67.

Mirjalili, S., Saremi, S., Mirjalili, S. M., & Coelho, L. dos S. (2016). Multi-objective grey wolf optimizer: A novel algorithm for multi-criterion optimization. *Expert Systems with Applications, 47*, 106–119.

Mirmirani, S., & Li, H. C. (2004). A comparison of VAR and neural networks with genetic algorithm in forecasting price of oil. In *Applications of artificial intelligence in finance and economics*. Emerald Group Publishing Limited.

Nelwamondo, F. V., Golding, D., & Marwala, T. (2013). A dynamic programming approach to missing data estimation using neural networks. *Information Sciences, 237*, 49–58.

Niu, H., Yang, Y., Zeng, L., & Li, Y. (2021). ELM-QR-based nonparametric probabilistic prediction method for wind power. *Energies, 14*(3), 701.

Oehmcke, S., Zielinski, O., & Kramer, O. (2016). kNN ensembles with penalized DTW for multivariate time series imputation. In *2016 International Joint Conference on Neural Networks (IJCNN)* (pp. 2774–2781). IEEE.

Pelikan, M., Goldberg, D. E., & Cantú-Paz, E. (1999). BOA: The Bayesian optimization algorithm. In *Proceedings of the genetic and evolutionary computation conference GECCO-99* (pp. 525–532). Citeseer.

Powers, J. T., Margossian, B., & Smith, B. A. (1991). Using a rule-based algorithm to disaggregate end-use load profiles from premise-level data. *IEEE Computer Applications in Power, 4*(2), 42–47. https://doi.org/10.1109/67.75875

Pradeepkumar, D., & Ravi, V. (2017). Forecasting financial time series volatility using Particle Swarm Optimization trained Quantile Regression Neural Network. *Applied Soft Computing, 58*, 35–52. https://doi.org/10.1016/j.asoc.2017.04.014

Qin, Q., Xie, K., He, H., Li, L., Chu, X., Wei, Y.-M., & Wu, T. (2019). An effective and robust decomposition-ensemble energy price forecasting paradigm with local linear prediction. *Energy Economics, 83*, 402–414.

Qingyang, L., Long, Y., Wang, J., Xiao, C., Song, Y., & Li, H. (2020). Application analysis of big data technology in energy internet. In *2020 International Conference on Intelligent Computing, Automation and Systems (ICICAS)* (pp. 160–163). IEEE.

Radovic, M., Ghalwash, M., Filipovic, N., & Obradovic, Z. (2017). Minimum redundancy maximum relevance feature selection approach for temporal gene expression data. *BMC Bioinformatics, 18*(1), 1–14.

Ramírez-Gallego, S., Lastra, I., Martínez-Rego, D., Bolón-Canedo, V., Benítez, J. M., Herrera, F., & Alonso-Betanzos, A. (2017). Fast-mRMR: Fast minimum redundancy maximum relevance algorithm for high-dimensional big data. *International Journal of Intelligent Systems, 32*(2), 134–152.

Ribeiro, M. H. M., & dos Santos, C. L. (2020). Ensemble approach based on bagging, boosting and stacking for short-term prediction in agribusiness time series. *Applied Soft Computing, 86*, 105837.

Rifkin, J. (2011). *The third industrial revolution: How lateral power is transforming energy, the economy, and the world*. Macmillan.

Roozbeh, N. A., Awasthi, A., & Bhuiyan, N. (2021). Industry 4.0 and demand forecasting of the energy supply chain: A literature review. *Computers and Industrial Engineering, 154*, 107128. https://doi.org/10.1016/j.cie.2021.107128

Sadorsky, P. (2021). A random forests approach to predicting clean energy stock prices. *Journal of Risk Financial Management, 14*(2), 48.

Sagiroglu, S., & Sinanc, D. (2013). Big data: A review. In *2013 International Conference on Collaboration Technologies and Systems (CTS)* (pp. 42–47). IEEE.

Schapire, R. E. (2003). The boosting approach to machine learning: An overview. In *Nonlinear estimation classification* (pp. 149–171).

Sedgwick, P. (2012). Pearson's correlation coefficient. *BMJ, 345.*

Shaw, S. R., Leeb, S. B., Norford, L. K., & Cox, R. W. (2008). Nonintrusive load monitoring and diagnostics in power systems. *IEEE Transactions on Instrumentation and Measurement, 57*(7), 1445–1454. https://doi.org/10.1109/TIM.2008.917179

Singh, S., & Majumdar, A. (2017). Deep sparse coding for non–intrusive load monitoring. *IEEE Transactions on Smart Grid, 9*(5), 4669–4678.

Singhal, D., & Swarup, K. S. (2011). Electricity price forecasting using artificial neural networks. *International Journal of Electrical Power Energy Systems, 33*(3), 550–555.

Sun, Q., Teng, F., & Zhang, H. (2017). Energy internet and its key control issues. *Acta Automatica Sinica, 43*(2), 176–194.

Sun, S., Wang, S., Zhang, G., & Zheng, J. (2018). A decomposition-clustering-ensemble learning approach for solar radiation forecasting. *Solar Energy, 163*, 189–199.

Tabatabaei, S. M., Dick, S., & Xu, W. (2017). Toward non-intrusive load monitoring via multi-label classification. *IEEE Transactions on Smart Grid, 8*(1), 26–40. https://doi.org/10.1109/TSG.2016.2584581

Teng, Z., Lv, J., & Guo, L. (2019). An improved hybrid grey wolf optimization algorithm. *Soft Computing, 23*(15), 6617–6631.

Ting, J. -A., Theodorou, E., & Schaal, S. (2007). A Kalman filter for robust outlier detection. In *2007 IEEE/RSJ International Conference on Intelligent Robots and Systems* (pp. 1514–1519). IEEE.

Ventosa, M., Baıllo, A., Ramos, A., & Rivier, M. (2005). Electricity market modeling trends. *Energy Policy, 33*(7), 897–913.

Wang, B., & Wang, J. (2020). Deep multi-hybrid forecasting system with random EWT extraction and variational learning rate algorithm for crude oil futures. *Expert Systems with Applications, 161*, 113686.

Wang, D., Tan, D., & Liu, L. (2018). Particle swarm optimization algorithm: An overview. *Soft Computing, 22*(2), 387–408.

Wang, J., Lei, C., & Guo, M. (2020). Daily natural gas price forecasting by a weighted hybrid data-driven model. *Journal of Petroleum Science and Engineering, 192*, 107240. https://doi.org/10.1016/j.petrol.2020.107240

Wang, J., Niu, T., Du, P., & Yang, W. (2020). Ensemble probabilistic prediction approach for modeling uncertainty in crude oil price. *Applied Soft Computing, 95*, 106509. https://doi.org/10.1016/j.asoc.2020.106509

Wang, J., Yang, W., Du, P., & Niu, T. (2020c). Outlier-robust hybrid electricity price forecasting model for electricity market management. *Journal of Cleaner Production, 249*, 119318.

Wang, J., Cao, J., Yuan, S., & Cheng, M. (2021). Short-term forecasting of natural gas prices by using a novel hybrid method based on a combination of the CEEMDAN-SE-and the PSO-ALS-optimized GRU network. *Energy, 233*, 121082. https://doi.org/10.1016/j.energy.2021.121082

Wang, Y., Tu, Y., Lu, S., Liu, Q., Ying, Y., & Zhou, X. (2020d). A review of the global energy internet and the suggestions to China. In *IOP Conference Series: Earth and Environmental Science* (Vol. 1 p. 012187). IOP Publishing.

Wang, Y., Zhao, Z., Huo, H., Zhang, D., Zhu, J., & Qi, C. (2020e). Research on key technologies and scenario application of emerging business under the energy internet. In *E3S Web of Conferences*. EDP Sciences.

Wang, Z., Yu, X., Mu, Y., Jia, H., Jiang, Q., & Wang, X. (2021). Peer-to-Peer energy trading strategy for energy balance service provider (EBSP) considering market elasticity in community microgrid. *Applied Energy, 303*, 117596. https://doi.org/10.1016/j.apenergy.2021.117596

Webb, M. G., & Ricketts, M. J. (1980). *The economics of energy*. Macmillan International Higher Education.

Wei, Y. -M., Fan, Y., Han, Z. -Y., & Zou, L. (2010). *Energy economics: Modeling and empirical analysis in China*. Center for Energy and Environmental Policy Research (CEEP).

Wei, Y.-M., & Liao, H. (2016). *Energy economics: Energy efficiency in China*. Springer.

Wu, Y., Wu, Y., Guerrero, J. M., & Vasquez, J. C. (2021). A comprehensive overview of framework for developing sustainable energy internet: From things-based energy network to services-based management system. *Renewable Sustainable Energy Reviews*, 150, 111409.

Xiao, W., Beichang, G., Tianlei, Z., Yingying, L., Zhang, Z., & Fan, Y. (2017). Management system research of large-scale new energy development based on energy internet. In *2017 IEEE Conference on Energy Internet and Energy System Integration (EI2)* (pp. 1–6). IEEE.

Xiao, X., Xiao, Y., Zhang, Y., Qiu, J., Zhang, J., & Yildirim, T. (2021). A fusion data preprocessing method and its application in complex industrial power consumption prediction. *Mechatronics, 77*, 102520. https://doi.org/10.1016/j.mechatronics.2021.102520

Xu, J., Huang, S., Shi, L., & Sharma, S. S. (2021). Trade conflicts and energy firms' market values: Evidence from China. *Energy Economics, 101*, 105434. https://doi.org/10.1016/j.eneco.2021.105434

Xu, Q., Jiang, C., & He, Y. (2015). An exponentially weighted quantile regression via SVM with application to estimating multiperiod VaR. *Statistical Methods and Applications, 25*(2), 285–320. https://doi.org/10.1007/s10260-015-0332-9

Yang, C. C., Soh, C. S., & Yap, V. V. (2014). Comparative study of event detection methods for non-intrusive appliance load monitoring. *Energy Procedia, 61*, 1840–1843. https://doi.org/10.1016/j.egypro.2014.12.225

Yang, J., Rahardja, S., & Fränti, P. (2021a). Mean-shift outlier detection and filtering. *Pattern Recognition, 115*, 107874.

Yang, W., Wang, J., Niu, T., & Du, P. (2020). A novel system for multi-step electricity price forecasting for electricity market management. *Applied Soft Computing, 88*, 106029.

Yang, X.-Y., Liu, Y.-X., & Xing, G.-T. (2021). Analysis of wind farm participation in the frequency regulation market considering wind power uncertainty. *International Journal of Electrical Power and Energy Systems, 130*, 106946. https://doi.org/10.1016/j.ijepes.2021.106946

Yanguas, P. P., Hauenstein, C., & Oei, P.-Y. (2021). The death valley of coal—Modelling COVID-19 recovery scenarios for steam coal markets. *Applied Energy, 288*, 116564. https://doi.org/10.1016/j.apenergy.2021.116564

Yao, Y., Gao, C., Chen, T., Yang, J., & Chen, S. (2021). Distributed electric energy trading model and strategy analysis based on prospect theory. *International Journal of Electrical Power and Energy Systems, 131*, 106865. https://doi.org/10.1016/j.ijepes.2021.106865

Yin, W. J., Ming, Z. F., Wen, T., & Zhang, C. (2022). Photovoltaic curve management using demand response with long and short-term memory. *Energy, 238*, 121729. https://doi.org/10.1016/j.energy.2021.121729

Yu, K. (2019). Energy cooperation under the belt and road initiative: Implications for global energy governance. *The Journal of World Investment and Trade, 20*(2–3), 243–258.

Yu, S., Sun, Y., Niu, X., & Zhao, C. (2010). Energy internet system based on distributed renewable energy generation. *Electric Power Automation Equipment, 5*, 104–108.

Zha, Y., Zhang, T., Huang, Z., Zhang, Y., Liu, B., & Huang, S. J. (2014). Analysis of energy internet key technologies. *Scientia Sinica Informationis, 44*(6), 702–713.

Zhang, C., Zhong, M., Wang, Z., Goddard, N., & Sutton, C. (2018a). Sequence-to-point learning with neural networks for non-intrusive load monitoring. In *Proceedings of the AAAI Conference on Artificial Intelligence* (Vol. 1).

Zhang, H., Yang, Y., Zhang, Y., He, Z., Yuan, W., Yang, Y., Qiu, W., & Li, L. (2021). A combined model based on SSA, neural networks, and LSSVM for short-term electric load and price forecasting. *Neural Computing Applications, 33*(2), 773–788.

Zhang, J.-L., Zhang, Y.-J., & Zhang, L. (2015). A novel hybrid method for crude oil price forecasting. *Energy Economics, 49*, 649–659.

Zhang, L., Zhou, C., Ma, M., & Liu, X. (2004). Solutions of multi-objective optimization problems based on particle swarm optimization. *Journal of Computer Research Development, 7*(41), 7.

Zhang, Y., Fang, C., Ren, H., Lu, Y., & Yang, X. (2018). Research on distributed energy system based on spatial structure analysis. *Energy Procedia, 152*, 9–14. https://doi.org/10.1016/j.egypro.2018.09.051

Zhao, H., & Lifeng, W. (2020). Forecasting the non-renewable energy consumption by an adjacent accumulation grey model. *Journal of Cleaner Production, 275*, 124113.

Zhao, L., Cheng, L., Wan, Y., Zhang, H., & Zhang, Z. (2015). A VAR-SVM model for crude oil price forecasting. *International Journal of Global Energy Issues, 38*(1–3), 126–144.

Zhihong, J., Jian, H., Wenzhou, L., Zhe, C., Ning, L., Siyuan, W., Xiao, Z., & Chang, L. (2018). Energy internet—A new driving force for sustainable urban development. *Energy Procedia, 152*, 1206–1211.

Zhiren, R., Bo, T., Longfeng, W., Hui, L., Yanfei, L., & Haiping, W. (2019). Non-intrusive load identification method based on integrated intelligence strategy. In *2019 25th International Conference on Automation and Computing (ICAC)* (pp. 1–6). IEEE.

Zhou, K., Fu, C., & Yang, S. (2016). Big data driven smart energy management: From big data to big insights. *Renewable Sustainable Energy Reviews, 56*, 215–225.

Zhou, K., Yang, S., & Shao, Z. (2016). Energy internet: The business perspective. *Applied Energy, 178*, 212–222.

Zhu, D., Ma, R., Zhang, S., Liu, J., & Sha, J. (2021). Energy big data application based on energy big data center. In *International Conference on Machine Learning and Big Data Analytics for IoT Security and Privacy* (pp. 511–516). Springer.

Zoha, A., Gluhak, A., Imran, M. A., & Rajasegarar, S. (2012). Non-intrusive load monitoring approaches for disaggregated energy sensing: A survey. *Sensors, 12*(12), 16838–16866.

Big Data Analysis of Energy Economics in Oil Market

2

2.1 Introduction

As industrial blood and raw material, crude oil occupies an important position in the global market of energy economy and is an important energy product. It has not only general commodity attributes, but also financial and political attributes. Crude oil price fluctuations will have an important impact on the economy and politics of the world (Barrales-Ruiz & Mohammed, 2021). The "Statistical Review of World Energy 2021" published by the British Petroleum Company plc stated that, 2020 has become a very special year in the world's energy history (BP), due to the outbreak of Coronavirus Disease 2019 (COVID-19).

- In 2020, global energy consumption fell 4.5% year-on-year, which was the largest drop since the end of World War II.
- In 2020, renewable energy, led by wind and solar energy, will continue to maintain its momentum of development.
- In 2020, fossil energy will still be the absolute hegemon in the field of energy consumption, accounting for as high as 83.1%.
- In 2020, global carbon emissions fell by 6.3% year-on-year.

With the decline in energy consumption, especially the unprecedented decline in oil demand, renewable energy has maintained a steady development momentum. The total installed capacity of solar (photovoltaic) and wind energy has increased by 238GW, which exceeds the new installed capacity in any previous year. The share of wind and solar power in the global power structure hit a record high at the same time, (O'Shaughnessy et al., 2021).

The "Statistical Review of World Energy" report shows that although world oil consumption has fallen sharply year-on-year, oil consumption still tops the list (31.2%). Among them, coal, natural gas, hydropower, renewable energy, and nuclear energy account for 27.2%, 24.7%, 6.9%, 5.7%, and 4.3% respectively.

© Science Press 2022

H. Liu et al., *Big Data in Energy Economics*, Management for Professionals,
https://doi.org/10.1007/978-981-16-8965-9_2

Fossil energy will still be the absolute hegemon in the energy consumption field in 2020, accounting for as high as 83.1%.

Crude oil is the main consumption energy, and its international price fluctuations generally impacts the economic development of various countries. Forecasting international crude oil prices with high accuracy is therefore a significant task (Cortazar et al., 2021; Costa et al., 2021).

According to different forecast horizons, crude oil price forecasts are divided into two categories, including short-term and long-term crude oil price forecasts. For the former, the predictable range is generally from a few days to several months. For the latter, the predictable range can reach one year or even several years. International crude oil pricing is affected by politics, economy, or some unexpected events. These uncertainties make the long-term forecast of crude oil pricing face great challenges (Costa et al., 2021).

To better grasp the dynamic changes of crude oil prices, statistical analysis methods and traditional machine learning methods have been introduced. These methods have achieved certain success in capturing the law of changes in oil prices. Due to the widespread use of machine learning, more and more forecasting models based on machine learning are gradually being applied in the field of energy price time series forecasting. There are many predictive models based on econometric techniques, including Linear Regression (LR) models (Shabri & Samsudin, 2017), Vector AutoRegressive (VAR) models (Allegret et al., 2015; Maghyereh, 2006), AutoRegressive (AR) models (Salisu & Oloko, 2015), Generalized AutoRegressive Conditional Heteroskedasticity (GARCH) models (Wang et al., 2016), AutoRegressive Integrated Moving Average (ARIMA) models (Wang et al., 2020; Zhang & Wang, 2015) and Artificial Neural Networks (ANNs) (Baruník & Malinska, 2016), etc.

Unfortunately, traditional machine learning models and statistical models for predicting crude oil prices ignore the impact of certain historical values on the future values during the fitting process. To a certain extent, these models omit some of the dynamic characteristics of historical crude oil prices, which makes it difficult for them to obtain a completely logical sequence of crude oil price changes (Busari & Lim, 2021).

In subsequent research, deep learning prediction models are promoted for predicting crude oil price (Huang & Deng, 2021). The artificial intelligence technology enables the model to better deal with the non-stationary dynamic characteristics of crude oil time series. The mining of deeper features is able to avoid the impact of certain sudden political and economic events on the accuracy of crude oil price forecasts (Zhang et al., 2021).

In this chapter, two deep learning models, Bi-directional Long Short-Term Memory (BiLSTM) network and Deep Belief Network (DBN) are used to perform big data multi-step forecasting for West Texas Intermediate (WTI) Spot Prices, Europe Brent Spot Prices, and New York Mercantile Exchange (NYMEX) Crude Oil Future prices.

- The experiments compare the 10-day short-term prediction performance of BiLSTM and DBN in crude oil futures and spot prices.
- The BiLSTM and DBN can effectively capture the changing laws of crude oil futures and spot prices.
- The experiments confirm that the predictions of WTI and Brent crude oil prices are useful as a guide to the econometric analysis for the future crude oil market.

2.2 Influencing Factors Analysis of Oil Prices

As an important industrial raw material, crude oil has the attributes of a general commodity. Due to the complexity of crude oil assets, the international crude oil market is a typical complex nonlinear system. The crude oil prices fluctuation is the result of many intertwined and comprehensive effects of risk factors. BP's "Statistical Review of World Energy 2021" gives the historical trend of crude oil (BP 2021).

The statistical review shows that the price fluctuation is affected by the basic factors of supply and demand in the long run. In addition, global emergencies can also cause crude oil prices to fluctuate. For example, the Pennsylvanian oil boom from 1860 to 1880, the Yom Kippur war in 1973, the Iranian revolution in 1979, the Asian financial crisis in 1997, the global financial crisis in 2008, and the COVID-19 in 2020, all impacted greatly on the price. These emergencies have profoundly changed the pattern of crude oil supply and demand (Gong et al., 2021).

Recently, under the influence of COVID-19, various countries have implemented blockades, curfews, and other measures. In addition, the tourism industry has also been severely affected by COVID-19, and the transportation industry related to it has also been severely affected. Air passenger volume, freight volume, railway passenger volume, and freight volume have fallen sharply, which has led to a rapid decline in the demand for transportation fuel. It can be seen that global emergencies will affect the supply and demand relationship of crude oil, and thus affect the price (BP 2021).

Since entering the twenty-first century, with the continuous evolution of the global financial market, crude oil futures have occupied an increasingly prominent position in the international bulk commodity market. The financial attributes of crude oil as a non-renewable resource have gradually become prominent. For a long time, the supply and demand relationship, as a core factor, determines the price of crude oil. The oil trading market is involved. The import and export of crude oil and changes in crude oil reserves in various regions of the world make crude oil prices highly non-linear and non-stationary.

To conduct a more scientific and effective econometric analysis of crude oil prices, the WTI Spot Prices, Europe Brent Spot Prices, and NYMEX's Crude Oil Future prices C1–C4 are used as the input variables of the model. In addition, the deep learning models are utilized to carry out crude oil price forecasting modeling.

2.2.1 Data Description of Crude Oil Prices Influencing Factors

In the study of this chapter, observations of 10 variables for 7647 days from February 8, 1991 to July 23, 2021 are used. Table 2.1 describes in detail the horizon, unit, dataset description and data availability of the influence factors studied.

In the case study of this chapter, a total of 7647 data from February 8, 1991 to July 23, 2021 are used to train and evaluate the BiLSTM and DBN models. This study uses daily data, and its time interval is 1 day. The WTI Spot Prices, Europe Brent Spot Prices and Crude Oil Future prices C1–C4 involved are divided into training sets and test sets. Among them, training data accounted for 70%, 1st to 5353rd is training data, test data accounted for 30%, 5354th to 7647th is test data. Figure 2.1 shows the historical crude oil prices data.

The figure shows that the crude oil price fluctuates up and down around 75, and the variation range is between −40 and 150. Crude oil futures and spot price series data are typical strong non-stationary time series data. The statistical parameters of these data are presented in Table 2.2.

To find out the relationship between these strong non-stationary wave characteristics, a correlation analysis is conducted.

2.2.2 Correlation Analysis of the Factors Affecting Crude Oil Prices

Many factors affect the price of crude oil. WTI crude oil and Brent crude oil are the most important benchmark oils for international crude oil prices. The relationship between WTI crude oil, Brent crude oil, and NYMEX crude oil futures prices is discussed in this chapter. The relationship between two sets of time series is usually measured by the correlation coefficient. The normalization method is used to pre-process WTI Spot Prices, Europe Brent Spot Prices and Crude Oil Future prices C1–C4 to obtain homogeneous data sets before performing correlation analysis.

The original time series data X_i input to the model is standardized to $X_{i(norm)}$. The normalization method is provided as follows:

$$X_{i(norm)} = \frac{X_i - X_{\min}}{X_{\max} - X_{\min}}, \quad i = 1, 2, 3, \ldots, 361 \tag{2.1}$$

where, $X_{\max}$ and $X_{\min}$ represent the maximum and minimum value of the original data set, respectively.

Pearson Correlation Coefficient (PCC) is originally used in the field of statistics to interpret the correlation between two variables. With the development of natural science, PCC has been widely used in many fields. Its expression is provided as

Table 2.1 Data description of crude oil influencing factors

Influence factors	Horizon	Unit	Dataset description	Data availability
WTI Spot Price	Daily	Dollars per barrel	Cushing, OK WTI Spot Price FOB https://www.eia.gov/dnav/pet/hist/RWTCD.htm	Freely available
Europe Brent Spot Price	Daily	Dollars per barrel	Europe Brent Spot Price FOB https://www.eia.gov/dnav/pet/hist/RBRTED.htm	Freely available
Crude Oil Proved Reserves	Annual	Million barrels	U.S. Crude Oil Proved Reserves https://www.eia.gov/dnav/pet/hist/LeafHandler.ashx?n=PET&s=RCRR01NUS_1&f=A	Freely available
Imports of Crude Oil	Weekly	Thousand barrels per day	Weekly U.S. Imports of Crude Oil https://www.eia.gov/dnav/pet/hist/LeafHandler.ashx?n=PET&s=WCRIMUS2&f=W	Freely available
Exports of Crude Oil	Weekly	Thousand barrels per day	Weekly U.S. Exports of Crude Oil https://www.eia.gov/dnav/pet/hist/LeafHandler.ashx?n=PET&s=WCREXUS2&f=W	Freely available
Crude Oil Future Contract 1	Daily	Dollars per barrel	Cushing, OK Crude Oil Future Contract 1, NYMEX https://www.eia.gov/dnav/pet/hist/RCLC1D.htm	Freely available
Crude Oil Future Contract 2	Daily	Dollars per barrel	Cushing, OK Crude Oil Future Contract 2, NYMEX https://www.eia.gov/dnav/pet/hist/RCLC2D.htm	Freely available

(continued)

Table 2.1 (continued)

Influence factors	Horizon	Unit	Dataset description	Data availability
Crude Oil Future Contract 3	Daily	Dollars per barrel	Cushing, OK Crude Oil Future Contract 3, NYMEX https://www.eia.gov/dnav/pet/hist/RCLC3D.htm	Freely available
Crude Oil Future Contract 4	Daily	Dollars per barrel	Cushing, OK Crude Oil Future Contract 4, NYMEX https://www.eia.gov/dnav/pet/hist/RCLC4D.htm	Freely available

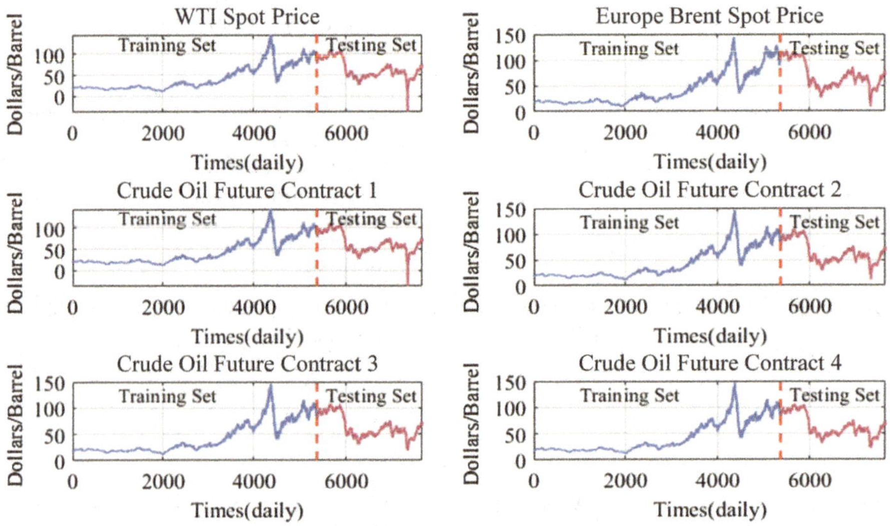

Fig. 2.1 The historical prices data from February 8, 1991 to July 23, 2021

Table 2.2 The statistical parameters of the crude oil dataset

Minimum (%)	Maximum (%)	Mean (%)	Standard deviation	Median	Quantile
−36.9800	145.3100	48.6648	28.9162	44.1500	16.1285
9.1000	143.9500	50.3746	32.3229	44.7100	14.7055
−37.6300	145.2900	48.7262	28.9259	44.4300	16.2470
11.0200	145.8600	48.9116	29.0168	45.1600	16.4500
11.3300	146.1300	49.0176	29.0976	45.8000	16.5685
11.6100	146.4300	49.0609	29.1614	46.2500	16.6670

follows (Benesty et al., 2009):

$$PCC = \frac{\sum_{t=1}^{N}\left((X_t - \overline{X})(Y_t - \overline{Y})\right)}{\sqrt{\left(\sum_{t=1}^{N}(X_t - \overline{X})^2 \sum_{t=1}^{N}(Y_t - \overline{Y})^2\right)}} \tag{2.2}$$

where, X and Y represent different time series data matrices that are input to the model $-1 \le PCC \le 1$. If $PCC = 0$, X and Y do not have a linear correlation. If $PCC > 0$, X and Y have a positive linear correlation. If $PCC < 0$, X and Y have a negative linear correlation. The closer the PCC is to ± 1, the higher the correlation between the two variables.

In the research of this chapter, the predictive variables that have a significant relationship with crude oil price fluctuations are analyzed. The Pearson correlation analysis is adopted to evaluate the correlation between crude oil futures prices and spot prices. Figure 2.2 shows the correlation coefficients between six variables including WTI crude oil prices, and the values range from 0 to 1.

Figure 2.2 shows that the correlation coefficients among the six variables of WTI Spot Prices, Europe Brent Spot Prices, and Crude Oil Future prices C1–C4 are all greater than 0.95, which indicates a significant correlation. The BiLSTM and DBN deep learning models are used to predict the fluctuation trend of all the six variables.

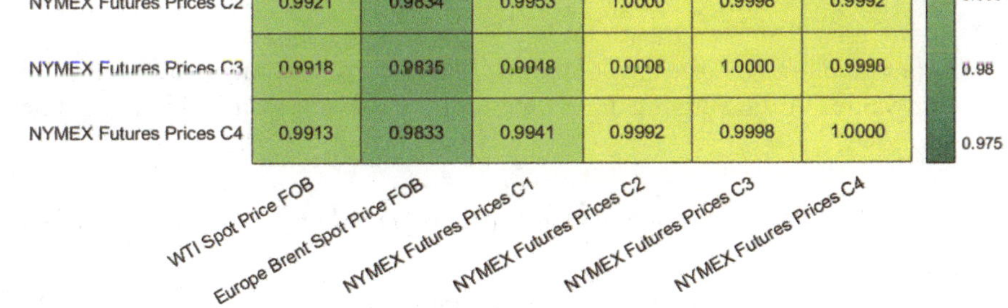

Fig. 2.2 Correlation analysis of crude oil futures and spot prices

2.3 Big Data Forecasting of Oil Prices

2.3.1 Base Forecasting Models

The performance of the BiLSTM and DBN for predicting the crude oil price are compared in this chapter. The algorithm theories involved in this process are provided as follows.

2.3.1.1 Bi-Directional Long Short-Term Memory

LSTM cell consists of three gates, namely input gate i_t, forget gate f_t, and output gate o_t. For each LSTM cell, the calculation can be organized as follows (Hochreiter & Schmidhuber, 1997):

(a) Forget gate f_t. By calculating the previous output and the current input, the output value of the forget gate can be obtained. The expression is provided as follows:

$$f_t = \sigma\left(W_f \times \left[h_{t-1}, x_t\right] + b_f\right) \tag{2.3}$$

(b) Input gate i_t. The previous output and the current input are input into the input gate, and the output of the input gate and the candidate state can be calculated. The expression is given as follows:

$$i_t = \sigma\left(W_i \times \left[h_{t-1}, x_t\right] + b_i\right) \tag{2.4}$$

$$\widehat{C}_t = \tanh\left(W_C \times \left[h_{t-1}, x_t\right] + b_C\right) \tag{2.5}$$

(c) Update the current cell state. The current cell state is obtained by the product of the output of the forget gate and the previous cell state plus the product of the two outputs of the input gate. The expression is provided as follows:

$$C_t = f_t \times C_{t-1} + i_t \times \widehat{C}_t \tag{2.6}$$

(d) Output gate o_t. The previous output h_{t-1} and current input x_t are used as the input values of the output gate at time t, and the expression is finally obtained as follows:

$$o_t = \sigma\left(W_o \times \left[h_{t-1}, x_t\right] + b_o\right) \tag{2.7}$$

$$h_t = o_t \times \tanh(C_t) \tag{2.8}$$

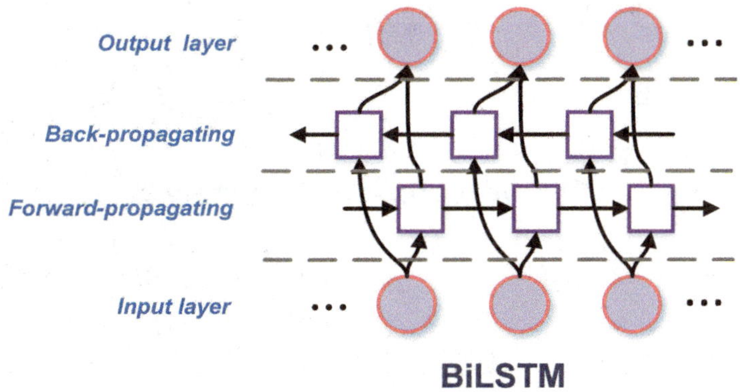

Fig. 2.3 Model structure of BiLSTM neural network

where σ is sigmoid activation function, h_{t-1} is the previous hidden state, W_o is the weight, x_t is the current input vector, b is the bias vector, i_t is the input gate, $\hat{C}_t$ is the candidate state, tanh() is the hyperbolic tangent activation function, C_t is the state at time t, o_t is the output gate.

BiLSTM is a combination of two LSTMs in different directions. Each direction has independent parameters. It can fit forward and backward data and connect the prediction results in series (Wang et al., 2017). Compared with the one-way fitting of LSTM, BiLSTM adds inverse LSTM, which allows it to capture more information (Liu et al., 2021). The diagram of BiLSTM is presented in Fig. 2.3.

BiLSTM forward calculation hidden vector is denoted by $\vec{h}$, reverse calculation hidden vector is denoted by $\overleftarrow{h}$, then the expression of the final output y_t is provided as follows:

$$h_t = \alpha\vec{h} + \beta\overleftarrow{h} \tag{2.9}$$

$$y_t = \sigma(h_t) \tag{2.10}$$

where α and β are constants, and $\alpha + \beta = 1$.

2.3.1.2 Deep Belief Network

By stacking several Restricted Boltzmann Machines (RBM), a DBN model can be obtained. A back propagation neural network is then used for fine-tuning. The structure of DBN is presented in Fig. 2.4 (Liu et al., 2021).

DBN can be obtained by stacking the hidden layer of the upper RBM as the visible layer of the next RBM. The data is input from the visible layer of the lowest RBM for bottom-up unsupervised learning. The top RBM outputs the last feature value. Then these features are input into Back Propagation Neural Network (BPNN), and the RBM fine-tuning layer by layer is realized through the back propagation of BPNN (Hua et al., 2015).

Fig. 2.4 Model structure of DBN neural network

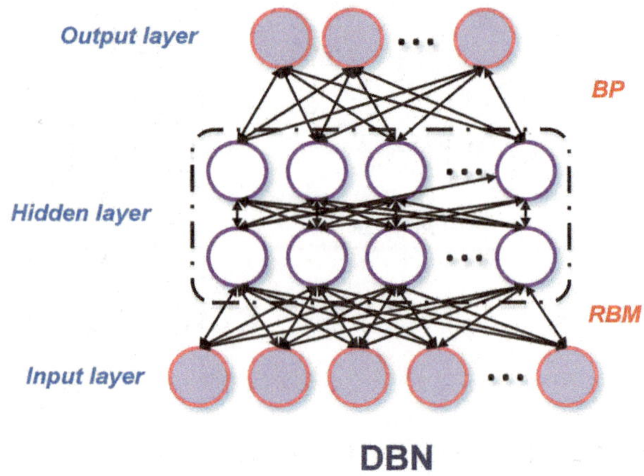

For RBM, the visible layer of neurons is denoted as $v = \{v_1, v_2, \ldots, v_n\}$, the hidden layer neuron is expressed as $h = h\{h_1, h_2, \ldots, h_m\}$, the energy function of RBM is provided and expressed as follows:

$$E(v, h) = -\sum_{i=1}^{n} a_i v_i - \sum_{j=1}^{m} b_j h_j - \sum_{i=1}^{n}\sum_{j=1}^{m} w_{ij} v_i h_j \tag{2.11}$$

where a_i is the bias in visible layer, b_j is the bias in hidden layer, and w_{ij} is the connection weight.

The joint probability distribution function between the visible and hidden layer is expressed as follows:

$$P(v, h) = \frac{1}{\sum_{i=1}^{m} \sum_{j=1}^{n} e^{-E(v,h)}} e^{E(v,h)} \tag{2.12}$$

The conditional probability distribution functions of visible layer v and hidden layer h are expressed as follows:

$$p(h_j = 1|v) = sigmoid\left(b_j + \sum_{i=1}^{n} w_{ij} v_i\right) \tag{2.13}$$

$$p(v_i = 1|h) = sigmoid\left(a_i + \sum_{j=1}^{m} w_{ij} h_j\right) \tag{2.14}$$

Through the contrastive divergence algorithm, the update of its weights and biases can be expressed as follows:

$$\Delta w_{ij} = \eta\left(\langle v_i h_j \rangle_{data} - \langle v_i h_j \rangle_{recon}\right) \tag{2.15}$$

$$\Delta a_i = \eta\left(\langle v_i \rangle_{data} - \langle v_i \rangle_{recon}\right) \tag{2.16}$$

$$\Delta b_j = \eta\left(\langle h_j \rangle_{data} - \langle h_j \rangle_{recon}\right) \tag{2.17}$$

where η is the learning rate.

Through Eqs. (2.15) to (2.17), the update and optimization of RBM can be completed.

2.3.2 Crude Oil Futures and Spot Prices Time Series Forecasting Model

To reflect the entire crude oil price prediction modeling process, its technical framework diagram is provided as shown in Fig. 2.5.

2.3.2.1 The Modeling Steps of the Proposed Model

The modeling steps of the proposed model are provided as follows:

(a) Correlation analysis between crude oil futures price and spot price. The PCC is adopted to evaluate the correlation of the six variables of WTI Spot Prices, Europe Brent Spot Prices, and Crude Oil Future prices C1–C4 in the US Energy Information Administration data set. Highly correlated factors are screened out as the model input for the next stage.

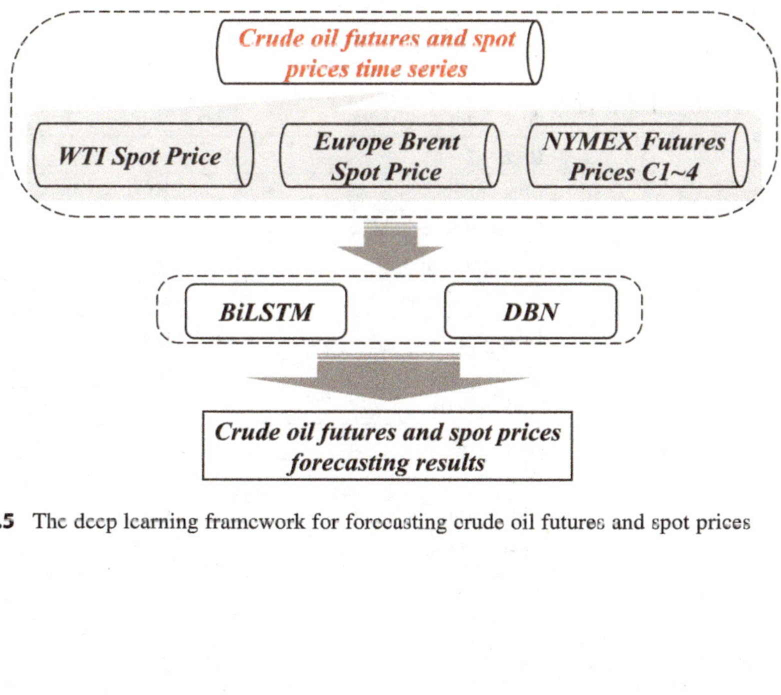

Fig. 2.5 The deep learning framework for forecasting crude oil futures and spot prices

(b) Data set division. The data set is divided into training set (70%) and test set (30%). The 1st to 5353rd data samples are in the training set, and the 5354th to 7647th data samples are in the test data.
(c) Model input. The time series data of these six variables are respectively input into the BiLSTM and DBN deep learning models. The training set is used during the model training process.
(d) Model output. The test data is then input into the models, and the performance metrics presented in Sect. 2.3.3 are adopted to evaluate the performance.

2.3.2.2 The Algorithm Parameters Settings of the Models

The algorithm parameters of the models are provided in Table 2.3.

2.3.3 Performance Metrics

Three evaluation metrics are used to evaluate the errors of models, including Mean Absolute Error (MAE), Mean Absolute Percentage Error (MAPE), and Root Mean Square Error (RMSE). The smaller the metrics, the higher the accuracy. The equations of the MAE, MAPE, and RMSE are given as follows:

$$MAE = \left(\sum_{t=1}^{N_Y} \left|Y_t - \widehat{Y}_t\right|\right) / N_Y \tag{2.18}$$

$$RMSE = \sqrt{\frac{1}{N_Y} \sum_{t=1}^{N_Y} \left(Y_t - \widehat{Y}_t\right)^2} \tag{2.19}$$

Table 2.3 The parameters of the models

Algorithm parameter	Value
BiLSTM	
Max epochs	30
Gradient threshold	1
L2 regularization	0
BiLSTM-layer	300
Learning rate	0.01
DBN	
Max epochs	30
Batch size	1
Activation function	sigm
Momentum factor	0.0
Learning rate	0.01

$$MAPE = \left(\sum_{t=1}^{N_Y} \left| (Y_t - \widehat{Y}_t)/Y_t \right| \right) / N_Y \tag{2.20}$$

where $\widehat{Y}$ is the prediction, Y is the real data, and N_Y is the number of the data.

2.3.4 Results and Discussions

In the experiments and discussions in this chapter, to find a suitable basic predictor, the performance of BiLSTM and DBN deep learning models in crude oil spot price and futures price prediction are compared. The multi-step forecasting results are listed in Table 2.4, where the smallest errors for different forecasting steps are marked in bold. Figure 2.6 shows the 10-step forecasting results. In the parallel graph of 1-step forecasting results, the forecasting value and actual value are close. For the convenience of observation, the displayed graph is selected as 10-step forecasting results. Figure 2.7 shows the stacking diagrams of 1-step to 10-step error indices under different prices series data.

Summarizing the experimental results of Table 2.4 and Figs. 2.6 and 2.7, the following analysis and conclusions can be obtained:

(a) Both BiLSTM and DBN can better capture the trends of WTI Spot Prices, Europe Brent Spot Prices, and Crude Oil Future prices C1–C4. Figure 2.6 shows that BiLSTM and DBN can fit the changing trend of the original crude oil price time series with nice performance. Even when the fluctuation amplitude is large, the fit of the prediction curve is very good. This phenomenon may be because the deep learning network has deeper structures so that the nonlinear fluctuation characteristics can be modeled well, thereby solving complex data-driven problems.

(b) In most cases, DBN performs better than BiLSTM. When the steps number is 1-step, all prediction results of DBN are better than BiLSTM. For 1-step forecasting results of the WTI spot price, Europe Brent Spot Price, Crude Oil Future Contract 1-4, the MAEs of BiLSTM are 1.4337, 2.5305, 1.3634, 1.2757, 2.1137, and 1.5376 (dollars per barrel). For the 1-step forecasting results of the WTI spot price, Europe Brent Spot Price, Crude Oil Future Contract 1-4, the MAEs of DBN are 1.0810, 1.0057, 1.0315, 0.9110, 0.9385, and 0.8912 (dollars per barrel). The error stacking chart shown in Fig. 2.7 demonstrates that when predicting the data of Europe Brent Spot Price and Crude Oil Future Contract 3-4, the MAEs, MAPEs, and RMSEs of DBN are significantly smaller than BiLSTM. This phenomenon may be because BiLSTM has a certain gradient disappearance problem, which makes BiLSTM a bit tricky when dealing with larger-scale time series data. DBN realizes bottom-up unsupervised training by stacking multiple RBMs, which can efficiently

Table 2.4 Multi-step forecasting results of crude oil futures and spot prices

Lead time	BiLSTM			DBN		
	MAE	RMSE	MAPE	MAE	RMSE	MAPE
	WTI Spot Price			WTI Spot Price		
1-day	1.4337	2.2685	2.7152	**1.0810**	**1.9316**	**2.1744**
5-day	2.6522	3.7684	4.9423	**2.6295**	**3.5028**	**4.9008**
10-day	**4.2354**	5.8317	**7.7799**	4.3592	**5.3940**	7.9282
	Europe Brent Spot Price			Europe Brent Spot Price		
1-day	2.5305	3.2417	5.2169	**1.0057**	**1.3487**	**1.7663**
5-day	4.9083	6.1796	10.1559	**2.4026**	**3.0507**	**4.2427**
10-day	7.7661	9.5905	15.9294	**3.7946**	**4.7147**	**6.6916**
	Crude Oil Future Contract 1			Crude Oil Future Contract 1		
1-day	1.3634	2.1674	2.5803	**1.0315**	**1.9423**	**2.0857**
5-day	**2.3682**	**3.3543**	**4.4876**	2.4392	3.4847	4.6860
10-day	**3.5027**	**4.8042**	**6.6858**	3.9133	5.3907	7.4232
	Crude Oil Future Contract 2			Crude Oil Future Contract 1		
1-day	1.2757	1.6918	2.3131	**0.9110**	**1.2682**	**1.6847**
5-day	2.1999	2.8658	3.9911	**2.0203**	**2.6698**	**3.7099**
10-day	3.2151	4.1704	5.8596	**2.9761**	**3.9648**	**5.5230**
	Crude Oil Future Contract 3			Crude Oil Future Contract 3		
1-day	2.1137	2.6632	4.0650	**0.9385**	**1.2947**	**1.6596**
5-day	4.0113	4.9211	7.7341	**2.5401**	**3.2816**	**4.3439**
10-day	6.4089	7.7011	12.3061	**4.4305**	**5.6539**	**7.4463**
	Crude Oil Future Contract 4			Crude Oil Future Contract 4		
1-day	1.5376	2.0097	2.8312	**0.8912**	**1.2061**	**1.5760**
5-day	2.7974	3.5788	5.2217	**2.1076**	**2.6621**	**3.7468**
10-day	4.3907	5.5205	8.2673	**3.4028**	**4.1267**	**6.1210**

Remark The units of all other performance indicators except MAPE in the table are 'Dollars per barrel', and the unit of MAPE is '%'

learn the hidden layer. At the same time, the supervised tuning process overcomes the difficult problem of traditional deep neural network training, so that this situation can be improved.

(c) Even when the prediction steps number is big, DBN can achieve satisfactory prediction accuracy. For the 10-step forecasting results of the WTI spot price, Europe Brent Spot Price, Crude Oil Future Contract 1-4, the MAEs of DBN are 4.3592, 3.7946, 3.9133, 2.9761, 4.4305, and 3.4028 (dollars per barrel). This phenomenon may be because, different from BPNN, DBN is not easy to fall into local optimum under the influence of randomly initialized weights. DBN's efficient hidden layer learning and supervised tuning process make it have excellent performance on prediction tasks.

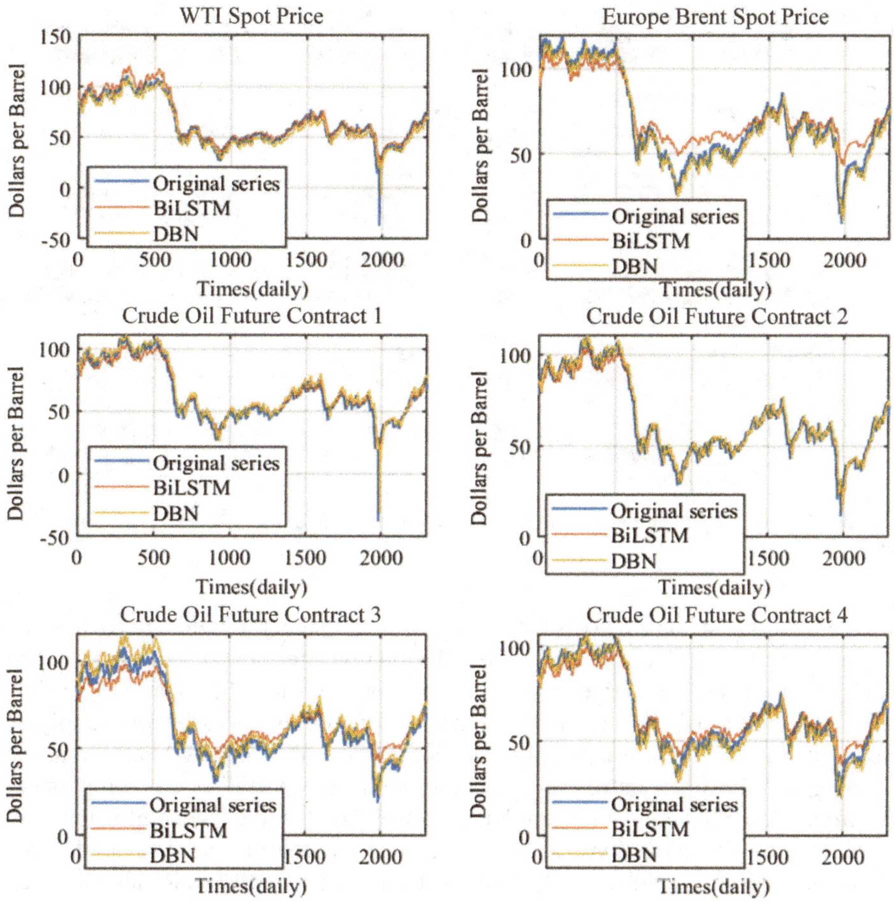

Fig. 2.6 The 10-step forecasting results of crude oil futures and spot prices data

2.4 Econometric Analysis of Oil Prices

2.4.1 Energy Economic Analysis of Crude Oil Market

Crude oil, as a vital strategic resource, is closely related to developing the world economy and reshaping the international structure. The supply-side and demand-side analyses for crude oil market are given as follows.

2.4.1.1 International Crude Oil Price Fluctuations on the Supply Side

(a) Organization of the Petroleum Exporting Countries (OPEC) dominates the global crude oil supply market.

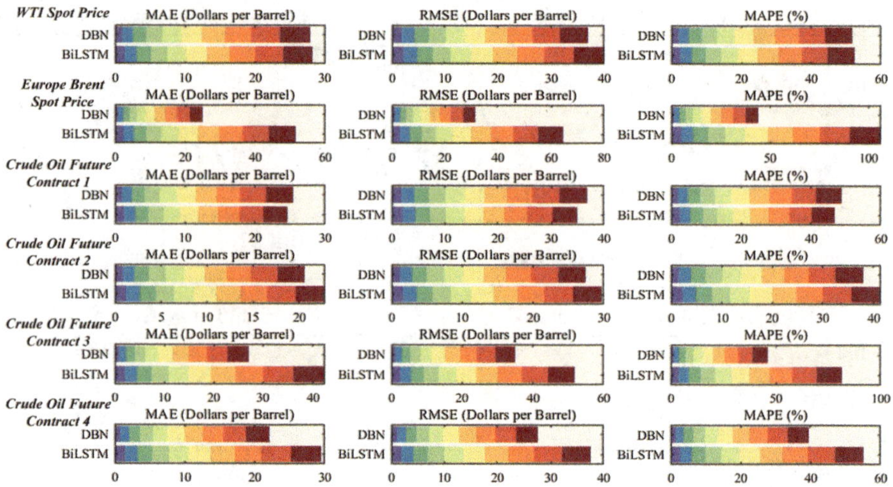

Fig. 2.7 The stacking diagrams of 1-step to 10-step error indices under different prices time series data

Affected by the sudden COVID-19, the global demand for liquid fuels has not yet fully recovered. The current crude oil situation presents an oversupply. The rebalancing of the oil market is determined by the game between the world's major oil producers. The redundant capacity of "OPEC+" crude oil supply is about900 × 10^4 bbl/d, and it has absolute control over the global crude oil market. To safeguard the interests of oil-producing countries, "OPEC+" may implement the production restriction plan until the end of 2022. OPEC generally tends to lag the release of production capacity behind the growth of demand. In addition, production growth in the United States and other non-OPEC countries has been limited by COVID-19, energy policies, extreme weather, and operating conditions of regional oil and gas companies. The international crude oil market will continue to maintain a situation where supply capacity exceeds demand. The OPEC still controls the main redundant production capacity of global crude oil and dominates the international crude oil market (Markit, 2021; WM 2021).

(b) Exploration and development investment is hovering at a low level, and growth in oil production capacity has slowed significantly.

Since 2014, due to the impact of low oil prices, oil and gas companies around the world have drastically reduced their upstream investment. In 2014, the upstream investment was about 780 billion U.S. dollars, and in 2016 it fell to 430 billion U.S. dollars, a decrease of about 45%. From 2016 to 2019, upstream investment rebounded slightly to 480 billion U.S. dollars. In 2020, the upstream investment will drop by about 30% compared with 2019, to about 340 billion U.S. dollars, which is still less than 50% of the 2014 peak. The upstream investment in 2021 is expected to be slightly higher than that in 2020, at about 350 billion U.S. dollars (IEA 2021).

In recent years, large international oil companies have successively proposed carbon emission reduction and carbon neutral goals. These companies have set an energy transition path to reduce the upstream oil and gas industry. Enterprises expand investment in renewable energy and transform into integrated energy companies. As the upstream investment in oil and gas continues to linger at a low level, the energy transformation happened in oil industry has accelerated the loss of upstream investment, the growth rate of oil production capacity has slowed significantly. It is expected that the future may cause a shortage of global crude oil supply.

(c) The United States broke the global crude oil supply pattern, becoming the world's largest producer.

Data from the U.S. Energy Information Administration (EIA) and OPEC show that, the U.S. has continuously expanded its crude oil production and has become the world's largest producer in recent years. At present, the United States has formed an important international crude oil supply force with Saudi Arabia and Russia, and its influence on the international crude oil prices trend is gradually increasing. The United States has become the world's largest producer for the following three reasons. First, the U.S. shale oil production capacity has been greatly increased, which has led to continuous improvement in shale oil production efficiency and continuous rise in drilling data. Second, the United States has become increasingly prominent in the global crude oil supply structure, and its control over the pricing power of the crude oil market has been significantly enhanced. Third, the United States has intervened in the crude oil production of countries in the Middle East by exerting political pressure and other means, which has also affected international crude oil pricing to a certain extent.

2.4.1.2 International Crude Oil Price Fluctuations on the Demand Side

(a) The crude oil market is hit by the rapid development of clean energy.

The American Energy Information Association stated that the global energy consumption structure will undergo significant changes. In the next ten or even thirty years, the consumption of oil will not change significantly, but the consumption proportion will decrease. The share of clean and renewable energy will rise sharply, thus impacting the crude oil market.

(b) With repeated outbreaks of COVID-19, the mutant virus has wreaked havoc across the globe.

The global oil industry has undoubtedly been severely affected by the outbreak and continuation of COVID-19 (Sharif et al., 2020). Due to the sudden drop and rebound of oil demand, global oil inventory prices fluctuated violently. Earlier, professionals believed that global oil demand would return to pre-epidemic levels within a year or two. But now it seems that this view is too optimistic. The policymakers' views on energy security, awareness of the urgency of climate change, policies, technology, and investment for energy will also be affected by COVID-19.

Statistics from the World Health Organization (WHO) show that the second round of COVID-19 ravaged the world in early 2021, with the epidemic center mainly in Europe and the Americas. Under the influence of the mutant virus, the world has been hit hard by the third round of the more violent epidemic starting from mid-to-late February 2021. The epidemic centers are mainly in South Asia, Southeast Asia, Latin America, and Europe. From May to June, due to the widespread vaccination, the overall development of the global epidemic slowed down. At the end of June, the delta novel coronavirus, which can cause a decline in the protective effect of vaccines, began to raging around the world. It has led to the restart of blockade measures in many places, and the development of the epidemic is once again facing severe challenges. Currently, COVID-19 is evolving in the direction of combating vaccines, which brings greater uncertainty to the development of the epidemic.

(c) The global economy has recovered unevenly, and the political and economic structure and development trends have changed significantly.

The great uncertainty brought about by the development of the epidemic has affected the progress of the global economic recovery, leading to obvious regional differences (Ozili & Arun, 2020). In the first half of 2021, Asia and North America will recover their economies at the fastest speed, with China and the United States driving global economic growth. In contrast, the recovery in Europe, the Middle East, Africa and Latin America has been relatively slow. It is expected that the global economy will generally improve in the future, but it will still be affected by the epidemic, with greater uncertainty.

Developing countries will gradually recover from COVID-19 with difficulty. The World Bank believes that the world's major economies will quickly recover from the epidemic in 2021. The economic output of many developing countries will not return to pre-epidemic levels until 2022 or 2023, and some vulnerable economies will be long-term facing economic difficulties. The global economy will continue to be driven by China and the United States. The trend of Sino-US relations will have a major impact on the global political and economic structure and development trends.

(d) The demand for crude oil continues to grow linearly, and the peak time may accelerate.

According to the research results of IHS Markit and the International Energy Agency (IEA), the global demand for liquid fuels (including petroleum, condensate, biodiesel and processing gains, etc.) is expected to gradually recover. The main driving force of the recovery comes from the travel and consumption of people in China, the United States and European countries. The recurrence of the epidemic will reduce the demand in Southeast Asia, South Asia, Latin America and other regions, but after the epidemic is over, there will be a sharp increase in demand, and even retaliatory growth (Jefferson, 2020).

Due to the profound changes in human production and lifestyles caused by COVID-19, coupled with the impact of carbon emission reduction and carbon peaking targets proposed by various countries, the growth rate of global liquid fuel demand may further slowdown. Demand for liquid fuels could peak by 2025, up from the agency's consensus forecast of 2030–2035.

2.4.2 Big Data Prediction Technology

Crude oil is the blood of the national economy and an important strategic resource. Predictive modeling of its price has also become a popular topic in energy economy. There are many popular traditional mathematical statistics models and econometric models, including ARIMA (Zhao & Wang, 2014), Random Walk (RW) (Yu et al., 2017), GARCH (Ahmed & Shabri, 2014), VAR (Ramyar & Kianfar, 2019). These traditional models perform well in dealing with linear relationships, but they are difficult to deal with the complex nonlinear connotations of financial time series. Subsequently, the rapid development of artificial intelligence technology gave birth to many neural networks. The models based on deep learning perform even better in the task of strong non-stationary time series prediction.

This chapter selects samples of crude oil futures prices and spot prices during the 30-year period from February 1991 to July 2021 for research. The crude oil price prediction performance of BiLSTM and DBN are compared. The experimental results show that the crude oil prices output by the two data-driven deep learning models can reproduce the curve fluctuation trend of the original crude oil prices with high details. The effectiveness of the BiLSTM and DBN models is verified using the data set of the U.S. Energy Information Administration, and the model gives full play to its characteristics of using historical data for memory. MAE, RMSE, MAPE and other mainstream error evaluation index data show the feasibility of the model in processing nonlinear financial time series data.

The research work in this chapter examines the influence of variables such as WTI Spot Prices, Europe Brent Spot Prices and Crude Oil Future prices C1–C4 on the forecast of crude oil price fluctuations. The PCC shows that there is a significant correlation between these factors, and its coefficient value exceeds 0.95.

This work provides a machine learning modeling basis for policy making in the international energy market.

2.4.3 Policies and Recommendations

(a) The government correctly studies and judges the impact of oil prices on the economy, and introduces control measures in due course.

High oil prices will negatively impact the macroeconomics of some countries in the world, including increased expenditure on imported crude oil, increased oil reserve costs, increased production costs of enterprises, increased inflationary pressures, and decreased Gross Domestic Product (GDP). It is recommended that government departments correctly study and judge the impact of oil prices on the macroeconomics, and introduce relevant industrial control measures in due course.

(b) The enterprises continue to promote cost reduction and efficiency enhancement, and explore new profit growth points.

High oil prices will push up production costs and have a negative impact on non-oil companies. The most affected companies are mainly concentrated in chemical, aviation, shipping, logistics, tourism and other industries and related industrial chains. It is recommended that relevant non-oil companies stick to the bottom-line thinking, continue to reduce costs and increase efficiency. Increase support for new businesses and new formats, and gradually cultivate new profit growth methods.

(c) The structure of overseas oil and gas assets should be optimized and the transformation of petroleum energy should be promoted.

High oil prices have more advantages than disadvantages for oil companies as a whole, while low oil prices have brought greater operating risks to oil companies. It is recommended that oil companies grasp the relatively high oil price window, increase upstream reserves and production, and strengthen the optimization of overseas high-cost oil and gas asset structures.

The enterprise professional service sector should expand market growth and cultivate new growth methods. The enterprise refining and chemical sector has accelerated the elimination of backward refining and chemical production capacity, grasped the rhythm and trend of oil price changes, fast production and sales, and kept low inventories. Enterprises actively promote energy transformation, accelerate the layout of the renewable energy industry, and continue to increase capital investment in new energy.

2.5 Conclusions

This chapter analyzes the international crude oil market with examples of machine learning modeling, and the following conclusions can be drawn:

(a) Crude oil prices have an important impact on the international market. Crude oil is the primary energy fuel around the world. Its global prices fluctuation will affect the economic development of various countries, involving commodity prices, trade and investment, production and consumption, import and export, and other aspects. Exploring the mechanism and reasons of crude oil price fluctuations is significant to the maintenance of international economic stability.
(b) Sudden global events and monetary policy can significantly impact crude oil prices. The repeated attacks of COVID-19 have brought severe shocks to the global oil industry, and the uneven recovery of the global economy has largely affected the pricing of crude oil. The sharp decline and rebound of oil demand caused sharp fluctuations in crude oil prices. Developing countries will gradually recover from COVID-19 with difficulty. The global economy will continue to be driven by the policies of China and the United States. The trend of Sino-US relations will have a major impact on the global political and economic structure and development trends.
(c) Changes in the supply and demand relationship affect the price of crude oil almost decisively. OPEC's crude oil dominance allows it to have absolute control over the global crude oil market, and the game of oil-producing countries determines the supply of crude oil in the market. The "OPEC+" production restriction plan has affected the supply side of crude oil, while the development of COVID-19 and new clean energy has affected the demand side. The fluctuations in supply and demand dominate the changes in crude oil prices.
(d) Big data forecasting technology has guiding significance for energy policy regulation. The influencing factors of oil price fluctuations are complex, and the non-linear, uncertain, and dynamic characteristics of the fluctuations are significant. Accurately predicting the fluctuations of crude oil prices is a challenging task. Oil price volatility is of great significance for calculating the value at risk and for investors to conduct risk management. At the same time, accurate oil price fluctuation forecasts can help government officials specify appropriate policies to avoid the impact of sudden changes in oil prices on economic development and social stability to a certain extent.
(e) The deep learning model can better capture the fluctuation law of crude oil price data and track the ups and downs. Even for the peaks and valleys with large fluctuations, the prediction curve fits well. The number of hidden layers in the deep learning network enables it to learn non-linear fluctuation characteristics well, thereby solving complex data-driven problems. Compared with traditional mathematical and statistical models and econometric models, the better memory mechanism of deep learning models can help it better meet the forecasting requirements of the financial market.
(f) Different forecasting models have different performances on different prediction steps and different data sets. BiLSTM predicts WTI Spot Price and Crude Oil Future Contract 1 better than DBN, when the number of prediction steps is high. BiLSTM's two-way combination structure enables it to capture more information contained in sequence data. When DBN predicts Europe Brent

Spot Prices and Crude Oil Future prices C2–C4, its prediction effect is significantly better than BiLSTM. The reason for this phenomenon may be that BiLSTM has a certain gradient disappearance problem, which makes BiLSTM a bit tricky when dealing with larger-scale time series data. DBN realizes bottom-up unsupervised training by stacking multiple RBMs, which can efficiently learn the hidden layer. At the same time, the supervised tuning process overcomes the difficulty of traditional deep neural network training, thereby improving the situation.

References

Ahmed, R. A., & Shabri, A. B. (2014). Daily crude oil price forecasting model using arima, generalized autoregressive conditional heteroscedastic and support vector machines. *American Journal of Applied Sciences, 11*(3), 425.

Allegret, J.-P., Mignon, V., & Sallenave, A. (2015). Oil price shocks and global imbalances: Lessons from a model with trade and financial interdependencies. *Economic Modelling, 49*, 232–247.

Barrales-Ruiz, J., & Mohammed, M. (2021). Financial regimes and oil prices. *Resources Policy, 74*, 102299. https://doi.org/10.1016/j.resourpol.2021.102299

Baruník, J., & Malinska, B. (2016). Forecasting the term structure of crude oil futures prices with neural networks. *Applied Energy, 164*, 366–379.

Benesty, J., Chen, J., Huang, Y., & Cohen, I. (2009). Pearson correlation coefficient. In *Noise reduction in speech processing* (pp. 1–4). Springer.

BP British Petroleum Company. (2021). Statistical Review of World Energy 2021. https://www.bp.com/en/global/corporate/energy-economics/statistical-review-of-world-energy.html. Accessed 2 September 2021.

Busari, G. A., & Lim, D. H. (2021). Crude oil price prediction: A comparison between AdaBoost-LSTM and AdaBoost-GRU for improving forecasting performance. *Computers & Chemical Engineering*, 107513. https://doi.org/10.1016/j.compchemeng.2021.107513

Cortazar, G., Ortega, H., & Valencia, C. (2021). How good are analyst forecasts of oil prices? *Energy Economics, 102*, 105500. https://doi.org/10.1016/j.eneco.2021.105500

Costa, A. B. R., Ferreira, P. C. G., Gaglianone, W. P., Guillén, O. T. C., Issler, J. V., & Lin, Y. (2021). Machine learning and oil price point and density forecasting. *Energy Economics, 102*, 105494. https://doi.org/10.1016/j.eneco.2021.105494

Gong, X.-L., Liu, J.-M., Xiong, X., & Zhang, W. (2021). The dynamic effects of international oil price shocks on economic fluctuation. *Resources Policy, 74*, 102304. https://doi.org/10.1016/j.resourpol.2021.102304

Hochreiter, S., & Schmidhuber, J. (1997). Long short-term memory. *Neural Computation, 9*(8), 1735–1780.

Hua, Y., Guo, J., & Zhao, H. (2015). Deep belief networks and deep learning. In *Proceedings of 2015 International Conference on Intelligent Computing and Internet of Things* (pp. 1–4). IEEE.

Huang, Y., & Deng, Y. (2021). A new crude oil price forecasting model based on variational mode decomposition. *Knowledge-Based Systems, 213*, 106669. https://doi.org/10.1016/j.knosys.2020.106669

IEA International Energy Agency. (2021). Oil 2021: Analysis and forecast to 2026 [EB/OL].

Jefferson, M. (2020). A crude future? COVID-19s challenges for oil demand, supply and prices. *Energy Research Social Science, 68*, 101669.

Liu, H., Yang, R., Wang, T., & Zhang, L. (2021). A hybrid neural network model for short-term wind speed forecasting based on decomposition, multi-learner ensemble, and adaptive multiple error corrections. *Renewable Energy, 165*, 573–594. https://doi.org/10.1016/j.renene.2020.11.002

Maghyereh, A. (2006). Oil price shocks and emerging stock markets: A generalized VAR approach. In *Global stock markets and portfolio management* (pp. 55–68). Springer.

Markit, I. H. S. (2021). *Global crude oil markets price long-term outlook fundamentals: 2nd quarter 2021*. IHS Markit.

O'Shaughnessy, E., Heeter, J., Shah, C., & Koebrich, S. (2021). Corporate acceleration of the renewable energy transition and implications for electric grids. *Renewable and Sustainable Energy Reviews, 146*, 111160. https://doi.org/10.1016/j.rser.2021.111160

Ozili, P. K, & Arun, T. (2020). Spillover of COVID-19: Impact on the global economy. Available at SSRN 3562570.

Ramyar, S., & Kianfar, F. (2019). Forecasting crude oil prices: A comparison between artificial neural networks and vector autoregressive models. *Computational Economics, 53*(2), 743–761.

Salisu, A. A., & Oloko, T. F. (2015). Modeling oil price–US stock nexus: A VARMA–BEKK–AGARCH approach. *Energy Economics, 50*, 1–12.

Shabri, A., & Samsudin, R. (2017). Hybridizing wavelet and multiple linear regression model for crude oil price forecasting. In *Proceedings of the International Conference on Computing, Mathematics and Statistics (iCMS 2015)* (pp. 157–164). Springer.

Sharif, A., Aloui, C., & Yarovaya, L. (2020). COVID-19 pandemic, oil prices, stock market, geopolitical risk and policy uncertainty nexus in the US economy: Fresh evidence from the wavelet-based approach. *International Review of Financial Analysis, 70*, 101496.

Wang, J., Hu, F., & Li, L. (2017). Deep bi-directional long short-term memory model for short-term traffic flow prediction. In *International Conference on Neural Information Processing* (pp. 306–316). Springer.

Wang, J., Zhou, H., Hong, T., Li, X., & Wang, S. (2020). A multi-granularity heterogeneous combination approach to crude oil price forecasting. *Energy Economics, 91*, 104790. https://doi.org/10.1016/j.eneco.2020.104790

Wang, Y., Wu, C., & Yang, L. (2016). Forecasting crude oil market volatility: A Markov switching multifractal volatility approach. *International Journal of Forecasting, 32*(1), 1–9.

WM Wood Mackenzie. (2021). *Macro oils long-term outlook h1 2021*. Wood Mackenzie.

Yu, L., Zhang, X., & Wang, S. (2017). Assessing potentiality of support vector machine method in crude oil price forecasting. *EURASIA Journal of Mathematics, Science and Technology Education, 13*(12), 7893–7904.

Zhang, T., Tang, Z., Wu, J., Du, X., & Chen, K. (2021). Multi-step-ahead crude oil price forecasting based on two-layer decomposition technique and extreme learning machine optimized by the particle swarm optimization algorithm. *Energy, 229*, 120797. https://doi.org/10.1016/j.energy.2021.120797

Zhang, Y.-J., & Wang, J. (2015). Exploring the WTI crude oil price bubble process using the Markov regime switching model. *Physica A: Statistical Mechanics and Its Applications, 421*, 377–387.

Zhao, C., & Wang, B. (2014). Forecasting crude oil price with an autoregressive integrated moving average (ARIMA) model. In *Fuzzy information and engineering and operations research and management* (pp. 275–286). Springer.

Big Data Analysis of Energy Economics in Coal Market

3

3.1 Introduction

As basic energy and chemical raw material, coal is called "industrial food and black gold", and occupies a strategic position in national economic development. Changes in coal prices will not only affect the production and operation of the coal industry, but also affect prices, investment, consumption, and the country's imports and exports. In addition, changes in coal prices will also affect related industries and even national economic development and social stability.

Coal is a commodity, and its price is affected by national macro control, government supervision, and the international coal market. External factors such as climate temperature changes and the transformation of energy consumption structure will also cause fluctuations in coal prices. Coal prices have a greater impact on Gross Domestic Product (GDP), international energy industry trends, and energy market. Studying the modeling and forecasting of coal price fluctuations is not only an important part of policy making, commodity pricing, asset allocation, and risk management, but also an important task of the market economy. Accurate coal price forecasting means grasping market opportunities, while also conducive to saving material opportunity costs, and even promoting the transformation and development to happen in the energy industry. Therefore, forecasting coal prices has also become a popular topic in energy economy (Krzemień et al., 2015).

Modeling research for coal price forecasting is in full swing, and its forecasting models mainly have three types, including: (a) statistical models (Dong et al., 2010; Xue & Sriboonchitta, 2014), (b) machine learning models (Herrera et al., 2019; Matyjaszek et al., 2019), and (c) hybrid models (Wang & Wang, 2019).

Lago et al. (2018) pointed out that traditional statistical models as linear predictors tend to perform poorly in processing high-frequency data. These models include exponential smoothing, moving average, AutoRegressive Integrated Moving Average (ARIMA), and so on. Tseng et al. (2018) pointed out that traditional statistical techniques are more susceptible to the influence of objective factors

© Science Press 2022

H. Liu et al., *Big Data in Energy Economics*, Management for Professionals,
https://doi.org/10.1007/978-981-16-8965-9_3

and have greater limitations, which makes it difficult for most statistical analysis methods to adapt to high-precision forecasting. The development of artificial intelligence technology gave birth to a batch of machine learning models. These AI models have been developed for predicting coal price and have achieved satisfactory results. However, the classic Multi-Layer Perception (MLP), Radial Basis Function (RBF) neural network, and Recurrent Neural Network (RNN) are limited to a single hidden layer, which leads to their disadvantages in multi-step prediction (Lago et al., 2018). In contrast, a hybrid model composed of decomposition and integration methods can obtain better prediction accuracy than a single model (Qiu et al., 2016).

Alameer et al. (2020) found that the powerful ability to capture complex nonlinear features of the hybrid modeling model makes it more suitable for solving the complex problem in coal price forecasting. At the same time, incorporating multiple variables into the input of the prediction model can enhance the coal price prediction capability of the model.

In the case of this chapter, a multi-factor decomposition integrated hybrid model is proposed to perform big data multi-step forecasting of coal prices for econometric analysis of the coal market.

- In this study, a Decomposition - Adaptive Boosting (AdaBoost). MRT—Multi-Factor (MF)—Deep Belief Network (DBN) hybrid model based on deep learning has been developed. Experiments have shown that the model can effectively predict coal prices in the future for many months.
- This study compares the impact of multiple decomposition methods on the model's predictive ability.
- AdaBoost.MRT obtains stronger and more robust prediction performance by integrating multiple weak DBN predictors.
- This work confirmed that Crude oil, Iron ore, Gold (UK), Silver (UK), Copper London Metal Exchange (LME), Natural Gas (US), AUD&EUR Exchange rates, AUD&CNY Exchange rates, AUD&USD Exchange rates and other factors can help the model better to capture the changing laws of coal prices.

3.2 Influencing Factors Analysis of Coal Prices

The factors that affect coal prices are always complex and changeable, and some factors even have different influencing capabilities at different time periods. Studying the reasons for the fluctuation of coal prices is significant for effective coal prices forecasting.

Coal, oil and gas, as the three basic sources of energy, have a certain correlation in price. Ji et al. (2014) used panel co-integration test and Vector AutoRegressive (VAR) model to study the long-term equilibrium relationship and short-term dynamic relationship between the prices. Li et al. (2017) explored the relationship between the prices of European natural gas and the prices of coal imports. It is found that there exists a positive long-term relationship between the two prices from 1994 to 2014. The rise in crude oil prices in recent years has also promoted the use of coal in some industries. As an important primary energy source, the relationship between the prices of coal, crude oil, and natural gas must be comprehensively considered (Chiappini et al., 2019).

It is worth mentioning that the prices of gold, silver, copper, iron, etc. will also affect the trend of coal prices to a certain extent (Haque et al., 2017). Alameer et al. (2019) introduced the impact of fluctuations in oil prices and coal prices on the prices of metals such as copper, aluminum, nickel, lead, tin, gold, zinc, silver, platinum, and palladium. Some scholars have found that the soaring prices of oil and coal have increased metal prices through the impact of inflation (Cologni & Manera, 2008). (Bildirici & Sonustun, 2018) studied the impact on oil exporting countries caused by of oil and gold prices. It is extremely important to the economic growth of countries around the world. There are also studies documenting the relationship between the crude oil prices and the prices of gold, silver, copper, and other metals (Singhal et al., 2019). Therefore, the dynamic changes of metal prices should also be incorporated into coal price forecasting.

Studies have shown that changes in currency exchange rates are also a powerful factor affecting the prices of import and export commodities. Under normal circumstances, international coal import and export trade is settled in U.S. dollars. The Free on Board (FOB) price of exported coal is not only affected by market supply and demand factors, but also by changes in the exchange rate between the currency of the exporting country and the U.S. dollar. Most of the time, the currency exchange rate is the main factor affecting coal prices.

To predict coal prices more scientifically and effectively, Coal (Australia), Crude oil, Iron ore, Gold (UK), Silver (UK), Copper (LME), Natural Gas (US), AUD&EUR Exchange rates, AUD&CNY Exchange rates, and AUD&USD Exchange rates are used as input variables of the model to perform multivariate price forecasts.

3.2.1 Data Description of Coal Prices Influencing Factors

In the research of this chapter, observations covering 361 months from June 1991 to June 2021 for 10 variables are used. Table 3.1 details the horizon, unit, dataset website, and availability of the data used.

In the case study of this chapter, a total of 361 data from June 1991 to June 2021 are used to train and evaluate the multi-factor coal prices hybrid forecasting model. This study uses monthly data, and its time interval is one month. The datasets are divided into training sets and test sets. Among them, training data accounted for 70%, and the 1st to 250th samples are training data. Test data accounted for 30%, and the 251st to 361st samples are test data. Figure 3.1 shows the historical coal prices data.

Table 3.1 Data description of coal price influencing factors

Influence factors	Horizon	Unit	Dataset description	Data availability
Coal (Australia)	Monthly	US Dollars per metric ton	Coal, Australian thermal coal Monthly Price https://www.indexmundi.com/commodities/?commodity=coal-australian&months=360	Freely available
Crude oil	Monthly	US Dollars per barrel	Crude Oil (petroleum) Monthly Price https://www.indexmundi.com/commodities/?commodity=crude-oil&months=360	Freely available
Iron ore	Monthly	US Dollars per dry metric ton	Iron Ore Monthly Price https://www.indexmundi.com/commodities/?commodity=iron-ore&months=360	Freely available
Gold (UK)	Monthly	US Dollars per troy ounce	Gold Monthly Price https://www.indexmundi.com/commodities/?commodity=gold&months=360	Freely available
Silver (UK)	Monthly	US Dollars per troy ounce	Silver Monthly Price https://www.indexmundi.com/commodities/?commodity=silver&months=360	Freely available
Copper (LME)	Monthly	US Dollars per metric ton	Copper, grade A cathode Monthly Price https://www.indexmundi.com/commodities/?commodity=copper&months=360	Freely available
Natural Gas (US)	Monthly	US Dollars per million metric British thermal unit	Natural Gas Monthly Price https://www.indexmundi.com/commodities/?commodity=natural-gas&months=360	Freely available
AUD&EUR Exchange rates	Monthly	–	AUD-Australia [Australian dollar/$AU]—EUR-Euroland [Euro/€] https://fxtop.com/en/historical-exchange-rates.php?A=1&C1=AUD&C2=EUR&TR=1&DD1=&MM1=&YYYY1=&B=1&P=&I=1&DD2=31&MM2=07&YYYY2=2021&btnOK=Go%21	Freely available
AUD&CNY Exchange rates	Monthly	–	AUD-Australia [Australian dollar/$AU]—CNY-China [Chinese yuan renminbi (RMB)/¥] https://fxtop.com/en/historical-exchange-rates.php?A=1&C1=AUD&C2=CNY&TR=1&DD1=&MM1=&YYYY1=&B=1&P=&I=1&DD2=31&MM2=07&YYYY2=2021&btnOK=Go%21	Freely available
AUD&USD Exchange rates	Monthly	–	AUD-Australia [Australian dollar/$AU]—USD-United States [US dollar/$] https://fxtop.com/en/historical-exchange-rates.php?A=1&C1=AUD&C2=USD&TR=1&DD1=&MM1=&YYYY1=&B=1&P=&I=1&DD2=31&MM2=07&YYYY2=2021&btnOK=Go%21	Freely available

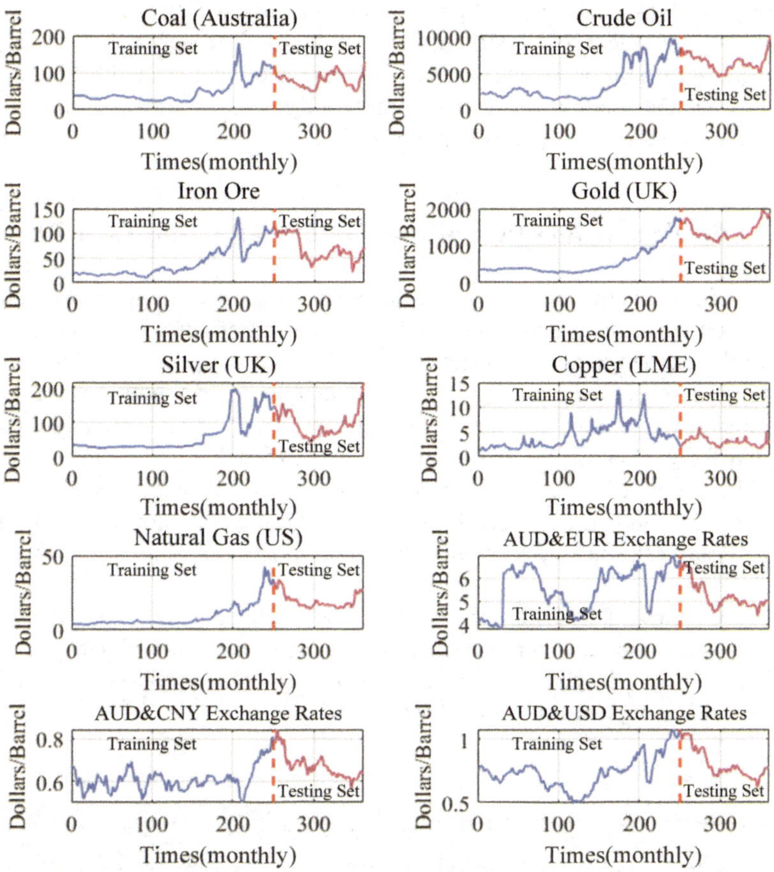

Fig. 3.1 The historical prices data from June 1991 to June 2021

Table 3.2 The statistical parameters of the coal price and other related factors dataset

Minimum	Maximum	Mean	Standard deviation	Skewness	Kurtosis
22.2500	180.0000	59.7596	31.0975	52.2100	25.1000
1377.2800	10,161.9700	4524.6719	2514.9851	4471.7900	1568.5275
10.4100	132.8300	49.1985	30.9495	43.8600	14.6085
256.0800	1968.6300	809.0694	513.5260	598.1900	275.5530
26.4700	214.4300	70.8791	48.5297	59.7800	27.5900
1.1900	13.5200	3.7575	2.1547	2.9700	1.6200
3.6500	42.7000	12.4479	8.6588	10.4400	4.1100
3.7794	6.9535	5.4524	0.8394	5.4305	4.0794
0.4976	0.8447	0.6284	0.0637	0.6179	0.5422
0.4998	1.0767	0.7585	0.1288	0.7492	0.5443

The figure shows that the coal price and other related factors fluctuate greatly, which belong to the typical strong non-stationary time series data. Table 3.2 presents the statistical parameters of the coal price and other related factors dataset.

To find out the relationship between these strong non-stationary wave characteristics, a correlation analysis is conducted.

3.2.2 Correlation Analysis of the Factors Affecting Coal Prices

The price of coal is affected by many factors. This section will explore the relationship between coal prices and prices of crude oil, natural gas, gold, silver, copper, iron, and the exchange rate of currencies.

In the research of this chapter, it aims to analyze the predictive variables that are significantly related to coal price fluctuations. The Pearson correlation analysis is adopted for evaluating the correlation between model input variables and coal prices. The calculation methods for normalization and Pearson correlation coefficient can be found in Sect. 2.2.2 of Chap. 2. Figure 3.2 shows the correlation coefficient between ten variables (including coal price), and the value range is [0, 1].

Figure 3.2 shows that except for AUD&EUR Exchange Rates and AUD&CNY Exchange Rates, the correlation coefficients between the other variables and coal prices all exceed 0.7, which indicates a strong correlation. These variables are therefore used as input to the hybrid model.

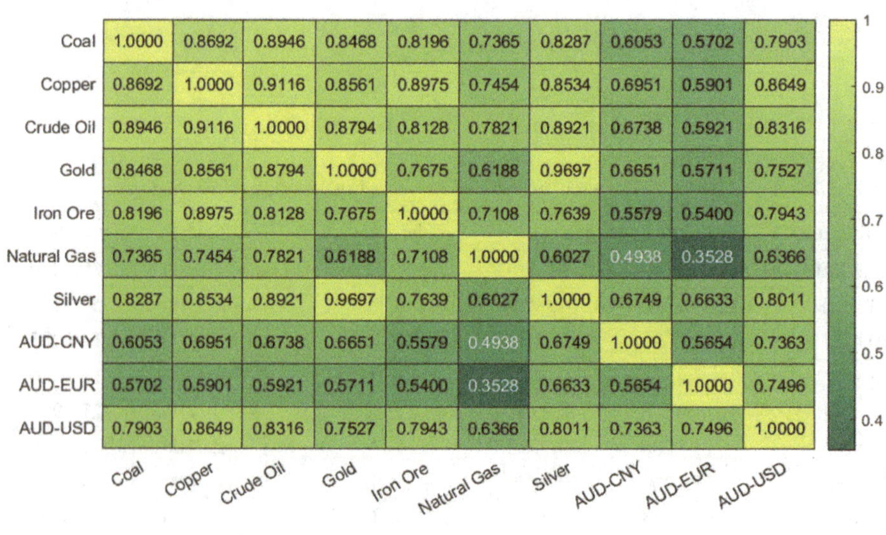

Fig. 3.2 Correlation analysis of the factors affecting coal price

3.3 Big Data Forecasting of Coal Prices

3.3.1 The Components of the Proposed Model

This chapter proposes a multi-factor coal price deep learning model based on decomposition method and integration technology. The methods involved in the proposed model include Pearson correlation analysis, time series decomposition, AdaBoost.MRT integration, and DBN.

3.3.1.1 Decomposition Methods

Empirical Mode Decomposition

Empirical Mode Decomposition (EMD) algorithms are generally applicable to the analysis of linear, nonlinear, stationary, and non-stationary signals. EMD can decompose the original time series signal x_t into a series of Intrinsic Mode Functions (IMF) and a residual component r (Huang et al., 1998). Signals with local characteristics at different time scales can be found in the IMFs. The analysis of the original signal can be completed by performing feature extraction or filtering processing on the IMF components (Boudraa & Cexus, 2007).

Generally, for a given time series signal x_t, all extreme points are determined first, and then the upper and lower envelopes $x_{\max(t)}$ and $x_{\min(t)}$ are obtained by interpolation and fitting for the maximum and minimum points, and the average value is m_t. A new time series h_t is obtained by calculating the difference between x_t and m_t. The expression is provided as follows:

$$h_t = x_t - m_t \tag{3.1}$$

If h_t satisfies the two conditions of the IMF component, one is that the number of zeros and extreme points of the signal are the same or differs by one at most, and the other is that the signal is zero mean, then h_t is the IMF component. Otherwise, instead of x_t, h_t is used to continue the screening and guiding to meet the two conditions.

Each IMF component is removed from the original time series, and the above screening steps are repeated until monotone sequence or constant sequence R_n is obtained, then the EMD decomposition process is finished. The expression of the EMD decomposition of the original time series x_t is provided as follows:

$$x_t = \sum_{i=1}^{n} IMF_i + R_n \tag{3.2}$$

Improved Complete Ensemble Empirical Mode Decomposition with Adaptive Noise

Improved Complete Ensemble Empirical Mode Decomposition with Adaptive Noise (ICEEMDAN), improved from EMD and CEEMDAN, provides more physical information for the IMF by further reducing the residual noise in the

CEEMDAN mode and avoiding false modes. The ICEEMDAN algorithm estimates the local average value, not the pattern, and only uses the white noise pattern instead of the white noise itself (Torres et al., 2011).

White noise $E_{1[w(i)]}$ is first added to the original signal x_t, and the following expression can be obtained:

$$x_t^{(i)} = x_t + \beta_0 E_1\left[w^{(i)}\right] \tag{3.3}$$

where $w(i)$ represents the i-th white noise to be added. In addition, the calculation of modal component IMF_1 is given as follows:

$$\hat{d}_1 = x - r_1 = x - \frac{1}{I}\sum_{i=1}^{I} M\left[X^{(i)}\right] \tag{3.4}$$

where $M()$ is the function used to generate local mean. The calculation of modal components IMF_2 is given as follows:

$$\hat{d}_2 = r_1 - r_2 = r_1 - \frac{1}{I}\sum_{i=1}^{I} M\left\{r_1 + \beta_1 E_2\left[w^{(i)}\right]\right\} \tag{3.5}$$

By analogy, the value of the k-th modal component IMF_k can be calculated, and the expression is given as follows:

$$\hat{d}_k = r_{k-1} - r_k = r_{k-1} - \frac{1}{I}\sum_{i=1}^{I} M\left\{r_{k-1} + \beta_{k-1} E_k\left[w^{(i)}\right]\right\} \tag{3.6}$$

where $k = 2, 3, \ldots, N$.

Wavelet Decomposition

The Wavelet Decomposition (WD) overcomes the defect that the Fourier transform aims to be localized in frequency. WD is a transform analysis method with both time (Morlet et al., 1982; Parsons et al., 2000). Wavelet decomposition can be divided into continuous wavelet decomposition and discrete wavelet decomposition. Through the WD method, the time series data can be decomposed into approximate parts and detailed parts. Then the information of different time and frequency of the original signal can be formed into a new time series array through the reconstruction method. The specific expression of WD is provided as follows:

$$Wf(a, v) = \left\langle f(t), \psi_{a,v}(t)\right\rangle = |a|^{-1/2}\int_R f(t)\overline{\psi\left(\frac{t-v}{a}\right)}\mathrm{d}t \tag{3.7}$$

where $a(a > 0)$ and v are the expansion factor and translation factor, respectively, $f(t)$ is the analysis signal function, and $\psi_{a,v}(t)$ is called the basic wavelet.

Wavelet Packet Decomposition

The wavelet packet is an extension of the wavelet concept. Different from the wavelet transform that only decomposes the low-frequency signal each time, Wavelet Packet Decomposition (WPD) can decompose the high-frequency part in more detail, taking full account of the time–frequency characteristics of the signal. The relationship of WPD can be expressed as follows (Gokhale & Khanduja, 2010):

$$F = D_1 D_2 + D_1 A_2 + A_1 A_2 + A_1 D_2 \tag{3.8}$$

Define the orthogonal scaling function $\varphi(x)$ and its corresponding wavelet function $\psi(x)$, assuming that $h(k)$ and $g(k)$ are the low-pass and high-pass filter coefficients, respectively, and $h(k)$ and $g(k)$ are the conjugate filter coefficients, respectively.

$$\begin{cases} \varphi(x) = \sqrt{2} \sum\limits_{k \in z} h(k)\varphi(2x - k) \\ \psi(x) = \sqrt{2} \sum\limits_{k \in z} g(k)\varphi(2x - k) \end{cases} \tag{3.9}$$

Here, $\mu_0 = \varphi(x)$, $\mu_1 = \psi(x)$, then:

$$\begin{cases} \mu_{2n}(x) = \sqrt{2} \sum\limits_{k \in z} h(k)\mu_n(2x - k) \\ \mu_{2n+1}(x) = \sqrt{2} \sum\limits_{k \in z} g(k)\mu_n(2x - k) \end{cases} \tag{3.10}$$

Assuming that subspace U_j^n is the closure space of function $\mu_n(x)$, subspace U_j^{2n} is the closure space of function $\mu_{2n}(x)$, and $g_j^n \in U_j^n$, then $g_j^n(x)$ can be expressed as follows:

$$g_j^n = \sum_l C_l^{j,n} \mu_n\left(2^j x - 1\right) \tag{3.11}$$

The WPD can be expressed as follows:

$$\begin{cases} C_l^{j,n} = \sum\limits_k h(k - 2l) C_k^{j+1,n} \\ C_l^{j,2n+1} = \sum\limits_k g(k - 2l) C_k^{j+1,n} \end{cases} \tag{3.12}$$

Decompose the wavelet packet and perform the inverse calculation to obtain the mathematical expression of WPD reconstruction, as follows:

$$C_l^{j+1,n} = \sum \left[h(l - 2k) C_l^{j,2n} + g(l - 2k) C_l^{j,2n+1} \right] \tag{3.13}$$

Maximal Overlap Discrete Wavelet Transform
Different from classic Dynamic Time Warping (DWT), Maximal Overlap Discrete Wavelet Transform (MODWT) is a non-orthogonal wavelet transform with high redundancy (Zhu et al., 2014). It still has properties such as smoothness, and there is no requirement for sample size. The relationship between wavelet filter $\left\{\hat{h}_l\right\}$ and scale filter $\{\hat{g}_l\}$ in MODWT and the wavelet filter and scale filter of DWT (Percival & Walden, 2000) are as follows:

$$\hat{h}_l = h_l/\sqrt{2} \tag{3.14}$$

$$\hat{g}_l = g_l/\sqrt{2} \tag{3.15}$$

In addition, they also need to meet conditions:

$$\sum_{l=0}^{L-1} \hat{h}_l = 0, \sum_{l=0}^{L-1} \hat{h}_l^2 = 1/2, \sum_{l=0}^{L-1} \hat{h}_l \hat{h}_{l+2k} = \sum_{l=-\infty}^{\infty} \hat{h}_l \hat{h}_{n+2k} = 0 \tag{3.16}$$

$$\sum_{l=0}^{L-1} \hat{g}_l = 1, \sum_{l=0}^{L-1} \hat{g}_l^2 = 1/2, \sum_{l=0}^{L-1} \hat{g}_l \hat{g}_{l+2k} = \sum_{l=-\infty}^{\infty} \hat{g}_l \hat{g}_{n+2k} = 0, \sum_{l=-\infty}^{\infty} \hat{g}_l \hat{h}_{l+2n} = 0 \tag{3.17}$$

The wavelet coefficients (detail signals) and scale coefficients (smooth signals) at scale j can be obtained as the following expressions:

$$\hat{W}_{j,t} = \sum_{t=0}^{N-1} \hat{h}_{j,t}^o X_{t-l \bmod N} \tag{3.18}$$

$$\hat{V}_{j,t} = \sum_{t=0}^{N-1} \hat{g}_{j,t}^o X_{t-l \bmod N} \tag{3.19}$$

where $\left\{\hat{h}_{j,l}^o\right\}$ and $\left\{\hat{g}_{j,l}^o\right\}$ are the filters obtained by periodizing $\left\{\hat{h}_{j,l}\right\}$ and $\{\hat{g}_{j,l}\}$ with N as the period.

The following expressions can be obtained using Mallat's algorithm:

$$\hat{D}_{j,t} = \sum\nolimits_{l=0}^{N-1} \hat{h}_{j,l}^o \hat{W}_{j,t+l \bmod N} \tag{3.20}$$

$$\hat{S}_{j,t} = \sum\nolimits_{l=0}^{N-1} \hat{g}_{j,l}^o \hat{V}_{j,t+l \bmod N} \tag{3.21}$$

where $t = 0, 1, \ldots, N-1$, $\hat{D}_{j,t}$, and $\hat{S}_{j,t}$ are respectively the t-th detail signal and the t-th smooth signal decomposed in the j-th layer. $\left\{\hat{D}_j\right\}$ and $\left\{\hat{S}_j\right\}$ are the detail signal and smooth signal in the j-th layer, respectively, $X_t = \sum_{j=1}^{J_0} \hat{D}_j + \hat{S}_{J_0}$ can be obtained by the Mallat algorithm of multi-resolution analysis, and J_0 is the number of decomposition layers.

In general, a sequence $\{X_t\}$ can be decomposed into low-frequency and high-frequency signals through MODWT.

Maximal Overlap Discrete Wavelet Packet Transform

Maximal Overlap Discrete Wavelet Packet Transform (MODWPT) is an optimization and improvement of MODWT (Liu et al., 2020). It can further decompose high-frequency signals that are not decomposed by MODWT, thereby improving frequency resolution. The decomposition coefficient of MODWPT are as follows:

$$W_{j.n} = \left\{ W_{j,n,t}, t = 0, \ldots, N-1 \right\} \tag{3.22}$$

where j is the number of decomposition layers, and n is the frequency index that changes with j.

Similarly, the decomposition coefficient of MODWPT can be calculated as follows:

$$W_{j,n,t} = \sum_{l=0}^{L-1} r_{n,l} W_{j-1,[n/2],(t-2j-1) \bmod N} \tag{3.23}$$

where if n mod 4 = 0 or 3, then $r_{n,l} = \left\{ \hat{g}_l \right\}$, if n mod 4 = 1 or 2, then $r_{n,l} = \left\{ \hat{h}_l \right\}$.

Empirical Wavelet Transform

Empirical Wavelet Transform (EWT) is a signal decomposition method based on wavelet analysis, and is adaptive (Gilles, 2013). EWT builds a set of empirical wavelet filter banks based on the spectral characteristics of the signal itself, which is used to extract different modes of the signal. The empirical wavelet contains both empirical scale function and empirical wavelet function. Their calculation methods are provided as follows:

$$\phi_n(\omega) = \begin{cases} 1, |\omega| \le \omega_n - \tau_n \\ \cos\left[\frac{\pi}{2}\beta\left(\frac{1}{2\tau_n}(|\omega| - \omega_n + \tau_n)\right)\right], \omega_n - \tau_n \le |\omega| \le \omega_n + \tau_n \\ 0, otherwise \end{cases} \tag{3.24}$$

$\psi_n(\omega)$

$$= \begin{cases} 1, \omega_n + \tau_n \le |\omega| \le \omega_{n+1} - \tau_{n+1} \\ \cos\left[\frac{\pi}{2}\beta\left(\frac{1}{2\tau_{n+1}}(|\omega| - \omega_{n+1} + \tau_{n+1})\right)\right], \omega_{n+1} - \tau_{n+1} \le |\omega| \le \omega_{n+1} + \tau_{n+1} \\ \sin\left[\frac{\pi}{2}\beta\left(\frac{1}{2\tau_n}(|\omega| - \omega_n + \tau_n)\right)\right], \omega_n - \tau_n \le |\omega| \le \omega_n + \tau_n \\ 0, otherwise \end{cases} \tag{3.25}$$

where τ_n and $\beta(x)$ can be provided as follows:

$$\begin{cases} \tau_n = \gamma\omega_n, 0 < \gamma < 1, \gamma < \min\left\{\frac{\omega_{n+1} - \omega_n}{\omega_{n+1} + \omega_n}\right\} \\ \beta(x) = x^4(35 - 84x + 70x^2 - 20x^3) \end{cases} \tag{3.26}$$

The reconstructed original signal can be expressed as follows:

$$f(t) = W_f^{\varepsilon}(0, t) * \phi_1(t) + \sum_{n=1}^{N} W_f^{\varepsilon}(n, t) * \psi_n(t) \tag{3.27}$$

where $*$ is the convolution calculation, $W_f^{\varepsilon}(0, t)$ is the approximate coefficient after Fourier transform, $W_f^{\varepsilon}(n, t)$ is the detail coefficient after Fourier transform.

3.3.1.2 Deep Belief Network

The details of Deep Belief Network (DBN) can be found in Sect. 2.3.1 of Chap. 2.

3.3.1.3 AdaBoost.MRT

As a classic predictive ensemble algorithm, the Boosting algorithm can continuously update its weights for the same sample points and train multiple weak regression models. Finally, by integrating these weak models with different weights, a strong regression model can be formed. As an improved algorithm, AdaBoost.MRT can eliminate the singularity in the misclassification function by adjusting the multivariate output, and finally achieve the purpose of enhancing the noise robustness (Zang et al., 2018). Specifically, the steps for implementing AdaBoost are provided as follows (Hastie et al., 2009):

(a) Input training set.
(b) Initialize the weight distribution of the training data. If there are N samples, the same weight is assigned to each sample.
(c) After setting the initial state, the sample T with a weight distribution of D_m is used to train M samples in a loop, and the m-th training is set to obtain a weak learner.

$$B = y_m(x), m = 1, 2, 3, \ldots, M \tag{3.28}$$

The maximum error of the *m*-th weak learner on the training set is expressed as follows:

$$E_m = \max|y_i - y_m(x_i)| \tag{3.29}$$

Then the relative error of each sample is expressed as follows:

$$e_{mi} = \frac{|y_i - y_m(x_i)|}{E_m} \tag{3.30}$$

Finally, the error rate of the *m*-th weak learner is expressed as follows:

$$e_m = \sum_{i=1}^{m} w_{mi} e_{mi} \tag{3.31}$$

According to the calculation of e_m, the weight of the weak learner in the final strong regressor is expressed as follows:

$$\alpha_m = \frac{e_m}{1 - e_m} \tag{3.32}$$

(d) Update the weight distribution of each sample of the training set and apply them to the next iteration.

$$D_{m+1} = \left(w_{m+1,1}, w_{m+1,2}, \ldots, w_{m+1,N}\right) \tag{3.33}$$

(e) To prevent the AdaBoost algorithm from overfitting, the normalization term β needs to be added. After adding the normalization term, the weak learner iteration of the AdaBoost algorithm is expressed as follows:

$$Y_m(x) = Y_{m-1}(x) + \beta \alpha_m y_m(x) \tag{3.34}$$

where $0 < \beta < 1$, when β is small, to achieve the same learning effect, an increase in the number of iterations of the weak learner is necessary and expected, thereby improving the model's generalization ability.

(f) Get the final strong regressor.

$$Y_M(x) = \sum_{m=1}^{M} \left(\ln \frac{1}{\alpha_m}\right) g_m(x) \tag{3.35}$$

where $Y_M(X)$ is the prediction of the AdaBoost model, and $g_m(x)$ is the median of all $\alpha_m y_m \leq (x)$.

3.3.2 Multi-factor Coal Price Hybrid Forecasting Model

A clear reflection of the connection relationship between various modules in the proposed model are provided in Fig. 3.3.

3.3.2.1 The Modeling Steps of the Proposed Model

The modeling steps of the proposed hybrid model are given as follows:

(a) Correlation analysis of coal price and its influencing factors. The PCC is used to evaluate the correlation of Coal (Australia), Crude Oil, Iron Ore, Gold (UK), Silver (UK), Copper (LME), Natural Gas (US), AUD&EUR Exchange Rates, AUD&CNY Exchange Rates and AUD&USD Exchange Rates. Highly correlated factors are chosen as the model input for multi-factor modeling.

(b) Data set division. The data set is divided into two parts: training and test set. The training set accounts for 70%, and the 1st to 250th samples are the training data. The test set accounts for 30%, and the 251st to 361st samples are the test data.

(c) Data pre-processing. The data is preprocessed by using decomposition algorithms such as EMD, ICEEMDAN, WD, WPD, MODWT, MODWPT, EWT, etc., and the effects of different decomposition methods on model performance are compared.

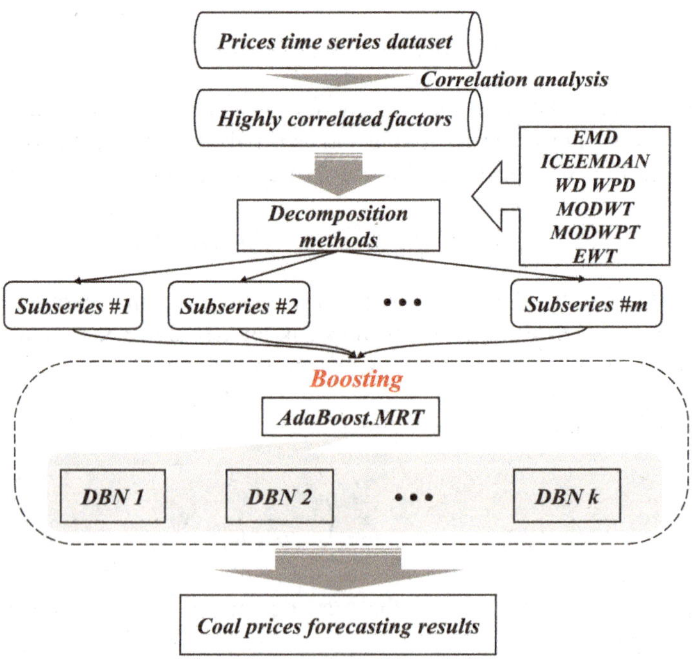

Fig. 3.3 The multi-factor hybrid model framework for coal prices forecasting

(d) Train weak predictors and build ensemble models. Establish k weak DBN predictors for the m series decomposed in step c, and realize the enhanced integration of weak predictors through the AdaBoost.MRT algorithm to obtain the strong predictor DBN. The final coal price prediction results under the influence of multiple factors are obtained by integrating the prediction results corresponding to each sub-predictor. Performance metrics are then used to evaluate the model.

3.3.2.2 The Algorithm Parameters Settings

Table 3.3 lists the DBN model and the parameter settings of the seven decomposition algorithms involved in this chapter.

3.3.3 Performance Metrics

Three evaluation metrics, namely Mean Absolute Error (MAE), Mean Absolute Percentage Error (MAPE), and Root Mean Square Error (RMSE) are used in this chapter, the equations and details can be found in Sect. 2.3.3 of Chap. 2.

3.3.4 Results and Discussions

In the experiment and discussion section of this chapter, in order to determine the most suitable decomposition algorithm under different prediction steps, the impact of different decomposition methods on forecasting performance is analyzed using control variable method. The factors input into the model are kept the same, the basic predictors used are all DBNs, and the integration method is AdaBoost.MRT. Experiments compare the hybrid model in the case of without decomposition methods, and with EMD, ICEEMDAN, WD, WPD, MODWT, MODWPT, and EWT decomposition methods. The multi-step forecasting results for coal time series data are listed in Table 3.4, where the smallest errors of different forecasting steps are marked in bold. Figure 3.4 shows the trend parallel graph between the model 3-step prediction results and the actual observation values. In the 1-step forecasting results, the forecasting results are close to the actual ones. For the convenience of observation, the displayed figure is selected as 3-step forecasting results. Figure 3.5 shows the stacking diagrams of 1-step to 10-step error indices under different decomposition methods (* represents the AdaBoost.MRT-MF).

Summarizing the experimental results shown in Table 3.4 and Figs. 3.4 and 3.5, the following analysis and conclusions can be obtained:

(a) The enhanced integration strategy of AdaBoost.MRT ensures the robustness of the prediction model, which makes the model have good stability during the prediction process and the prediction is more accurate. In the 1, 3, 5, 7, 10 day forecasting results, the MAEs of the DBN and AdaBoost.MRT-DBN

Table 3.3 The algorithm parameters of different methods

Algorithm parameter	Value
DBN	
Max epochs	100
Batch size	1
Activation function	sigm
Momentum factor	0.0
Learning rate	0.01
EMD	
Level of decomposition	8
Noise level	0.2
ICEEMDAN	
Level of decomposition	8
Ensemble size	500
Noise level	0.2
WD	
Level of decomposition	8
Mother wavelet	db10
WPD	
Level of decomposition	8
Mother wavelet	db10
MODWT	
Level of decomposition	8
Mother wavelet	Daubechies wavelet
MODWPT	
Level of decomposition	8
Mother wavelet	Daubechies wavelet
EWT	
Degree for the polynomial interpolation	6
Detection method	scalespace
Maximum number of bands	25
Sampling rate	1
Filter width	10

are 6.0654\$, 14.4465\$, 23.8740\$, 35.5341\$, 48.6052\$, 6.0088\$, 13.8431\$, 21.4142\$, 29.4493\$, 35.1194\$, respectively. The reason for this phenomenon may be that AdaBoost.MRT eliminates the singularity in the misclassification function by adjusting the multivariate output, and achieves enhanced noise robustness. At the same time, the enhanced integration of multiple weak predictors makes up for the shortcomings of a single predictor, and a more robust strong predictor has better prediction performance.

Table 3.4 Multi-step forecasting results of Coal (Australia) time series data

Model	Metrics	Horizons (monthly)				
		1-day	3-day	5-day	7-day	10-day
DBN	MAE	6.0654	14.4465	23.8740	35.5341	48.6052
	RMSE	7.7789	17.3276	27.5372	39.0443	52.2089
	MAPE	7.7177	18.3760	31.2380	47.0316	66.3590
AdaBoost.MRT-DBN	MAE	6.0088	13.8431	21.4142	29.4493	35.1194
	RMSE	7.7314	16.6866	24.9033	33.2596	40.3297
	MAPE	7.4453	18.3040	29.5494	41.7307	51.6429
MF-DBN	MAE	4.7482	11.9735	18.8209	26.2561	34.5910
	RMSE	6.3081	14.5889	21.9771	29.4644	37.9032
	MAPE	5.9298	15.2186	24.8141	35.2707	47.9149
*-DBN	MAE	4.4385	10.8627	15.6828	21.3029	28.2840
	RMSE	6.0248	13.2646	18.7842	25.0305	33.1167
	MAPE	5.5499	13.7603	20.6519	28.5067	38.4692
EMD-*-DBN	MAE	2.6526	4.1790	7.3966	11.6005	17.8777
	RMSE	3.7041	5.5912	8.9601	14.2782	21.7071
	MAPE	3.1839	5.2019	9.4781	14.8323	22.7701
ICEEMDAN-*-DBN	MAE	1.9350	5.1308	6.9313	8.4941	10.3594
	RMSE	5.2258	11.8754	12.0028	12.0567	13.0884
	MAPE	2.1498	5.8195	8.2447	10.5046	**13.4402**
WD-*-DBN	MAE	2.1699	4.7102	6.8097	8.3409	**10.2725**
	RMSE	2.9022	6.2982	8.6772	10.1741	**12.3898**
	MAPE	2.7770	6.2836	9.4298	11.2893	13.5353
WPD-*-DBN	MAE	1.7257	4.2253	6.7932	9.0897	12.7463
	RMSE	2.1922	5.9092	9.5743	13.2864	18.9300
	MAPE	2.1951	5.3604	8.4108	10.9402	14.6919
MODWT-*-DBN	MAE	**0.7589**	3.3932	**4.6620**	**6.3235**	12.0098
	RMSE	1.2517	4.4147	**6.0937**	**8.5190**	14.7333
	MAPE	**0.9358**	4.3672	**5.8991**	**7.9989**	15.9282
MODWPT-*-DBN	MAE	0.8116	**2.5055**	4.9486	7.6871	12.5271
	RMSE	**1.0687**	**3.1842**	6.2666	10.3211	17.4270
	MAPE	1.0526	**3.1805**	6.2304	9.3572	14.9176
EWT-*-DBN	MAE	1.4895	5.6878	10.0534	12.9378	14.3321
	RMSE	1.8480	7.1305	13.1323	17.0886	19.4952
	MAPE	1.9956	7.4280	12.5716	15.7056	16.9968

Remark The units of all other performance indicators except MAPE in the table are 'US Dollars per Metric Ton', and the unit of MAPE is '%', * represents the AdaBoost.MRT-MF

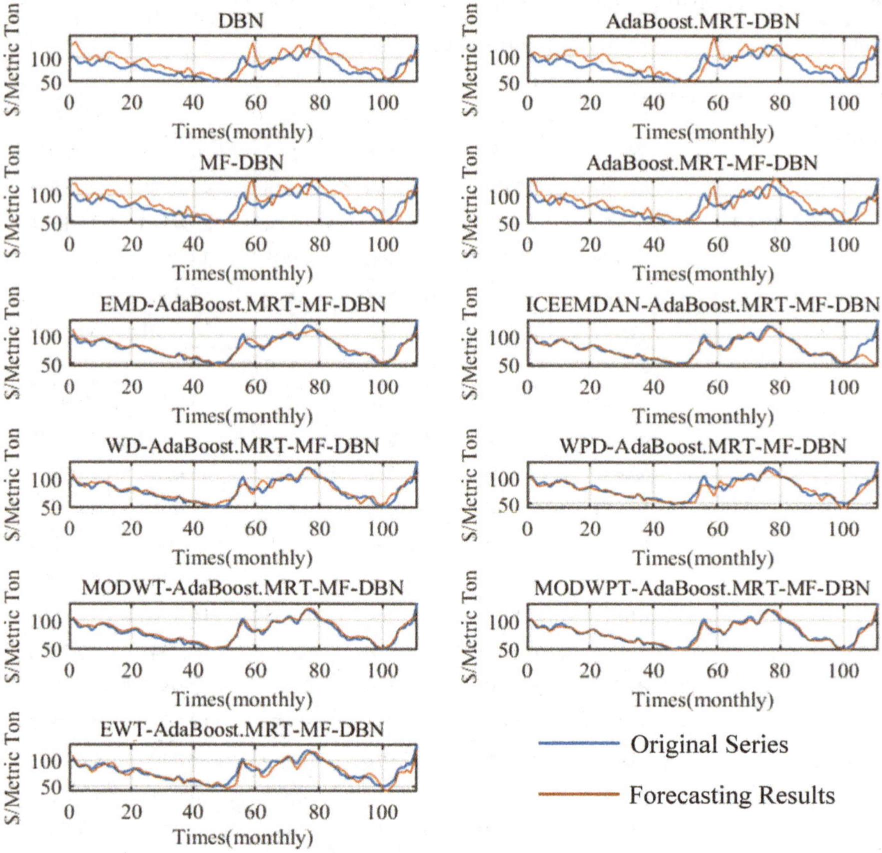

Fig. 3.4 The 3-step forecasting results of Coal (Australia) time series data under different models

(b) Compared with single factor forecasting, coal price forecasting under multiple factors is more reasonable, and it considers more components. In the 1, 3, 5, 7, 10 day forecasting results, the MAEs of the DBN and MF-DBN are 6.0654$, 14.4465$, 23.8740$, 35.5341$, 48.6052$, 4.7482$, 11.9735$, 18.8209$, 26.2561$, 34.5910$, respectively. It can be seen that factors such as fossil fuels, gold, silver, copper, iron, and currency exchange rates will affect the price of coal, which makes it volatile. The consideration of multiple factors makes the model training more adequate. Coal is the main predictive factor, and other high-correlation factors are used as auxiliary predictive factors, which makes the model easier to capture the trend of coal price changes. Among them, the screening process of Pearson correlation coefficient plays a very important role. It avoids the negative impact of low correlation factors on accuracy.

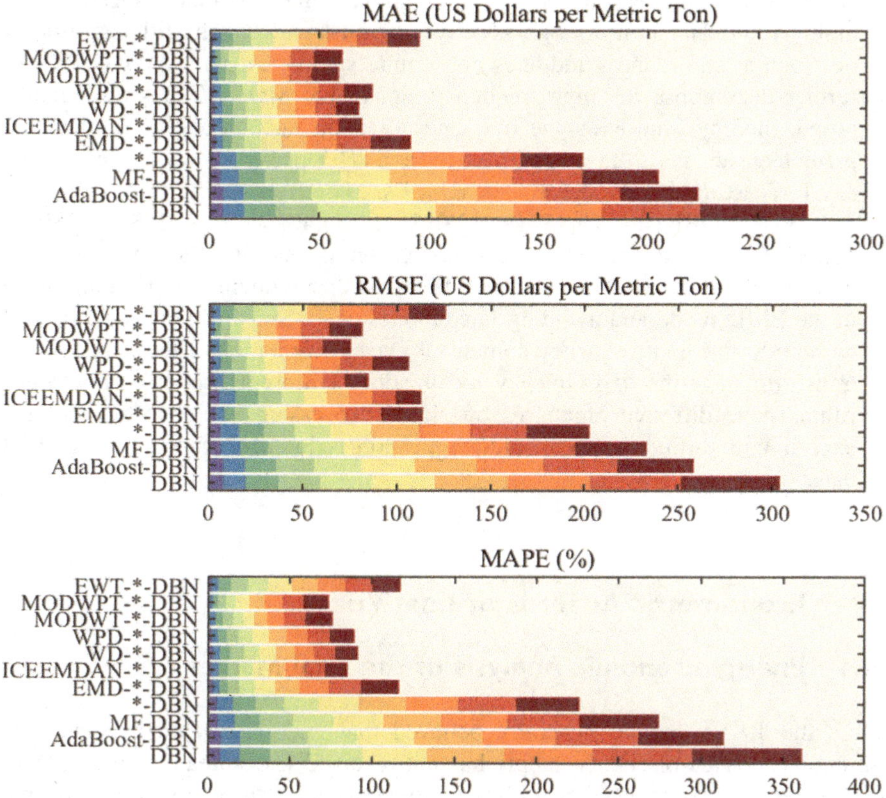

Fig. 3.5 The stacking diagrams of 1-step to 10-step error indices under different decomposition methods

(c) The decomposition method can effectively improve the accuracy. In the 1-day forecasting results, the MAEs of the *_DBN, EMD_*_DBN, ICEEMDAN_*_DBN, WD_*_DBN, WPD_*_DBN, MODWT_*_DBN, MODWPT_*_DBN, EWT_*_DBN are 4.4385$, 2.6526$, 1.9350$, 2.1699$, 1.7257$, 0.7589$, 0.8116$, 1.4895$, respectively. Evaluation metrics and the error stacking chart in Fig. 3.5, shows that after the decomposition process, the prediction accuracy has achieved a substantial increase. This shows that after the decomposition algorithm splits the original sequence into multiple stable sequences, it is easier for the predictor to learn the fluctuations of the data. Finally, the final model output is obtained by reversely superimposing the prediction results of each subseries.

(d) With different prediction steps, the decomposition algorithms showed their respective advantages in processing data features. When the predicted steps are small, MODWT and MODWPT perform best. In the 1, 3, 5, 7 day forecasting results, the MAEs of the MODWT_*_DBN and MODWPT_*_DBN are 0.7589$, 3.3932$, 4.6620$, 6.3235$, 0.8116$, 2.5055$, 4.9486$, 7.6871$, respectively. The reason for this phenomenon may be that MODWT is a highly redundant non-orthogonal wavelet transform, which still has properties such as smoothness and does not require sample capacity. MODWPT can further decompose the high frequency signal that MODWT has not decomposed, thereby improving the frequency division rate. In addition, when the predicted step is 10-day, ICEEMDAN and WD performed well. In the 10-day forecasting results, the MAEs, RMSEs, MAPEs of ICEEMDAN_*_DBN and WD_*_DBN are 10.3594$, 13.0884$, 13.4402%, 10.2725$, 12.3898$, 13.5353%, respectively. The possible reason is that ICEEMDAN provides more physical information for the IMF by further reducing the residual noise in the EMD mode and avoiding false modes. As a transform analysis method with both time and frequency domain, WD can overcome the defect of Fourier transform in terms of frequency localization. Different decomposition algorithms have different emphases on signal processing, which often requires experiments to find the most suitable pre-processing algorithm for the current data.

3.4 Econometric Analysis of Coal Prices

3.4.1 Energy Economic Analysis of the Coal Market

Before the 1970s, people generally believed that energy was an inexhaustible resource. The issue of energy supply has not received widespread attention. With the emergence of some unexpected international events in the energy crisis, people have begun to reflect on the issue of energy scarcity. Since the twenty-first century, issues such as the effective supply of energy, the efficient utilization of green energy, the safety of energy utilization, and the impact of energy price volatility on economic development have become the focus of research by governments and subject experts around the world (Sribna et al., 2019).

The International Energy Agency (IEA) "Global Coal Market Report (2018–2023)" pointed out that coal is still the core of the global energy system (IEA 2021). Due to the advantages of affordable price, abundant energy storage, and convenient transportation, coal is still the main source of energy in many countries. According to the report, global coal demand is predicted to continue to remain stable between 2018 and 2023. Among them, coal consumption in the European and American markets will decline, and this part of the decline will be offset by the increase in India and other Asian countries. As pointed out in the report, the contribution of coal in the global energy structure will be reduced from 27 to 25% in the next five years, and its main alternative energy sources will come from renewable energy and natural gas. Germany plans to close all coal power plants by 2038. This measure will be implemented in the UK and France in 2025 and 2021 respectively. Finland plans to ban coal completely by 2030. The proportion of coal consumption in China will drop to about 50% in 2030. In the same year,

China's coal consumption will exceed 4.2 billion tons. IEA believes that, as a high-carbon energy source, the sustainable development of coal in the future depends on Carbon Capture, Utilization, and Storage (CCUS). In the future development process, it is important to achieve clean mining and utilization of coal by firmly establishing the concept of clean energy development and continuously promoting the innovation and development of clean technology.

The coal industry has long been an important element in promoting the development of international finance (Song & Wang, 2019). Commodity prices, trade and investment, production and consumption, and imports and exports can all be greatly affected by fluctuations in coal prices. Furthermore, coal prices have even affected the economic development and social stability of many countries.

The external factors that affect coal prices can be summarized as coal supply and demand factors, national economic development status, coal alternative energy development status, currency issuance and inflation, relevant national policies, coal pricing mechanism, lack of external costs of coal mining, and the international coal market price etc. The internal factors affecting coal prices are summarized as corporate prices and circulation costs.

Long-term stable equilibrium and causal relationships can be found between coal prices and the macroeconomy. Specifically, coal price fluctuations have a certain impact on economic growth, and changes in economic growth will also cause coal prices to fluctuate, which indicates that there is a two-way guiding role between coal prices and economic growth. In addition, changes in coal prices will also impact the price level.

Macroeconomics also has a long-term and lasting impact on coal prices. Different macroeconomic variables have different effects on coal prices. Specifically, GDP has a long-term positive pulling effect on coal prices. The impact of industrial enterprise value added on coal prices is a weak negative feedback effect. The consumer price index has a positive and negative impact on coal prices, and money supply has a positive effect on coal prices. The overall coal price has a weak positive impact.

In general, changes in the market price of coal, the core of the global energy system, will have an important impact on all aspects of production and life.

3.4.2 Big Data Prediction Technology

Coal is so important that its price prediction modeling has also become a popular topic in global energy economy. This chapter selects a sample of coal prices and related factors during the 30 years from June 1991 to June 2021 for research. Compared with the most used traditional mathematical statistics and econometric forecasting methods, a multi-factor coal price hybrid ensemble forecasting model is proposed in this study.

The basic framework of the model includes multi-factor analysis, decomposition pre-processing, predictive modeling, and hybrid integration. The model can use time series data as a carrier to realize multiple outputs while performing multiple inputs. The proposed model shows excellent performance in forecasting international coal prices. The Australian thermal coal monthly price data set verifies the effectiveness of the proposed model in processing nonlinear data, and the model gives full play to its characteristics of using historical data for memory. The predicted coal price can reproduce the curve fluctuation trend of the original coal price with high detail.

The results show that under the combined effect of multiple high-correlation factors, the model can better capture the peaks and valleys of coal price changes. The optimized integration method based on AdaBoost.MRT increases the robustness of the model. The integrated strong predictor can better fit the overall change law of the original coal price data series. The decomposition algorithm disassembles the original time series data into multiple more stable series, which makes the model learning more fully and can more easily track the rise or fall of coal prices.

The research work in this chapter examines the influence of variables Crude Oil, Iron Ore, Gold (UK), Silver (UK), Copper (LME), Natural Gas (US), AUD&EUR Exchange Rates, AUD&CNY Exchange Rates and AUD&USD Exchange Rates on the fluctuation of coal prices. The results show that these factors are significantly correlated with coal prices. Taking the 1, 3, 5, 7, 10 day forecasting results as an example, the percentage improvement of the MAEs of MF-DBN versus DBN are 21.72%, 17.12%, 21.17%, 26.11%, and 28.83%, respectively. The prediction accuracy can be improved effectively by considering various factors comprehensively.

This work provides the basis of machine learning modeling for policy making in the international energy market.

3.4.3 Policies and Recommendations

The research in this chapter shows that there is a close connection between the fluctuation of coal prices and the macro economy. Fluctuations in coal prices will have a long-term impact on the macro economy (Wang et al., 2019). Conversely, changes in the macro economy will also have a certain impact on coal price fluctuations (Liu & Luan, 2015). To reduce the frequency of coal price fluctuations, stabilize commodity prices, and maintain the sustained and stable development of the national economy, some policy recommendations are given as follows:

(a) Country changes its development mode and maintains healthy economic development.

For areas rich in coal resources, their economic development often relies heavily on coal mining. This unhealthy method of economic development has brought great disadvantages. For example, the ecological problem of the production area, the collapse of the goaf. Seriously, it can even lead to distortions and imbalances in the economic structure. Excessive dependence on the coal resource industry will suppress the development of other industries. When faced with a sharp drop in coal prices and a sluggish market, or the exhaustion of coal after resources are over-exploited, the local economy will encounter huge difficulties. This phenomenon is called the "Resource Curse".

The economic growth rate of some countries with abundant natural resources is actually very slow. Spain in the seventeenth century, Russia in the nineteenth and twentieth centuries, Nigeria, Venezuela, and Gulf oil countries after World War II, were all countries with rich natural resources that had significantly slower economic growth than other countries over the same period (Sachs & Warner, 1995). On the contrary, some resource-poor countries, such as the Netherlands in the seventeenth century, Switzerland and Japan in the nineteenth century, and the Newly Industrialized Economies (NIE) after World War II, all experienced strong and rapid growth.

To avoid a similar "Resource Curse" phenomenon, corresponding countries need to introduce preferential policies to encourage resource-rich regions to change their development methods, and not to overuse a single energy resource. When coal prices fall sharply in the future, the impact on the economic structure and economic development of these places will be greatly reduced.

(b) Strengthen cooperation between countries and explore the new pattern of world energy development.

Under the dual background of the internationalization of resource flows and economic globalization, the volatility and impact of energy prices have not only affected the economic development of a country. It is no longer realistic to rely on a single country to solve the energy problem. This requires the cooperation of all countries around the world to solve it together. With the economic development of various countries, the energy issue has become imminent.

Countries should strengthen resource diplomacy and reduce diplomatic frictions in energy trade. The government should adhere to the principle of the supremacy of resources. On the premise of strengthening their core advantages, all regions have carried out diversified operations to reduce the impact of price fluctuations. In addition, the country also needs to restrict the import of low-quality coal, in order to reduce the environmental pollution, and reduce the impact of imported coal on coal prices. The state encourages enterprises to export coal through tax cuts and other methods to relieve some of the pressure of overcapacity. In this way, the market economy can be stabilized.

(c) Coal industry adheres to clean development and raises awareness of environmental protection.

Coal is the main energy source, and its continuous mining will aggravate ecological and environmental problems. In this new international situation, the coal industry needs to take the road of circular economy development based on environmental protection, green and comprehensive utilization of resources. Enterprises should speed up the development and use of wind energy, solar energy, and other clean energy sources. As low-carbon economy is the strategic direction of the future economic development, in order to improve the energy efficiency, coal enterprises must increase the transformation of technology. Actively apply new technologies and processes for emission reduction. Vigorously develop a circular economy and realize the comprehensive utilization and recycling of coal-related resources.

The coal industry must also rely on scientific and technological progress to promote cleaner production and comprehensive utilization. Reasonably develop and utilize coal resources and water resources to prevent ecological damage and environmental pollution. Coal mining should pay attention to water and soil conservation and reclamation and greening. Large areas of excavation damage caused during the production and construction of coal mines need to be treated in time. Adopt barriers, slope protection and land leveling measures to reduce soil erosion and increase vegetation coverage. Improve and protect water and soil resources and increase land production capacity. The power sector pays attention to the monitoring of sulfur emissions. According to the requirements of the environmental protection department, it is equipped with corresponding environmental protection facilities to meet the national emission standards. Strengthen the greening work along the railway. The surface of commercial coal in transportation is solidified to minimize environmental damage and pollution. Ultimately realize the unification of economic, social and environmental benefits.

(d) Optimizing the structural of coal industry and promoting industrial transformation and upgrading.

The storage conditions of coal resources in different regions are very different, which causes the development of coal mining enterprises to be extremely uneven. Many companies face the problem of low output and high consumption. For some enterprises in old mining areas, their competitiveness is relatively weak and transformation is difficult. It should abandon the development method that aims at quantity and speed, and shift to the goal of quality and efficiency.

The government should insist on optimizing the corporate structure to avoid vicious competition among coal companies. In order to avoid large fluctuations in the economy, the government needs to reduce the collapse or bankruptcy of a large number of energy companies caused by the rapid reduction of coal prices.

(e) Coal enterprises increase technological innovation and promote innovation in development concepts.

As a traditional labor-intensive industry, high-risk industry, and high-polluting industry, the coal industry has a relatively low proportion of high-quality talents. The innovation potential of enterprises is still huge, and its scientific and technological innovation system still needs to be improved and perfected. Coal enterprises must insist on innovation-driven, and then establish and improve technology, and innovate in investment, research and development, transformation and application mechanisms. The technological revolution is the driving force, and technological innovation must be the driving force of clean utilization of coal and the upgrading and transformation of coal industry.

Strengthen the idea innovation of the enterprise. Although the development of coal companies is inseparable from the support of national policies, coal companies need to recognize the reality and face increasingly severe tests. They cannot rely on the government blindly, but change from themselves. Coal enterprises should change from relying on government and policy support to relying on innovation and market competition. They need to firmly establish the concept of market economy, improve the quality of their own development, enhance the market competitiveness, to maintain the stable growth of social economy.

(f) Coal enterprises should strengthen basic management and implement lean management concepts.

Coal enterprises should strengthen the adjustment of infrastructure and promote lean management. How to achieve high-standard product quality with less investment, less resource consumption, and less expense events is the focus that coal companies need to explore. Produce more economic benefits, and continuously achieve the goal of enterprise development. Through scientific methods, the coal mining and utilization can be operated, monitored, and evaluated, to obtain satisfactory economic and social benefits.

3.5 Conclusions

This chapter analyzes the international coal market with examples of machine learning hybrid modeling, and the following conclusions can be drawn:

(a) Coal prices have an important effect on the stability of the international market. As an important element in promoting international finance development, the price fluctuation of the coal industry greatly impacts commodity prices, trade and investment, production and consumption, and import and export. Furthermore, coal prices have even affected the economic development and social stability of many countries. The work to explore the causes of coal price fluctuations highlights great significance.
(b) Big data forecasting technology plays a guiding role in energy policy regulation. Many factors affect coal prices, and data-driven artificial intelligence hybrid modeling can accurately capture coal price trends. Effective coal price forecasts can help policy implementers prepare for risk aversion in advance, thereby reducing the impact of coal price fluctuations on global economic development. It has important practical significance for maintaining sustained national economy development and social stability.

(c) A variety of factors can affect international coal prices. The output of the coal price prediction model considering many factors is closer to the real fluctuation. This chapter studies the influence of fossil fuels, gold, silver, copper, iron, and currency exchange rate on coal prices. The study found that the exchange rate between the US dollar and the Australian dollar greatly influences coal prices, and the price changes of crude oil, natural gas, gold, silver, copper, and iron also affect coal prices. With the combination of several highly correlated factors, the forecasting model can better capture the changing trend of coal prices.
(d) Compared with a single deep learning weak predictor, the optimized ensemble method based on AdaBoost.MRT increases the robustness of the hybrid model. The integrated strong predictor can better fit the original sequence. The hybrid model can also track extreme values such as peaks and valleys of coal prices. The experimental comparison demonstrates the effectiveness of the AdaBoost.MRT.
(e) The prediction accuracy can be effectively improved by the decomposition-based data pre-processing method. The decomposition algorithm disassembles the original data into multiple more stable series, and the subseries after the split has lower non-stationarity. This feature makes it easier to be learned by the model, and the deep information contained in it can be more fully mined. The research in this chapter compares the decomposition algorithms of EMD, ICEEMDAN, WD, WPD, MODWT, MODWPT, and EWT. The experimental results found that when the number of prediction steps is small, MODWT and MODWPT have the best effect on improving the accuracy, and when the steps are 10-day, ICEEMDAN and WD perform better. Research results in this chapter show that the data has different characteristics under different prediction steps, and experimental comparison is needed to determine the best decomposition algorithm under different prediction steps.

References

Alameer, Z., Abd Elaziz, M., Ewees, A. A., Ye, H., & Jianhua, Z. (2019). Forecasting copper prices using hybrid adaptive neuro-fuzzy inference system and genetic algorithms. *Natural Resources Research, 28*(4), 1385–1401.

Alameer, Z., Fathalla, A., Li, K., Ye, H., & Jianhua, Z. (2020). Multistep-ahead forecasting of coal prices using a hybrid deep learning model. *Resources Policy, 65*, 101588. https://doi.org/10.1016/j.resourpol.2020.101588

Bildirici, M. E., & Sonustun, F. O. (2018). The effects of oil and gold prices on oil-exporting countries. *Energy Strategy Reviews, 22*, 290–302.

Boudraa, A.-O., & Cexus, J.-C. (2007). EMD-based signal filtering. *IEEE Transactions on Instrumentation and Measurement, 56*(6), 2196–2202.

Chiappini, R., Jégourel, Y., & Raymond, P. (2019). Towards a worldwide integrated market? New evidence on the dynamics of US, European and Asian natural gas prices. *Energy Economics, 81*, 545–565.

Cologni, A., & Manera, M. (2008). Oil prices, inflation and interest rates in a structural cointegrated VAR model for the G-7 countries. *Energy Economics, 30*(3), 856–888.

Dong, B., Li, X., & Lin, B. (2010). Forecasting long-run coal price in China: A shifting trend time-series approach. *Review of Development Economics, 14*(3), 499–519.

Gilles, J. (2013). Empirical wavelet transform. *IEEE Transactions on Signal Processing, 61*(16), 3999–4010.

Gokhale, M. Y., & Khanduja, D. K. (2010). Time domain signal analysis using wavelet packet decomposition approach. *International Journal of Communications, Network and System Sciences, 3*(03), 321.

Haque, Md. A., Topal, E., & Lilford, E. (2017). Evaluation of a mining project under the joint effect of commodity price and exchange rate uncertainties using real options valuation. *The Engineering Economist, 62*(3), 231–253.

Hastie, T., Rosset, S., Zhu, J., & Zou, H. (2009). Multi-class adaboost. Statistics and its. *Interface, 2*(3), 349–360.

Herrera, G. P., Constantino, M., Tabak, B. M., Pistori, H., Su, J.-J., & Naranpanawa, A. (2019). Long-term forecast of energy commodities price using machine learning. *Energy, 179*, 214–221.

Huang, N. E., Shen, Z., Long, S. R., Wu, M. C., Shih, H. H., Zheng, Q., Yen, N.-C., Tung, C. C., & Liu, H. H. (1998). The empirical mode decomposition and the Hilbert spectrum for nonlinear and non-stationary time series analysis. *Proceedings of the Royal Society of London Series A: Mathematical, Physical and Engineering Sciences, 454*(1971), 903–995.

IEA International Energy Agency. (2021). Shaping a secure and sustainable energy future for all. https://www.iea.org/. Accessed 10 September 2021.

Ji, Q., Geng, J.-B., & Fan, Y. (2014). Separated influence of crude oil prices on regional natural gas import prices. *Energy Policy, 70*, 96–105.

Krzemień, A., Riesgo, F. P., Suárez, S. A., & Sánchez, L. F. (2015). Forecasting European thermal coal spot prices. *Journal of Sustainable Mining, 14*(4), 203–210. https://doi.org/10.1016/j.jsm.2016.04.002

Lago, J., De Ridder, F., & De Schutter, B. (2018). Forecasting spot electricity prices: Deep learning approaches and empirical comparison of traditional algorithms. *Applied Energy, 221*, 386–405.

Li, H., Chen, L., Wang, D., & Zhang, H. (2017). Analysis of the price correlation between the international natural gas and coal. *Energy Procedia, 142*, 3141–3146.

Liu, H., Yang, R., & Duan, Z. (2020). Wind speed forecasting using a new multi-factor fusion and multi-resolution ensemble model with real-time decomposition and adaptive error correction. *Energy Conversion and Management, 217*, 112995. https://doi.org/10.1016/j.enconman.2020.112995

Liu, X., & Luan, H. (2015). A brief analysis of the circumstances of the current coal industry and its trends in China. *International Journal of Mining Mineral Engineering, 6*(1), 87–96.

Matyjaszek, M., Fernández, P. R., Krzemień, A., Wodarski, K., & Valverde, G. F. (2019). Forecasting coking coal prices by means of ARIMA models and neural networks, considering the transgenic time series theory. *Resources Policy, 61*, 283–292.

Morlet, J., Arens, G., Fourgeau, E., & Glard, D. (1982). Wave propagation and sampling theory—Part I: Complex signal and scattering in multilayered media. *Geophysics, 47*(2), 203–221.

Parsons, S., Boonman, A. M., & Obrist, M. K. (2000). Advantages and disadvantages of techniques for transforming and analyzing chiropteran echolocation calls. *Journal of Mammalogy, 81*(4), 927–938.

Percival, D. B., & Walden, A. T. (2000). *Wavelet methods for time series analysis* (Vol. 4). Cambridge University Press.

Qiu, X., Suganthan, P. N., & Amaratunga, G. A. J. (2016). Electricity load demand time series forecasting with empirical mode decomposition based random vector functional link network. In *2016 IEEE International Conference on Systems, Man, and Cybernetics (SMC)* (pp. 001394–001399). IEEE.

Sachs, J. D., & Warner, A. (1995). *Natural resource abundance and economic growth.* National Bureau of Economic Research Cambridge

Singhal, S., Choudhary, S., & Biswal, P. C. (2019). Return and volatility linkages among International crude oil price, gold price, exchange rate and stock markets: Evidence from Mexico. *Resources Policy, 60*, 255–261.

Song, Y., & Wang, N. (2019). Exploring temporal and spatial evolution of global coal supply-demand and flow structure. *Energy, 168*, 1073–1080.

Sribna, Y., Trokhymets, O., Nosatov, I., & Kriukova, I. (2019). The globalization of the world coal market–contradictions and trends. In *E3S Web of Conferences* (p. 01044). EDP Sciences.

Torres, M. E., Colominas, M. A., Schlotthauer, G., & Flandrin, P. (2011). A complete ensemble empirical mode decomposition with adaptive noise. In *2011 IEEE International Conference on Acoustics, Speech and Signal Processing (ICASSP)* (pp. 4144–4147). IEEE.

Tseng, K.-K., Lin, R.F.-Y., Zhou, H., Kurniajaya, K. J., & Li, Q. (2018). Price prediction of e-commerce products through Internet sentiment analysis. *Electronic Commerce Research, 18*(1), 65–88.

Wang, B., & Wang, J. (2019). Energy futures prices forecasting by novel DPFWR neural network and DS-CID evaluation. *Neurocomputing, 338*, 1–15.

Wang, W., Li, Z., & Cheng, X. (2019). Evolution of the global coal trade network: A complex network analysis. *Resources Policy, 62*, 496–506.

Xue, G., & Sriboonchitta, S. (2014). Co-movement of prices of energy and agricultural commodities in biofuel Era: A period-GARCH Copula approach. In *Modeling dependence in econometrics* (pp. 505–519). Springer.

Zang, H., Cheng, L., Ding, T., Cheung, K. W., Liang, Z., Wei, Z., & Sun, G. (2018). Hybrid method for short-term photovoltaic power forecasting based on deep convolutional neural network. *IET Generation, Transmission and Distribution, 12*(20), 4557–4567.

Zhu, L., Wang, Y., & Fan, Q. (2014). MODWT-ARMA model for time series prediction. *Applied Mathematical Modelling, 38*(5–6), 1859–1865.

4 Big Data Analysis of Energy Economics in Wind Power Market

4.1 Introduction

Wind energy is one of the most commercially potential and most dynamic renewable energy sources. It is clean in use, low in cost, and inexhaustible. Wind power generation has the advantages of large room for growth in installed capacity, rapid cost reduction, renewable energy, wide distribution, and no pollution. While providing a stable power supply for economic growth, wind power can effectively alleviate air pollution, water pollution, and global warming issues. In the development of various new energy sources, wind power generation is a power generation method with relatively mature technology and conditions for large-scale development and commercial development. Wind power can reduce the large number of pollutants and carbon emissions generated by fossil fuel power generation. Large-scale promotion of wind power can make a positive contribution to energy conservation and emission reduction (Zhao et al., 2012). In the context of the increasingly serious global energy crisis and environmental crisis, wind energy resources have begun to receive widespread attention. The large-scale development of wind power generation provides a broad market space and prospects for the wind power generation equipment manufacturing industry.

According to the World Wind Energy Association, the world's installed capacity of wind power has grown at an average annual rate of about 30% in recent years (WWEA 2021). After a record-breaking 2020, the global wind power installation boom will cool off this year. Under the strong pull of China, global wind power construction will develop rapidly in 2020, but it will decline in 2021. Bloomberg New Energy Finance predicts that onshore wind power construction in Europe, the Middle East and Africa (EMEA), Latin America, and other Asia–Pacific markets will reach the highest level ever (BNEF 2021). The scale of onshore wind power construction in the United States will also be the same as last year's record. After 2025, onshore wind power construction in all regions except China has entered

© Science Press 2022

H. Liu et al., *Big Data in Energy Economics*, Management for Professionals,
https://doi.org/10.1007/978-981-16-8965-9_4

a plateau period, and the global wind power growth will be mainly driven by offshore wind power.

According to the statistics of the International Energy Agency (IEA), the carbon dioxide emissions of China mainly come from the combustion of coal (Downie, 2020). Therefore, to cope with carbon dioxide emission reduction, controlling the scale of coal power and the proportion of power generation will be an important starting point. It will have an important impact on future power supply structures. The big data analysis of energy economics in wind power market is essential to analyze the future demand development trend of wind power industry.

4.2 Multi-temporal and Spatial Scale Wind Power Big Data Forecasting

Wind power big data forecasting can deterministically improve the conversion efficiency and reduce the occurrence of wind abandonment. Compared with single-point forecasting, models based on spatial scale can further improve forecasting accuracy (Hu et al., 2021). Using spatial features to assist the model brings better performance and interpretability. Therefore, the section focuses on multi-temporal and spatial scale wind power forecasting models.

4.2.1 Description of Original Wind Dataset

To analyze spatial wind power forecasting models, datasets containing 415 wind turbines are utilized. They have a certain correlation on the spatial level. The data were downloaded from National Renewable Energy Laboratory (NREL). The resolution of wind power series is 5 min. Among the 415 wind turbines, four sites are selected as forecasting targets. The locations of these wind turbines are shown in Fig. 4.1.

Table 4.1 shows the statistics of wind power data in four target sites. After reaching the rated wind speed, the generated power of wind turbine keeps the rated output. Therefore, the maximum wind power values of four targets are 14 kW, 10 kW, 16 kW, and 16 kW, respectively. The minimum power demonstrates the actual wind speed is lower than the required value to start wind turbines. So, there is no wind power output. The standard deviations of four target sites are 5.5989 kW, 3.8933 kW, 6.3102 kW, and 5.9088 kW, respectively. The wind power in target site #2 is apparently more stable than other three targets.

The actual curves of wind power in four target sites are plotted in Fig. 4.2. Each wind power series contains 9216 samples in total. They are divided into three parts that serve different purposes. The 0–60% data form training set, which is utilized to train forecasting models. The 60–80% data from validation set, which is used to achieve feature selection results based on binary optimization algorithms. The last 20% data from test set, which is remained to evaluate the performance of models.

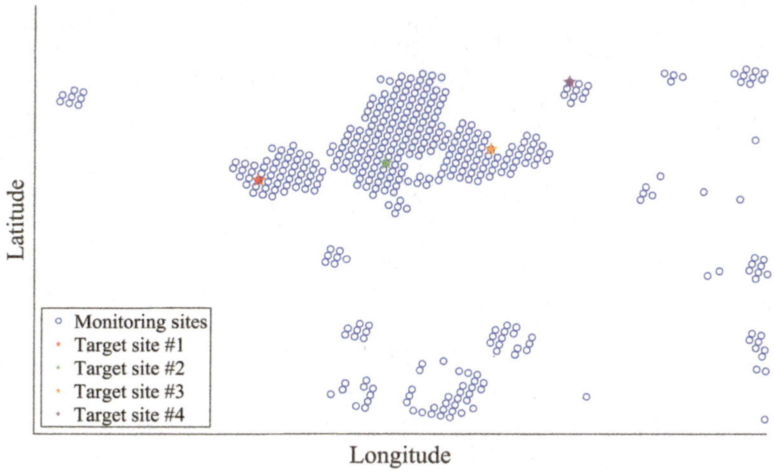

Fig. 4.1 Location of wind turbines in the dataset

Table 4.1 Statistics of wind power data in target sites

Target site	Maximum (kW)	Minimum (kW)	Mean (kW)	Standard deviation (kW)	Kurtosis
#1	14	0	7.8785	5.5989	1.3459
#2	10	0	6.1785	3.8933	1.4557
#3	16	0	9.4097	6.3102	1.4441
#4	16	0	7.5818	5.9088	1.5025

4.2.2 Framework of Wind Power Forecasting Models

Figure 4.3 shows the framework of spatial scale wind power forecasting models. The construction of models includes the following steps:

(a) The wind power dataset is separated into two groups, including target site and adjacent sites.
(b) The wind power series is divided into training set, validation set, and test set.
(c) Use mutual information to quantitatively assess the relevance of each site. Generate alternative spatial data by specific threshold.
(d) Perform spatial feature selection using binary optimization algorithms. The applied algorithms include Binary Particle Swarm Optimization (BPSO) (Khanesar et al., 2007), Binary Grey Wolf Optimization (BGWO) (Emary et al., 2016), Binary Differential Evolution (BDE) (Pampara et al., 2006), and Binary Harris Hawk Optimization (BHHO) (Too et al., 2019).
(e) Train Extreme Learning Machine (ELM) forecasting models based on spatial features.

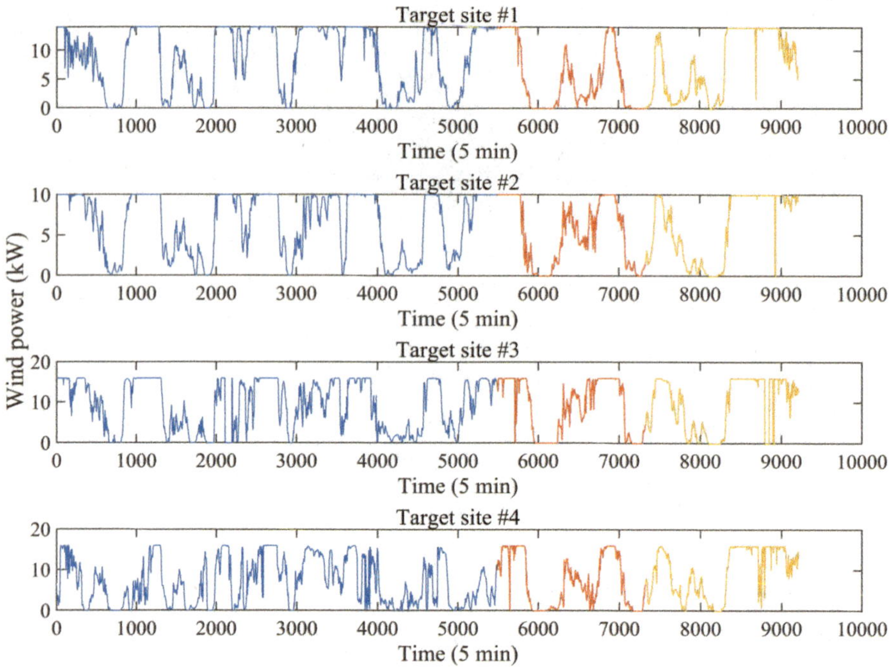

Fig. 4.2 Wind power data of four target sites

(f) Evaluate forecasting results by Mean Absolute Error (MAE) (Willmott & Matsuura, 2005), Root Mean Square Error (RMSE) (Chai & Draxler, 2014), and Pearson Correlation Coefficient (PCC) (Benesty et al., 2009).

4.2.3 Analysis of Wind Power Forecasting Models

4.2.3.1 Fitness Curve of Optimization Process

In the wind power forecasting models, spatial feature selection is achieved by different binary optimization algorithms. The convergence of these algorithms significantly affects the effectiveness of models. The binary optimization process includes 100 iterations. During 100 iterations, the fitness curves of four algorithms are shown in Fig. 4.4. It can be observed that:

(a) The fitness curves of all algorithms decrease during the optimization process. This proves that the comparison experiments are valid. Since MAE represents the error of the prediction model, the reduction of the error represents the improvement of the prediction performance. The MAE of forecasting models can be effectively reduced by selecting proper spatial features to train models.

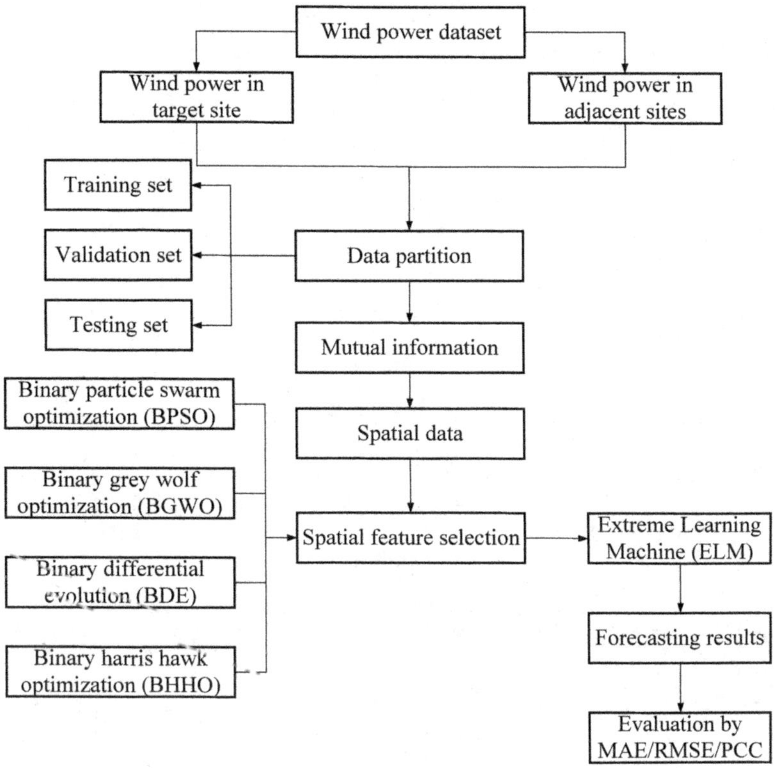

Fig. 4.3 Framework of spatial scale wind power forecasting models

(b) Regarding to the convergence performance of four algorithms, the BGWO algorithm is the best. It rapidly reduces fitness in the beginning iterations. Its fitness values for four target sites are significantly lower than other algorithms. It is determined by the algorithmic mechanism of BGWO itself, which converges to the optimal value more quickly. The results demonstrate that the BGWO may be more suitable for spatial feature selection of wind power data.

4.2.3.2 Selection of Spatial Features

The detailed selection results of spatial features for target sites #1–#4 are shown in Figs. 4.5, 4.6, 4.7 and 4.8. The alternative features contain wind power data of adjacent monitoring sites with 12 time lags. The selected features are labeled as 1, while the abandoned features are labeled as 0. After comparison of the selection results based on four algorithms, it can be observed that:

(a) The selection result of BDE is very poor, unable to find out features with strong characterization ability, but retains most of the features. The result is

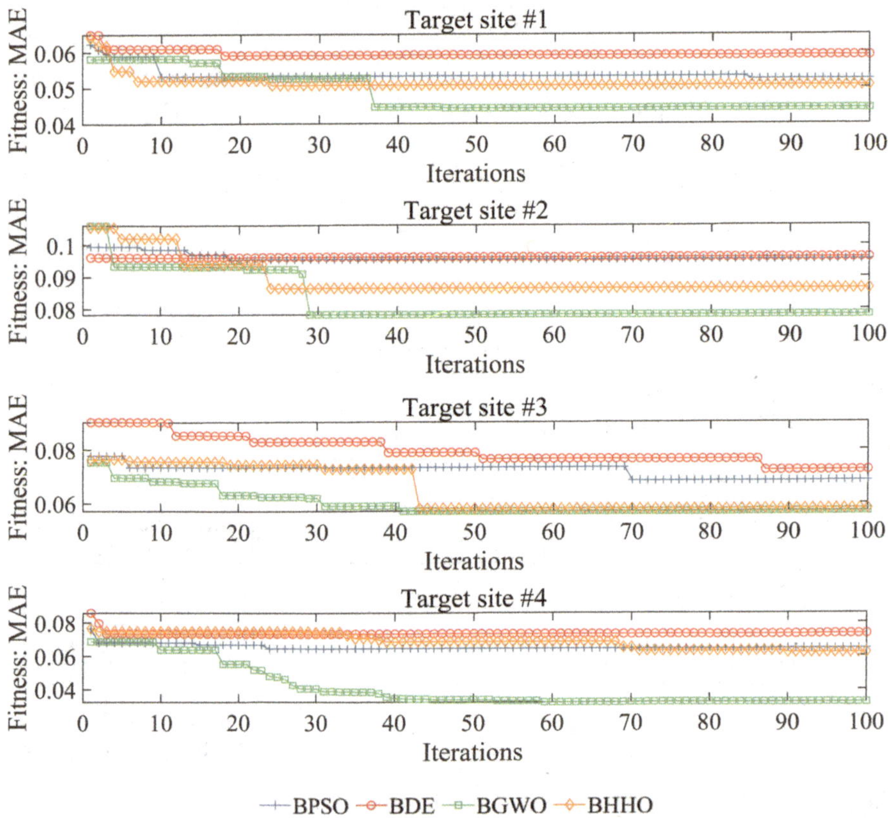

Fig. 4.4 Fitness curves of four optimization algorithms for spatial feature selection

a disaster for the forecasting model, because too much data may have strong redundancy, making it difficult for the model to learn useful information.

(b) In terms of the number of obtained features, the results of BPSO and BHHO are similar. It demonstrates that the two algorithms cannot properly select spatial features for wind power data.

(c) BGWO can select a few important features. No matter for which target site, the obtained features are far less than other algorithms. Especially for targets #4, only 4 features are selected in the end.

(d) Features of target sites or short-term time lags are more likely to be selected. The result makes sense because the data is more relevant to the future wind power of the target site.

4.2.3.3 Comparison of Wind Power Forecasting Results

To find out the best wind power forecasting model for spatial data, the forecasting results of ELM, BPSO-ELM, BDE-ELM, BGWO-ELM, and BHHO-ELM

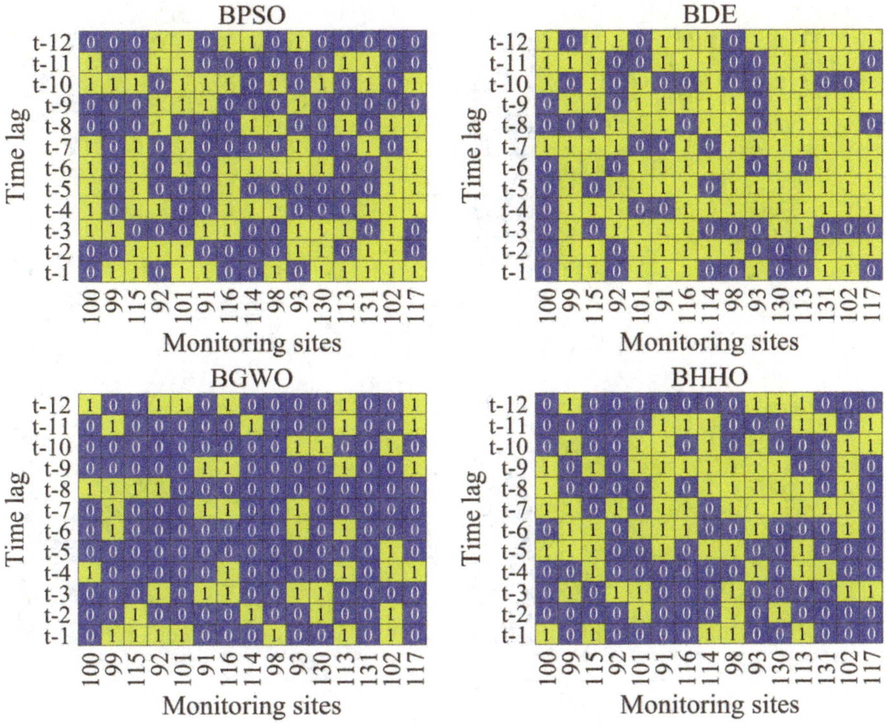

Fig. 4.5 Spatial feature selection results for target site #1

models are compared. Among these models, the ELM model absorbs all spatial data obtained by MI. The other four, using different optimization algorithms, are used for further spatial feature selection. Figures 4.9, 4.10, 4.11 and 4.12 show the 1-step wind power forecasting results of these models for target sites #1–#4, respectively. The top subfigure shows the overall prediction results of each model on the full test set. The bottom subfigure shows local prediction results for a specific time span. Based on Figs. 4.9, 4.10, 4.11 and 4.12, the following conclusions can be drawn:

(a) All the predicted results are in satisfactory agreement with the actual values. The 1-step ahead overall forecasting results follow the original wind power data in general.
(b) The BPSO-ELM and BPSO-BDE models generate outliers in some local forecasting results. For instance, in the 1-step forecasting results of target site #1, the original wind power data rapidly increases and reaches 15 kW in the 1330th sample. While the BPSO-ELM forecasting results show a significant lag and reach 15 kW after the 1340th sample. In the 1-step forecasting results of target site #2, the original wind power data remain the maximum in the

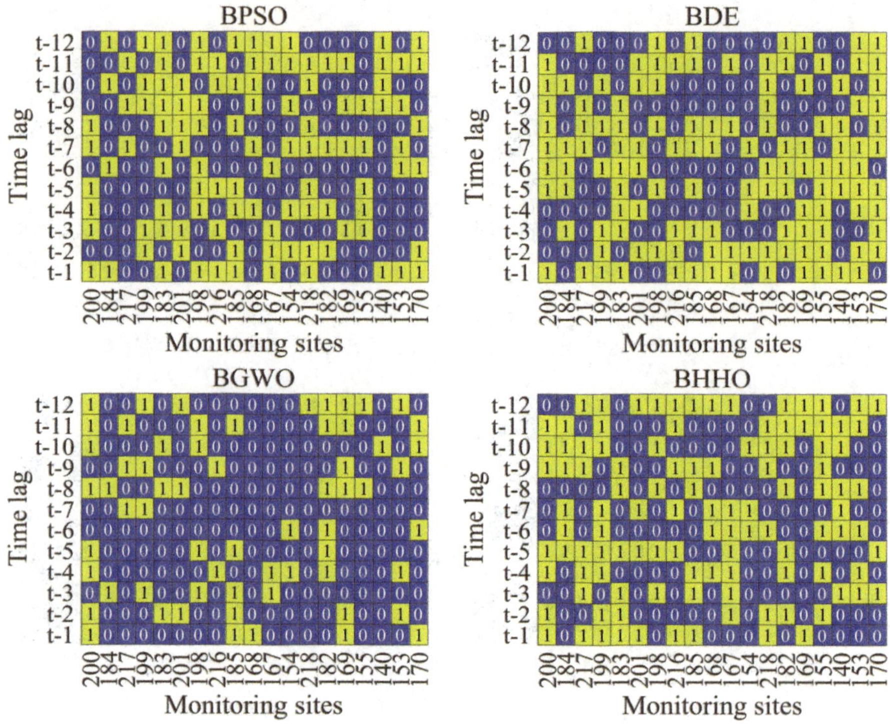

Fig. 4.6 Spatial feature selection results for target site #2

local subfigure. While the BDE-ELM forecasting results have a very large gap compared with the real value.

(c) The BGWO-ELM model has absolute advantage compared with other models. For instance, in the 1-step wind power forecasting results for target site #2, the results of BGWO-ELM model are the closest to the actual values. The result is determined by the characteristics of different algorithms. It can be concluded that the BGWO is more suitable for spatial feature selection of wind power data.

For a more detailed and quantitative analysis, the evaluation results of the forecasting results for target sites #1–#4 are listed in Tables 4.2, 4.3, 4.4 and 4.5, respectively. The utilized evaluation indices include MAE, RMSE, and PCC. According to Tables 4.2, 4.3, 4.4 and 4.5, it can be concluded that:

(a) Four binary optimization algorithms can generally guarantee better forecasting accuracy than the ELM model. For instance, in the 3-step forecasting results of target site #1, the MAE, RMSE, PCC of ELM model are 0.7384 kW, 1.2732 kW, and 0.9733, respectively. The MAE, RMSE, PCC of BPSO-ELM model are 0.7002 kW, 1.1644 kW, and 0.9774, respectively. The MAE, RMSE,

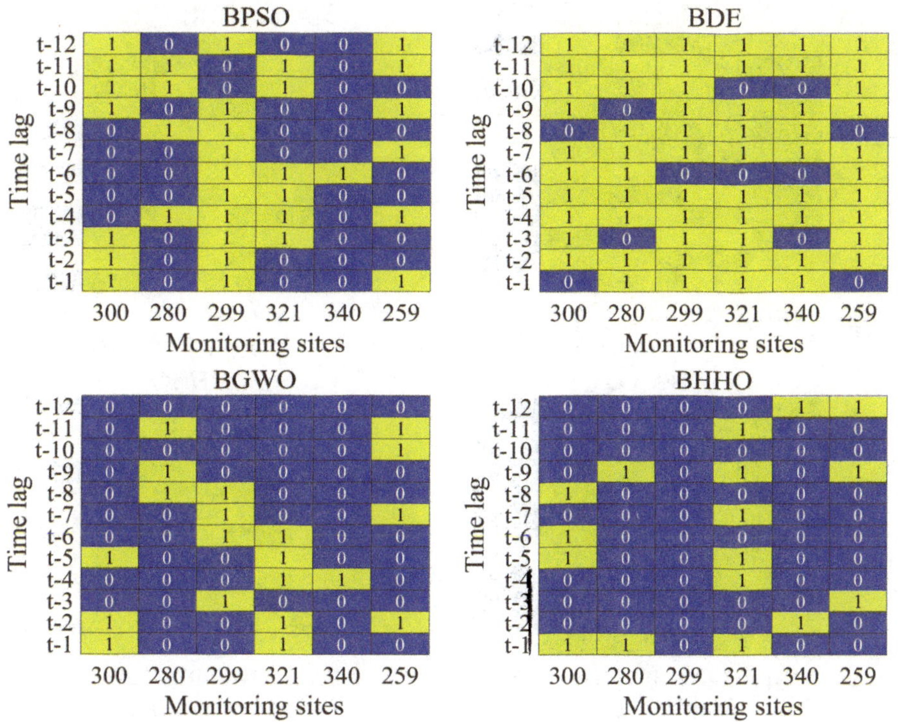

Fig. 4.7 Spatial feature selection results for target site #3

PCC of BDE-ELM model are 0.6859 kW, 1.1042 kW, and 0.9795, respectively. The MAE, RMSE, PCC of BGWO-ELM model are 0.6693 kW, 1.2097 kW, and 0.9760, respectively. The MAE, RMSE, PCC of BHHO-ELM model are 0.6837 kW, 1.1647 kW, and 0.9775, respectively. The MAE and RMSE values of BPSO-ELM, BDE-ELM, BGWO-ELM, and BHHO-ELM models are generally smaller than the ELM model. The PCC values of BPSO-ELM, BDE-ELM, BGWO-ELM, and BHHO-ELM models are generally larger than the ELM model.

(b) Among the four target sites, the wind power forecasting models show the worst accuracy in target site #3. The best forecasting model for target site #3 is the BHHO-ELM model. The MAE, RMSE, and the PCC of its 3-step forecasting results are 0.9544 kW, 1.8423 kW, and 0.9599, respectively. The best MAE, RMSE, and PCC of 3-step forecasting results for target site #4 are 0.4738 kW, 1.0325 kW, and 0.9847, respectively. The results are determined by the characteristics of dataset.

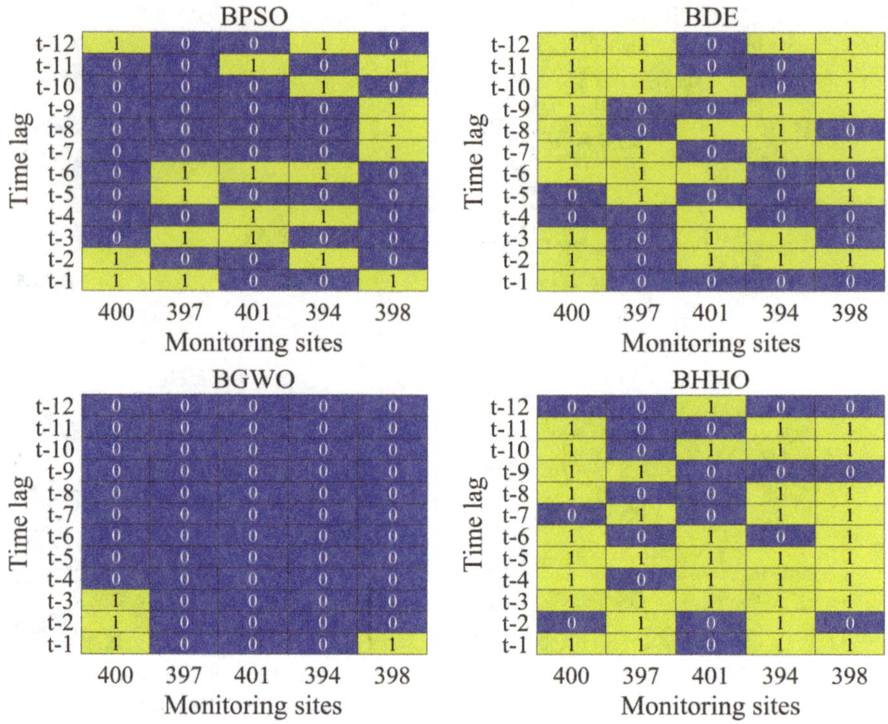

Fig. 4.8 Spatial feature selection results for target site #4

4.3 Conversion Efficiency of Wind Power Energy

The low utilization efficiency has seriously hindered the development of wind energy (Han et al., 2015). Therefore, it is very important to improve the utilization efficiency of wind energy. Wind power generation relies on wind turbines to convert wind energy into electrical energy. The wind energy utilization factor represents the ability of wind turbines to absorb natural wind, and is numerically equal to the conversion efficiency from wind energy to electricity, which is expressed by Cp. According to Betz theory, the maximum wind energy utilization factor of wind turbines is 0.593 (Ragheb & Ragheb, 2011). The size of the wind energy utilization coefficient is affected by the blade tip speed ratio and blade pitch angle (Zhao et al., 2015). Its specific expression is as follows (Ma et al., 2018):

$$C_p = \frac{P}{\frac{1}{2}\rho v^3 s}. \tag{4.1}$$

where ρ represents air density, P denotes actual shaft power obtained by wind turbine, s denotes sweep area of wind wheel, and v is the actual wind speed.

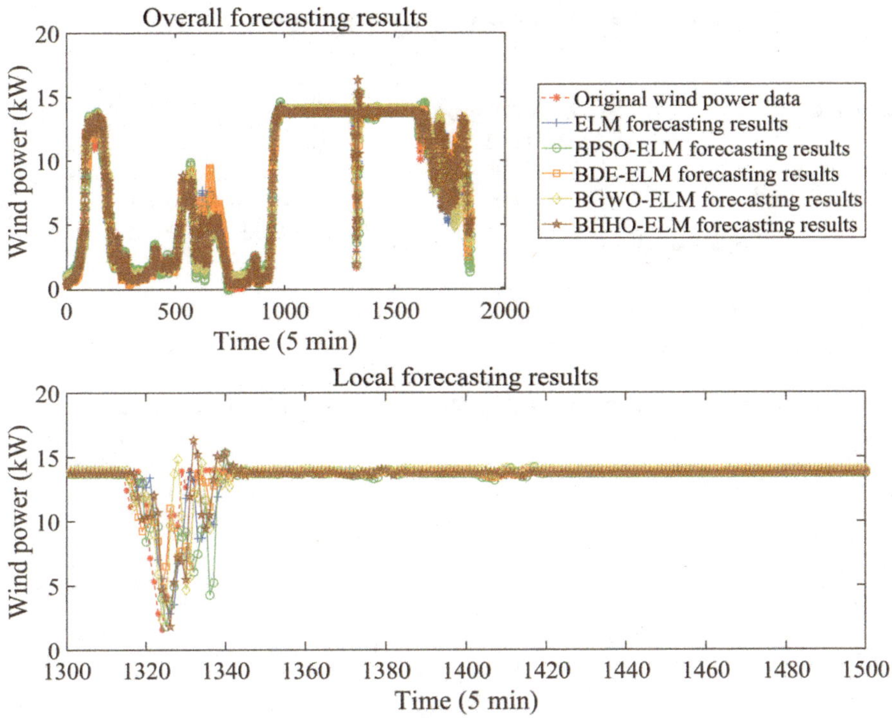

Fig. 4.9 The 1-step wind power forecasting results of different spatial selection methods for target site #1

When the wind wheel of the wind turbine is facing the wind direction, the wind wheel can get the most wind energy. To ensure that the wind wheel is facing the wind direction at all times, a yaw system is provided in the wind turbine. When the wind vane installed on the top of the nacelle detects that the wind wheel is not facing the wind direction, it will issue a yaw command. The nacelle and the wind wheel are rotated around the vertical axis of the tower through the yaw system to achieve the purpose of aligning the wind direction. The rotation speed of the wind wheel increases with the increase of wind speed. Too much wind speed will cause wind wheel speed to be too fast and the generator overload operation, all of which will cause operation accidents of the wind generator. To guarantee the safety of wind turbines, the wind turbines are equipped with speed-limiting safety devices to adjust the speed of the wind turbine rotors, so that they remain basically unchanged within a certain wind speed range, so that the wind turbines can operate under different wind conditions.

At the same time, the phenomenon of wind curtailment also affects the efficiency of wind energy conversion. Wind curtailment refers to the suspension of some wind farms because of erratic wind generation, inadequate capacity on the local grid, and ill-timed wind-farm construction plans, when wind turbines are

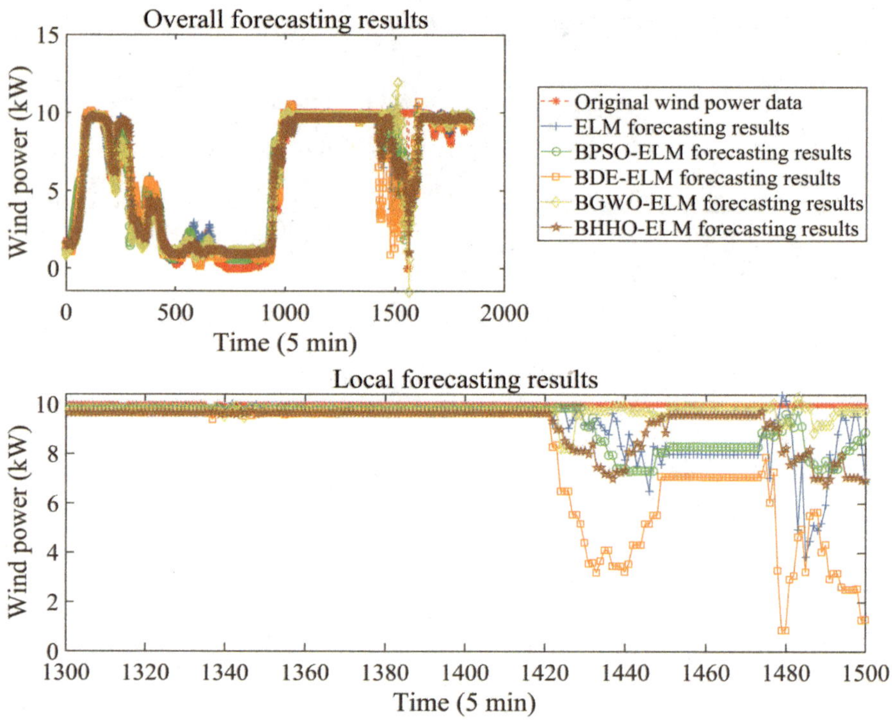

Fig. 4.10 The 1-step wind power forecasting results of different spatial selection methods for target site #2

under normal conditions in the early stage of wind power development. Since 2010, the phenomenon of wind curtailment began to appear. Since then, in 2011 and 2012, the national average wind curtailment rate has increased year by year. Since 2015, the problem of wind curtailment and power curtailment has become more serious. According to statistical data released by the National Energy Administration of China, the national average wind curtailment rate has reached 15% in 2015 (Lewis, 2016). Severe wind curtailment and power curtailment have brought about huge power loss and evaporation of social and economic benefits. In some years, annual installed capacity was almost offset to make up for the loss. In the long run, it is not just the hapless renewable-energy companies that will be overwhelmed by the financial pressure, but the entire industry chain as well.

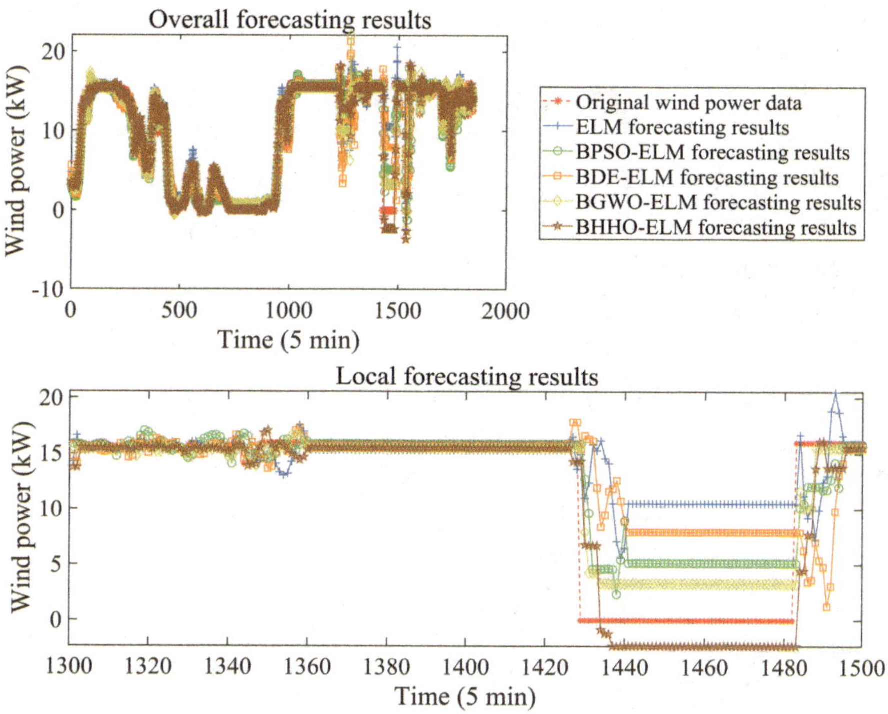

Fig. 4.11 The 1-step wind power forecasting results of different spatial selection methods for target site #3

4.4 Market Economy Analysis of Wind Power Application

4.4.1 Market Economy Analysis of Wind Power Application in China

In the section, the market economy of wind power application in China is analyzed. The data are obtained from Global Wind Energy Council (GWEC) (GWEC 2021) and National Energy Administration of China (NEAC).

In recent years, China has gradually released a series of price subsidies and wind-level pricing policies. In 2019, the state will actively promote the construction of unsubsidized and parity grid-connected projects for wind power, comprehensively promote the competitive allocation of wind power projects, establish a new mechanism for the consumption of renewable energy, and promote the market-oriented transaction of distributed renewable energy in combination with power reforms, etc. Comprehensively promote the high-quality development of renewable energy. As of 2019, the national wind curtailment rate has dropped. The wind curtailment rate of Xinjiang and Gansu both hit record lows.

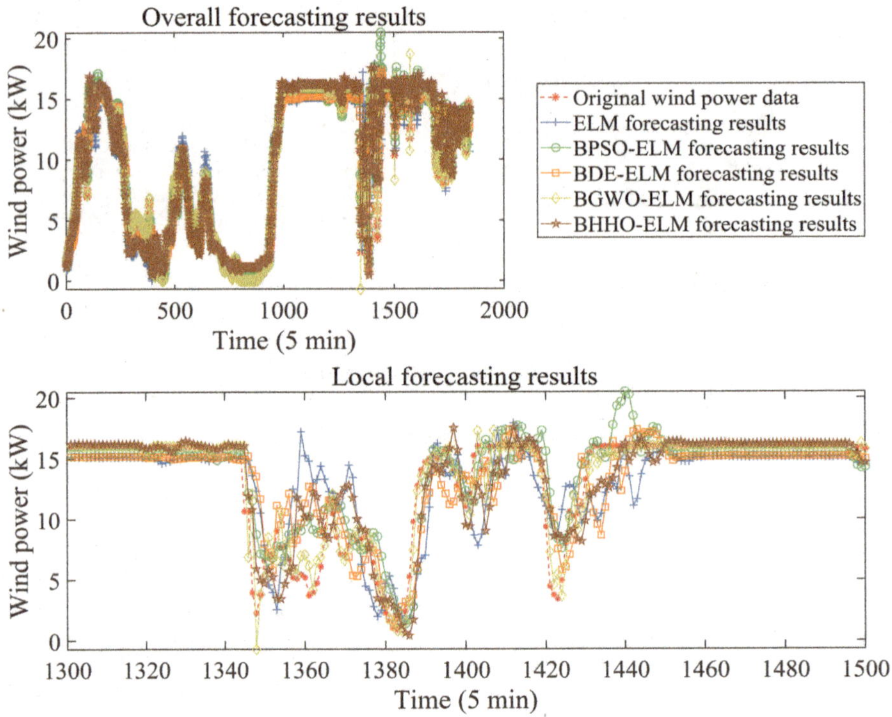

Fig. 4.12 The 1-step wind power forecasting results of different spatial selection methods for target site #4

From 2018 to 2020, North China has maintained a leading trend. Its installed capacity reached 59.76 million kW in 2019, a year-on-year growth rate of 11.2%. In 2019, in terms of growth rate, Central China ranked first with a growth rate of 41.76%, followed by South China with 26.71%. In the first half of 2020, South China surpassed Central China to become the first growth rate, with a growth rate of 12.25%; followed by East China, with a growth rate of 5.54%.

4.4.1.1 Increasing Installed Capacity

In 2019, China's installed wind power capacity reached 210.05 million kW, accounting for roughly 32.29% of the world's cumulative wind power installed capacity, an increase of approximately 1% point from the previous year. As of the first half of 2020, China's installed wind power capacity reached 216.75 million kW. From the perspective of the proportion, the proportion of cumulative installed capacity in China is showing an upward trend as a whole and is in the top position in the world.

Table 4.2 Evaluation of different wind power forecasting models for target site #1

Target site	Model	Step	MAE (kW)	RMSE (kW)	PCC
#1	ELM	1	0.6092	0.9966	0.9838
		2	0.6737	1.1344	0.9789
		3	0.7384	1.2732	0.9733
	BPSO-ELM	1	0.5907	0.9803	0.9841
		2	0.6432	1.0750	0.9809
		3	0.7002	1.1644	0.9774
	BDE-ELM	1	0.5553	0.8883	0.9869
		2	0.6163	0.9930	0.9835
		3	0.6859	1.1042	0.9795
	BGWO-ELM	1	0.5683	1.0311	0.9828
		2	0.6166	1.1148	0.9798
		3	0.6693	1.2097	0.9760
	BHHO-ELM	1	0.5693	0.9734	0.9846
		2	0.6257	1.0757	0.9810
		3	0.6837	1.1647	0.9775

Table 4.3 Evaluation of different wind power forecasting models for target site #2

Target site	Model	Step	MAE (kW)	RMSE (kW)	PCC
#2	ELM	1	0.8991	1.5386	0.9249
		2	0.9092	1.5454	0.9244
		3	0.9198	1.5493	0.9241
	BPSO-ELM	1	0.7048	1.2232	0.9544
		2	0.7057	1.2102	0.9554
		3	0.7095	1.2036	0.9560
	BDE-ELM	1	0.8920	1.7148	0.9126
		2	0.9171	1.7388	0.9098
		3	0.9433	1.7615	0.9071
	BGWO-ELM	1	0.7172	1.2827	0.9500
		2	0.7445	1.3279	0.9463
		3	0.7748	1.3795	0.9419
	BHHO-ELM	1	0.8738	1.3618	0.9438
		2	0.8849	1.3559	0.9442
		3	0.8970	1.3551	0.9443

Table 4.4 Evaluation of different wind power forecasting models for target site #3

Target site	Model	Step	MAE (kW)	RMSE (kW)	PCC
#3	ELM	1	1.5216	2.6778	0.9124
		2	1.6046	2.8011	0.9036
		3	1.6859	2.9165	0.8950
	BPSO-ELM	1	0.9048	1.7022	0.9655
		2	0.9943	1.8870	0.9574
		3	1.0966	2.0649	0.9487
	BDE-ELM	1	1.4688	2.6413	0.9145
		2	1.5597	2.7659	0.9058
		3	1.6455	2.8791	0.8976
	BGWO-ELM	1	0.8958	1.5729	0.9714
		2	0.9748	1.7276	0.9651
		3	1.0739	1.8960	0.9576
	BHHO-ELM	1	0.8161	1.4866	0.9743
		2	0.8801	1.6703	0.9673
		3	0.9544	1.8423	0.9599

Table 4.5 Evaluation of different wind power forecasting models for target site #4

Target site	Model	Step	MAE (kW)	RMSE (kW)	PCC
#4	ELM	1	1.2232	1.7097	0.9590
		2	1.2812	1.7952	0.9549
		3	1.3463	1.8816	0.9505
	BPSO-ELM	1	0.6421	0.9686	0.9866
		2	0.7553	1.1714	0.9804
		3	0.8726	1.3576	0.9736
	BDE-ELM	1	0.9893	1.3527	0.9749
		2	1.0716	1.4729	0.9702
		3	1.1507	1.5808	0.9658
	BGWO-ELM	1	0.1599	0.4320	0.9973
		2	0.3167	0.7573	0.9918
		3	0.4738	1.0325	0.9847
	BHHO-ELM	1	0.8874	1.3055	0.9760
		2	0.9611	1.4682	0.9694
		3	1.0269	1.6150	0.9628

4.4.1.2 Increasing Wind Power Generation

In the first half of 2008–2020, China saw an upward trend in electricity generation. The substantial increase in installed capacity has driven the rapid growth of power generation.

4.4.1.3 Accounting for an Essential Part of the World's New Wind Power Capacity

In 2019, the global new installed capacity of wind power was about 60.4 GW. In terms of global share, China's newly installed capacity is among the best in the world, which also shows China's strong support and emphasis on the development of the wind power industry.

4.4.1.4 Huge Development Potential of Offshore Wind Power

China's wind power industry mainly includes two types: onshore power generation and offshore power generation. Overall, offshore wind power is greener and more effective than onshore wind power. In recent years, as China's onshore wind power construction technology has become increasingly mature, the national wind power development policy has gradually tilted towards offshore power generation; in addition, offshore wind power resources are more extensive. On the coast of China's east coast, its exploitable wind energy resources are about 750 million kW. Not only the resource potential is huge, but also the means of development and utilization are mature, and the market is friendly. By the first half of 2020, it will increase to 6.99 million. Kilowatts, accounting for 3.22% of the total. It is expected that the market share of offshore wind power will further increase in the future.

4.4.2 Market Economy Analysis of Wind Power Application in America

Wind energy resources are greatly affected by topography. Most of the world's wind energy resources are concentrated in the coastal and open continental shrinking zones. From the distribution point of view, they are mainly distributed in North America, Asia, Latin America, and other places.

In the section, the market economy of wind power application in America is analyzed. The data are obtained from GWEC (GWEC 2021), U.S. Energy Information Administration (EIA 2021), U.S. Department of Energy (DE 2021), and International Renewable Energy Agency (IRENA) (IREA 2021).

According to a report by the U.S. Energy Information Administration, wind power generation in the United States increased from approximately 120 million MW in 2011 to 226 million MW in 2016, accounting for approximately 6% of the total U.S. power generation, showing strong growth. In 2016, wind power in the United States accounted for 6.2% of the national power generation, which was higher than the global average (4.6%). Vestas (43%), General Electric (42%), and Siemens (10%) account for 95% of the American wind turbine market. According to the report of the U.S. Department of Energy, in 2016, the U.S. invested 13 billion U.S. dollars in wind power and added 8.203 GW of new wind power installed capacity (90% of the new installed capacity was in the central states of the United States). Among the newly installed capacity is the 30 MW Rhode Island Wind Farm, the first offshore wind power plant in the United States. Another 48 MW of wind power has been decommissioned. At the end of 2016, the cumulative installed capacity reached 82.143 GW, ranking second in the world after China (cumulative installed capacity of 168.732 GW). The number of American wind power employees reached 102,000. In 2017, the wind power newly installed capacity was 6.3 GW, and the total was 87.5 GW; wind power reached 2542.5 trillion kWh (6.3%), an increase of 12% over 2016.

According to statistics from the IRENA, the installed capacity of wind energy in the United States increased from 24,651 to 87,544 MW, and the power generation increased from 55,696 to 254,250 GWh between 2008 and 2017.

According to GWEC forecasts, by 2024, there will be no utility-scale offshore wind power projects in North America. From 2025 to 2030, it is estimated that a total of 23 GW of offshore wind power projects will be built in the region. Although the Canadian coast has a high potential for technological development, the country's offshore wind power projects will not exceed 1 GW before 2030. At the state level, the East Coast industrial cluster is driving strong demand for offshore wind power. As of the first quarter of 2019, the announced total purchase target for offshore wind power reached 28.1 GW. GWEC predicts that the United States will build a total of 22.6 GW of offshore wind power projects by 2030.

4.4.3 Market Economy Analysis of Wind Power Application in Europe

In the section, the market economy of wind power application in Europe is analyzed. The data are obtained from Wind Europe (WE 2021). As early as the early 1990s, the European Union put forward plans and goals for vigorously developing wind power. So far, wind power has become the cheapest power generation mode in Europe. From the perspective of the regional development status of the European wind power industry, in 2020, Germany will have the largest wind power installed capacity and the Netherlands will have the largest new installed capacity; from the perspective of the proportion of wind power consumption in European countries, Denmark and Ireland will account for a relatively large amount of wind power consumption, respectively It is 48 and 38%.

From the perspective of wind power development prospects in European countries, the United Kingdom will be the focus of offshore wind power development from 2021 to 2025, and the United Kingdom will maintain an average annual installed capacity of 14.6 GW. In terms of onshore wind power development, Germany will be the focus of development, and Germany will Maintain an average annual installed capacity of 13 GW.

4.4.3.1 Cumulative Installed Wind Power Capacity in Europe Reaches Over 200 GW

Data show that from 2011 to 2020, the cumulative installed capacity of wind power in Europe has increased from 98 to 220 GW, and the annual new installed capacity has also continued to increase. In 2020, the cumulative installed capacity of wind power in Europe reaches over 200 GW.

4.4.3.2 Germany and Netherlands Have the Largest Installed Wind Power Capacities

In 2020, from the perspective of cumulative installed capacity, Germany's wind energy installed capacity share ranks first in Europe, reaching 29%, followed by Spain and the United Kingdom, with cumulative installed capacity shares of 12% and 11% respectively. From the perspective of new installed capacity, the Netherlands will have the largest new installed capacity in 2020, with a share of 14%; followed by Germany, Norway, and Spain, each with a share of more than 10%.

4.4.3.3 Denmark and Ireland Share the Largest Proportion of Wind Power Demand

From the perspective of wind power demand, in 2020, in the EU's electricity consumption structure, wind power will account for about 15%. Denmark is the world's leading wind power country, which started early in the field of wind power. It invested heavily in key areas such as research, development, certification, testing, and standards-setting of wind energy. From the perspective of the proportion of wind power consumption, Denmark and Ireland will account for relatively large wind power consumption, respectively 48%, 38%; followed by Germany, the United Kingdom, Portugal, Spain and Sweden. These countries account for about 20–30% of wind power consumption. In addition, these countries rank high in the European wind power installed capacity.

4.5 Conclusions

This chapter concentrates on the big data analysis of energy economics in wind power market. Several different spatial scale wind power big data forecasting models are proposed. Their forecasting performances have been thoroughly analyzed. It can be concluded that:

(a) Spatial data are helpful to the forecasting models. All the predicted results are in satisfactory agreement with the actual values. The 1-step ahead overall forecasting results follow the original wind power data in general. The MAE of forecasting models can be effectively reduced by selecting proper spatial features to train models, and the models with spatial data can sometimes outperform the single ELM model.

(b) Regarding the convergence performance of four different algorithms, the BGWO algorithm is the best. It rapidly reduces fitness in the beginning iterations. Its final fitness values for four target sites are significantly lower than other algorithms.
(c) The BGWO-ELM model has an absolute advantage compared with other models. The BGWO is more suitable for spatial feature selection of wind power data. However, the BPSO-ELM and BPSO-BDE models generate outliers in some local forecasting results.

Moreover, this chapter briefly introduces the conversion efficiency of wind power energy. The market economy of wind power applications is analyzed for China, America, and Europe. It can be found that with the increase of clean energy policy, the development of the wind power market shows a global trend of gradual expansion, and still has a tremendous development prospect. Major countries and regions have introduced industrial policies to encourage the development of wind power. In the future, Asia, North America, and Europe will continue to drive the growth of the wind power market to achieve the diversification of energy structure and carbon emission reduction targets.

References

Benesty, J., Chen, J., Huang, Y., & Cohen, I. (2009). Pearson correlation coefficient. In *Noise reduction in speech processing* (pp. 1–4). Springer.

BNEF Bloomberg New Energy Finance. (2021). New Energy Outlook 2021. https://about.bnef.com/new-energy-outlook/. Accessed 22 September 2021.

Chai, T., & Draxler, R. R. (2014). Root mean square error (RMSE) or mean absolute error (MAE)?—Arguments against avoiding RMSE in the literature. *Geoscientific Model Development, 7*(3), 1247–1250.

DE Department of Energy. (2021). https://www.energy.gov/. Accessed 22 September 2021.

Downie, C. (2020). Strategies for survival: The International Energy Agency's response to a new world. *Energy Policy, 141*, 111452.

EIA Energy Information Administration. (2021). https://www.eia.gov/. Accessed 22 September 2021.

Emary, E., Zawbaa, H. M., & Hassanien, A. E. (2016). Binary grey wolf optimization approaches for feature selection. *Neurocomputing, 172*, 371–381.

GWEC Global Wind Energy Council. (2021). Global Wind Report 2021. https://gwec.net/global-wind-report-2021/. Accessed 22 September 2021.

Han, W., Yan, P., Han, W., & He, Y. (2015). Design of wind turbines with shroud and lobed ejectors for efficient utilization of low-grade wind energy. *Energy, 89*, 687–701.

Hu, S., Xiang, Y., Zhang, H., Xie, S., Li, J., Gu, C., Sun, W., & Liu, J. (2021). Hybrid forecasting method for wind power integrating spatial correlation and corrected numerical weather prediction. *Applied Energy 293*, 116951.

IREA International Renewable Energy Agency. (2021). https://www.irena.org/. Accessed 22 September 2021.

Khanesar, M. A., Teshnehlab, M., Shoorehdeli, M. A. (2007). A novel binary particle swarm optimization. In *2007 Mediterranean Conference on Control and Automation* (pp. 1–6). IEEE.

Lewis, J. I. (2016). Wind energy in China: Getting more from wind farms. *Nature Energy, 1*(6), 1–2.

Ma, N., Lei, H., Han, Z., Zhou, D., Bao, Y., Zhang, K., Zhou, L., & Chen, C. (2018). Airfoil optimization to improve power performance of a high-solidity vertical axis wind turbine at a moderate tip speed ratio. *Energy, 150*, 236–252.

NEAC National Energy Administration of China. (2021). http://www.nea.gov.cn/. Accessed 22 September 2021.

NREL National Renewable Energy Laboratory. (2021). Wind Data and Tools. https://www.nrel.gov/research/data-tools.html. Accessed 13 September 2021.

Pampara, G., Engelbrecht, A. P., & Franken, N. (2006). Binary differential evolution. In *2006 IEEE International Conference on Evolutionary Computation* (pp. 1873–1879). IEEE.

Ragheb, M., & Ragheb, A. M. (2011). Wind turbines theory-the betz equation and optimal rotor tip speed ratio. *Fundamental and Advanced Topics in Wind Power, 1*(1), 19–38.

Too, J., Abdullah, A. R., & Saad, N. M. (2019). A new quadratic binary Harris hawk optimization for feature selection. *Electronics, 8*(10), 1130.

WE Wind Europe. (2021). https://windeurope.org/intelligence-platform/statistics/. Accessed 23 September 2021.

Willmott, C. J., & Matsuura, K. (2005). Advantages of the mean absolute error (MAE) over the root mean square error (RMSE) in assessing average model performance. *Climate Research, 30*(1), 79–82.

WWEA World Wind Energy Association. (2021). https://wwindea.org/. Accessed 22 September 2021.

Zhao, C., Luo, J., Xie, S., & Li, H. (2015). Experiment validation of vertical axis wind turbine control system based on wind energy utilization coefficient characteristics. In *2015 IEEE International Conference on Cyber Technology in Automation, Control, and Intelligent Systems (CYBER)* (pp. 1640–1644). IEEE.

Zhao, X., Wang, F., & Wang, M. (2012). Large-scale utilization of wind power in China: Obstacles of conflict between market and planning. *Energy Policy, 48*, 222–232.

Big Data Analysis of Energy Economics in Photovoltaic Power Generation Market

5

5.1 Introduction

Photovoltaic (PV) power generation can provide clean energy with renewable capabilities, which can help to reduce the consumption of non-renewable energy. The development of PV contributes to the effective reduction of greenhouse gas emissions and the mitigation of global climate change. The worldwide installed capacity of PV power generation systems in 2019 is 111.6GW, as illustrated in "Trends in PV applications 2020" released by IEA PVPS (Masson & Kaizuka, 2020). By the end of 2019, the global cumulative installed PV capacity was 623.2GW, of which 70% was installed in the past five years. The cumulative and annual installed capacity from 2011 to 2019 are shown in Fig. 5.1, which shows that the PV installation capacity has been increasing for several consecutive years.

Due to the vast amount and wide distribution of solar energy, it still has great development potential in the future. However, limited by natural and meteorological conditions, PV power generation systems have the characteristics of discontinuity, instability, and volatility. These characteristics lead to output fluctuations, which dramatically affects and threatens stability and safety when the PV system is connected to a power grid on a large scale. The connection of PV power will substantially impact the existing power grid and endanger its power quality. A simplified diagram of PV power integration is presented in Fig. 5.2.

5.2 Big Data Forecasting of Photovoltaic Power Generation

As with other renewable energy sources, photovoltaic power also suffers the drawbacks of being heavily affected by environmental conditions. The intermittent and fluctuation of the PV power may cause severe challenges for the grid, threatening its safe operation. Therefore, accurate short-time PV forecasting is one of the crucial tasks for PV power integration, which can lead to an increase in utilization

© Science Press 2022

H. Liu et al., *Big Data in Energy Economics*, Management for Professionals,
https://doi.org/10.1007/978-981-16-8965-9_5

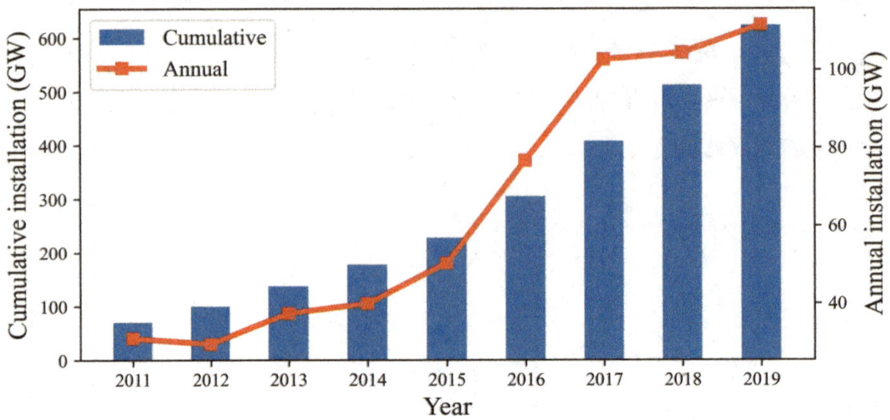

Fig. 5.1 The cumulative and annual PV installations (2011–2019)

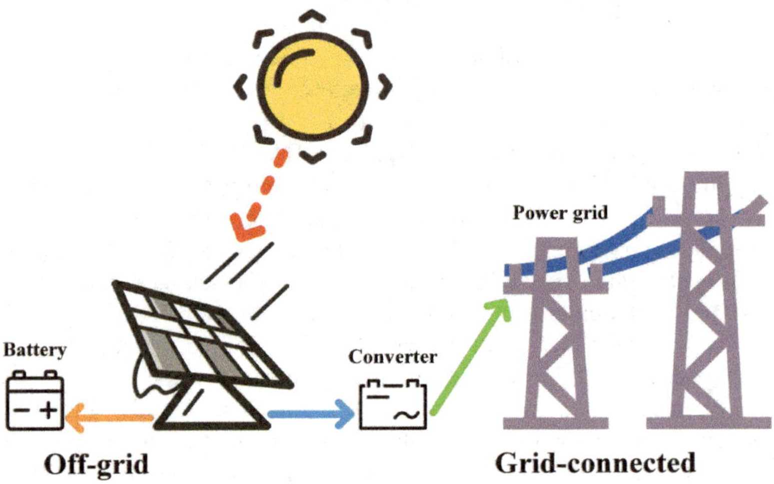

Fig. 5.2 Diagram of PV power system

efficiency and the guarantee of the safety and stability of the power grid (Zhen et al., 2021). Based on the accurately forecasted PV power, the dispatch of the electricity system can be appropriately scheduled, reducing the safety risks and absorbing as much PV power as possible.

5.2.1 Big Data Processing Engines

Apache Spark is a computing engine for big data processing and can be used to build large-scale, low-latency big data processing programs. Spark uses a task

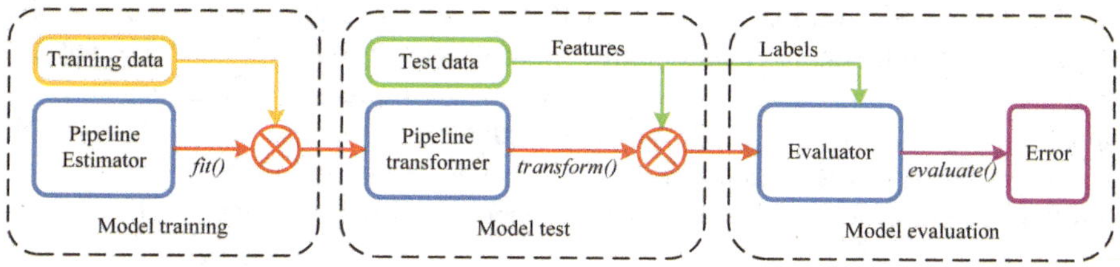

Fig. 5.3 Machine learning pipeline in Spark

scheduling mechanism based on a Directed Acyclic Graph (DAG) and implements in-memory calculations, effectively reducing disk reading and writing costs. Its running speed is higher than that of the Hadoop framework. The Spark ecosystem provides a machine learning library, MLlib, which includes regression algorithms (Syed et al., 2020). Using the regression model makes it possible to realize the parallel and distributed calculation of the big data of PV power generation and then realize the efficient, rapid, and accurate PV power generation forecast.

Spark MLlib provides a machine learning pipeline, which allows each step in the machine learning to be connected by a unified interface and embedded in a unified pipeline model. With the help of the pipeline model, data pre-processing, feature engineering, and machine learning algorithms can all be connected to form a continuous machine learning pipeline. Figure 5.3 shows the training, test, and evaluation process of a machine learning pipeline. It is worth noting that the pipeline has embedded all the data processing procedures involved in the entire machine learning task.

5.2.2 Forecasting Strategy and Methods

Multiple variables may influence the output power in forecasting PV power, such as the output history, time, solar azimuth, solar elevation, wind speed, temperature, humidity, etc. Therefore, to accurately forecast the PV power, both univariate and multivariate forecasting strategies can be used. In univariate forecasting, only the historical power output is used as a feature to train the model, while in multivariate forecasting, multiple variables can be input for training the model.

The prediction of photovoltaic power generation can adopt an indirect way. First, predict the solar radiation, and then obtain the prediction of the output power according to the corresponding conversion equations. In the direct prediction method, relevant models are used to predict PV output directly, without the need for predicting radiation first.

In the PV time series forecasting, there is no power output in the nighttime, and only outputs at daytime are meaningful and can be used for training forecasting models. This way, consecutive time series is not available, and generally used forecasting strategies and methods cannot be directly applied here. Therefore, the collected data needs to be processed first to address this drawback.

First, the classic training/test split method needs to be modified. Traditionally, suppose the original data is a consecutive time series without period or zero values. In that case, the time series is split into two parts, representing training and test samples (or into three parts, if validation samples are required). When dealing with the PV power, periodicity and multiple zero values exist, the approach mentioned above is not perfectly valid (it can still be operational but can result in large error metrics).

In dealing with PV power, the data collected at nighttime should be removed explicitly or implicitly, and the following two kinds of approaches can be applied.

The implicit approach is that, by averaging or summing all the samples in one day, one week, or one month, (i.e., down-sampling the PV data), a consecutive time series can then be constructed to fit into the general time series forecasting frameworks. This kind of forecasting helps in long-term forecasting. However, only low-resolution forecasting can be obtained by doing this, and the number of samples for forecasting models drops rapidly. In this chapter, this approach is called time series forecasting by aggregation.

The explicit approach is simply to delete all samples at night and keep only the daytime samples. This approach may be more flexible if forecasting with higher resolution is required. In this approach, two forecasting strategies are available. The first strategy is that the historical values of variables that happened at the same moment of one day are used to forecast the same moment in the next day. In this kind of forecasting, the forecasting models can be modeled with multiple inputs and outputs corresponding to all moments in daytime, or a number of different models can be constructed separately for different moments. The second strategy is that the models still take the classic methods like in the consecutive time series forecasting. But, the samples of one day must be restrained precisely in one day because the last several values of the previous day are not the adjacent history of the first several values of the next day (when the zero outputs at night are removed). In this chapter, this approach is called time series forecasting by filtration.

In conclusion, the main forecasting approaches and strategies in the PV power generation systems are presented in Fig. 5.4.

5.2.3 Forecasting Models

Recurrent Neural Networks (RNNs) are very suitable for solving tasks with time dependence due to the inherent recurrent structure, which allows the information learned from the previous moment to be selectively passed to the next moment. This chapter introduces two variants of the RNN to forecast the PV power, namely, Long-Short Term Memory (LSTM) network and Gated Recurrent Unit (GRU) network, both of which show superior performance in dealing with the time series forecasting tasks. In addition, a classic neural network model, Multilayer Perceptron (MLP) network, which has no recurrent structure, is also introduced for comparison.

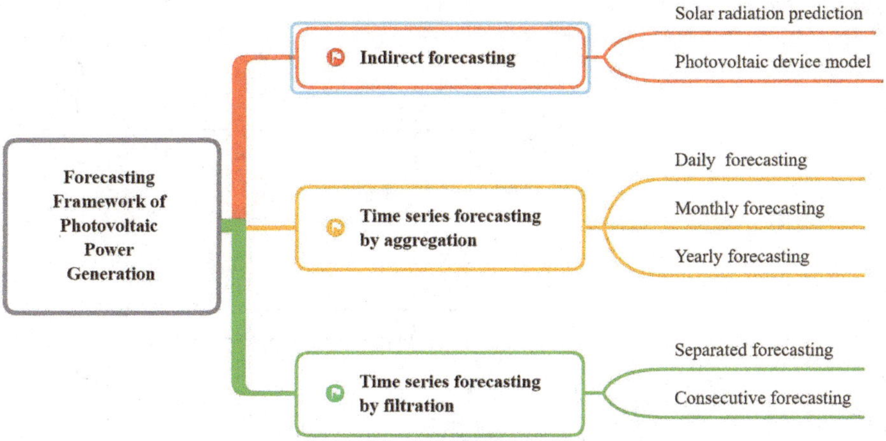

Fig. 5.4 Forecasting framework of PV power generation

5.2.3.1 Long-Short Term Memory Network

The details of LSTM network can be found in Sect. 2.3.1 of Chap. 2.

5.2.3.2 Gated Recurrent Units Network

Gated Recurrent Units (GRU) network is a modified and improved version of LSTM (Cho et al., 2014). GRU can also learn long-term dependencies by addressing the disappearance and explosion of gradient. However, only two gate control structures can be found in GRU, the update and reset gate. The cell state and hidden state, which can be found in LSTM cell are mixed here in GRU. Therefore, compared with LSTM, it has a faster calculation speed due to fewer parameters. The performance of GRU is similar to LSTM on most tasks, and it shows superior performance on some specific tasks. The structure of GRU cell is presented in Fig. 5.5.

The equations for each step of the information flow inside the GRU cell are given as follows (Chen et al., 2019):

$$
\begin{aligned}
z_t &= \sigma\left(W_z \times \left[h_{t-1}, x_t\right] + b_z\right) \\
r_t &= \sigma\left(W_r \times \left[h_{t-1}, x_t\right] + b_r\right) \\
\tilde{h}_t &= \tanh\left(W_{\tilde{h}} \times \left[r_t * h_{t-1}, x_t\right] + b_{\tilde{h}}\right) \\
h_t &= (1 - z_t) * h_{t-1} + z_t * \tilde{h}_t
\end{aligned}
\tag{5.1}
$$

where x_t is the input at the current moment. h_{t-1} and h_t are the output at the previous and current moment, respectively. W_z, W_r and $W_{\tilde{h}}$ are the connection weights. b_z, b_r and $b_{\tilde{h}}$ are the biases.

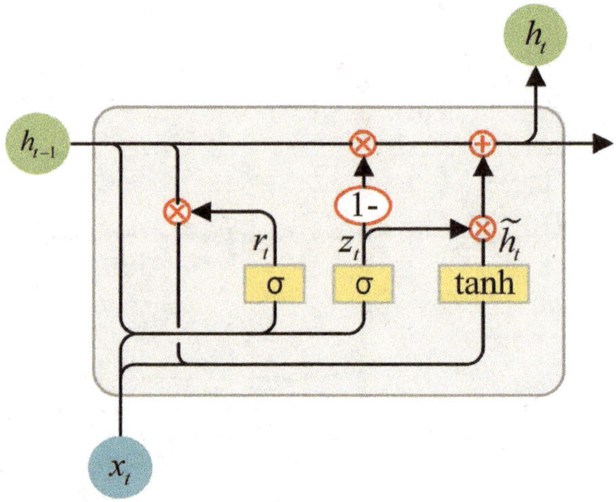

Fig. 5.5 Structure of GRU cell

5.2.3.3 Multilayer Perceptron Network

MLP network is a model that can solve nonlinear separable problems (Gardner & Dorling, 1998). An input layer, several hidden layers, and an output layer can be found in the structure of MLP. The acceptance of input features is processed in the input layer, and learning of the input features is processed in the hidden and output layer. The output layer is also responsible for outputting the result. MLP contains many neurons and is a multilayer feedforward neural network, which can effectively fit nonlinear functions. The structure of an MLP with a single hidden layer is presented in Fig. 5.6.

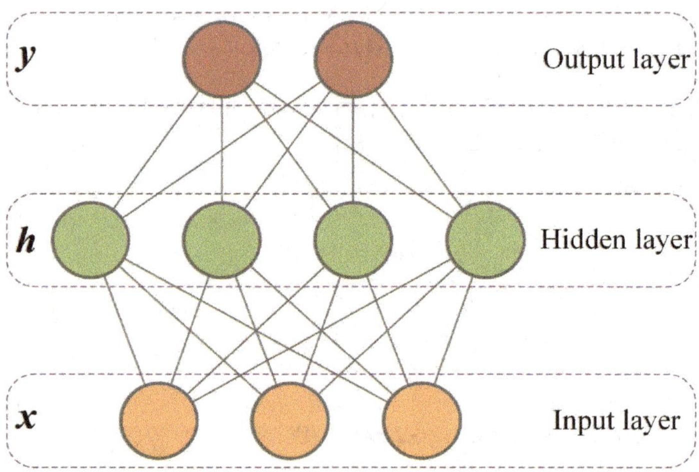

Fig. 5.6 Structure of an MLP

The equation for the forward information flow in the hidden layer is presented as follows:

$$H_j = f\left(\sum_{i=1}^{n} w_{ij}x_i - b_j\right) j = 1, 2, \cdots, l \tag{5.2}$$

where H_j is the output of the jth neuron. f is activation function, l is the number of neurons, n is the dimension of input, w_{ij} represents the weight from input i to output j, x_i is the ith element of the input, and b_j is the bias of the jth neuron.

5.3 Photovoltaic Power Consumption by Small and Medium-Sized Users

Photovoltaic power consumption by small and medium-sized users generally belongs to off-grid PV power. This type of PV power output is directly consumed or stored by users without being integrated into the large-scale power grid.

5.3.1 Dataset Description

The dataset is collected from a PV power generation device with automatic tracking ability, and the sampling interval of the dataset is 1 min. In this experiment, 27 days of data samples are selected, and the data is downsampled with a new interval of 10 min to smooth the original data.

The data collected in the first 18 days are used to construct training set, and the data collected in the next nine days are used to test the model performance. The output of the PV device within one day is visualized in detail in Fig. 5.7, which demonstrates that the power curve has large fluctuations but follows a trend of higher power at noon and lower power in the morning and evening. The whole dataset is shown in Fig. 5.8.

As shown in Fig. 5.8, during the time range covered by the dataset, the output of PV power in some days is relatively low, which is usually caused by bad weather conditions. In addition, there are lots of zeros during night, which are unfavorable for building forecasting models. Therefore, only the data whose output is greater than 5.0% of the rated output power of the PV device is retained for the construction of training and test sets. Several statistics of the dataset are calculated as shown in Table 5.1.

5.3.2 Experiments

5.3.2.1 Model Training

The training process is monitored by an early stopping mechanism and learning rate scheduler. If the validation loss did not drop in several epochs, the learning

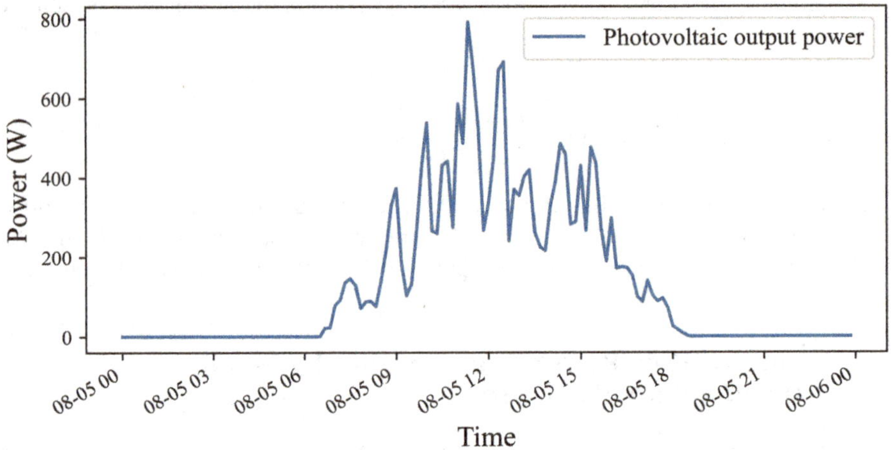

Fig. 5.7 PV power output within one day

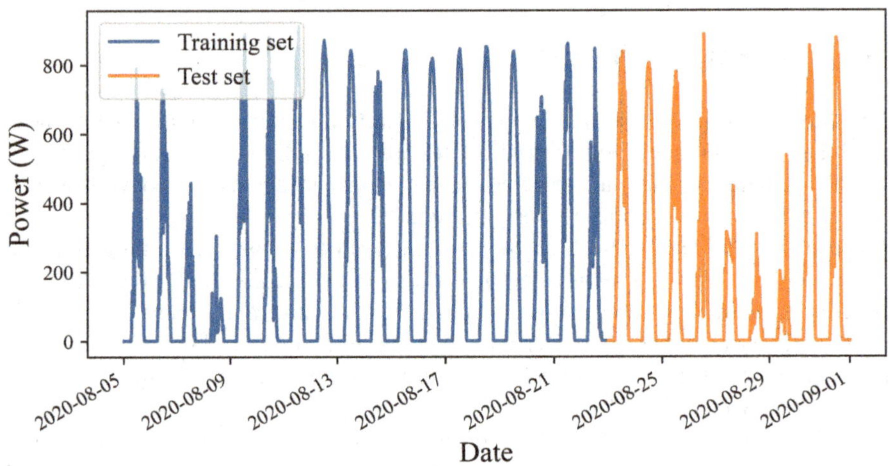

Fig. 5.8 Dataset visualization and splits

Table 5.1 Statistics of the dataset

Maximum (W)	Minimum (W)	Mean (W)	Standard deviation (W)	Kurtosis
912.218	0.000	192.801	275.638	−0.056

rate would automatically decline. If the model performance is still not improved in another number of epochs, the training process will be stopped in advance. A large number is assigned to the maximal training epochs. In this way, both under-fitting and over-fitting can be avoided.

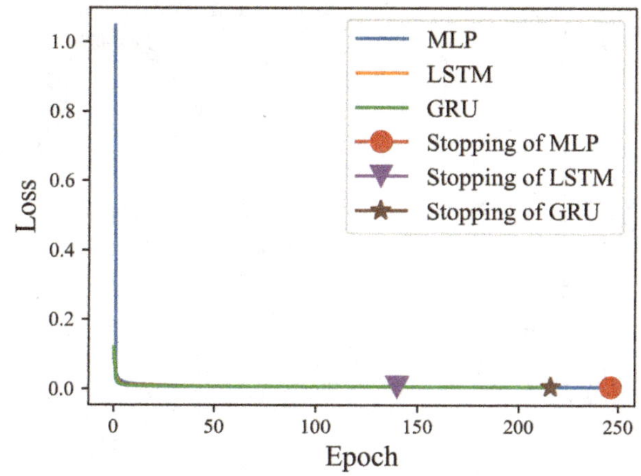

Fig. 5.9 Loss curve of the training

Training losses of different models are visualized in Fig. 5.9. The figure shows that the training of MLP is the last to be terminated. Although RNN models have strong structural complexity and time complexity, they are terminated after fewer iterations, proving the strong learning ability of the RNN models. It is worth noting that, since the duration of one epoch for training RNN is usually longer than MLP, the training time of the LSTM and GRU is not necessarily less than that of the MLP.

5.3.2.2 Model Evaluation

The forecasting results are presented in Fig. 5.10. The red line represents the lowest value for valid power output in the figure, and only the power outputs above the threshold are predicted. In addition, the predicted values below 0 are converted to 0 to fit the actual situation. The figure shows that all the models can capture the variation trend of PV power generation, and the models can adapt to PV power generation forecasts under different weather conditions.

The detailed forecasting for the first day in the test set is presented in Fig. 5.11,

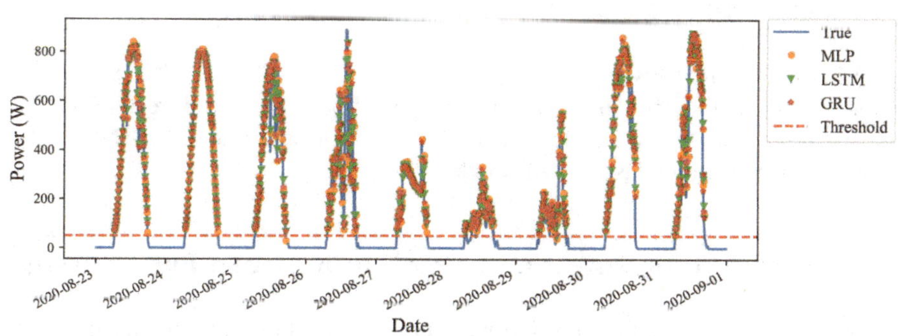

Fig. 5.10 Forecasting of the models

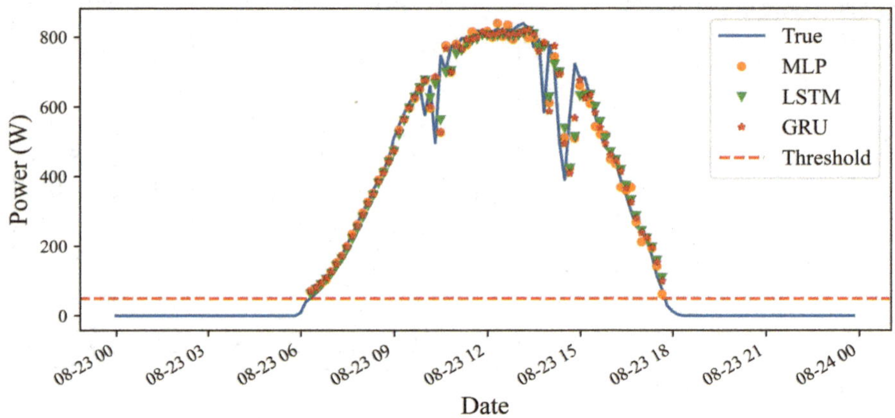

Fig. 5.11 Forecasting of the models for the first day in the test set

Table 5.2 Error metrics of the forecasting

Model	RMSE (W)	MAE (W)	MAPE (%)
MLP	86.3531	50.4825	21.1671
LSTM	**85.1260**	50.2734	22.1002
GRU	85.3089	**48.9445**	**20.9905**

Remark The value in bold represents the smallest value in each column

which shows that the prediction performance is still excellent in detail.

Three error metrics, namely RMSE, MAE, and MAPE, are adopted for error evaluation (Shcherbakov et al., 2013), and the details of the metrics can be found in Sect. 2.3.3 of Chap. 2. The error metrics of the three models are shown in Table 5.2. It is worth noting that, to eliminate the performance difference caused by random initialization of neural networks, the data in the table are the average errors of the results of multiple experiments.

The table shows that the best RMSE, MAE, and MAPE values are all obtained by the RNN models.

5.3.2.3 Results Analysis

In the experiment, one-step forecasting is performed on a high-resolution PV power dataset. All the models demonstrate nice performance, and the RNN-based models outperform the classic MLP model.

5.4 Photovoltaic Power Consumption in Urban Public Areas

Photovoltaic power consumption in urban public areas is usually directly from the urban power grid, which is first generated somewhere else and then integrated into the power grid. This type of PV power involves PV power integration technology and belongs to the grid-connected systems.

5.4.1 Dataset Description

The dataset is collected from a PV power plant, and the time range covers from January 1, 2014, to December 31, 2017. The time interval between two data samples is one day. The dataset contains 1461 samples in total. The samples collected in the first three years are used to construct the training set, and the samples in the last year are used as test data. There are missing values in the original dataset. Therefore, the dataset is first being completed using interpolation methods. This dataset is visualized in Fig. 5.12.

As shown in Fig. 5.12, the periodicity of the PV power output is very significant, large power outputs concentrate in summer, and outputs in winter are lower. However, there are many fluctuations in the power generation data, and these fluctuations are generally lower than the tendency of the whole time series data. The occasional bad weather conditions may cause these fluctuations. Therefore, if higher forecasting accuracy is required, more information about the surrounding weather status should also be provided to the forecasting models. The statistics of the PV dataset are presented in Table 5.3.

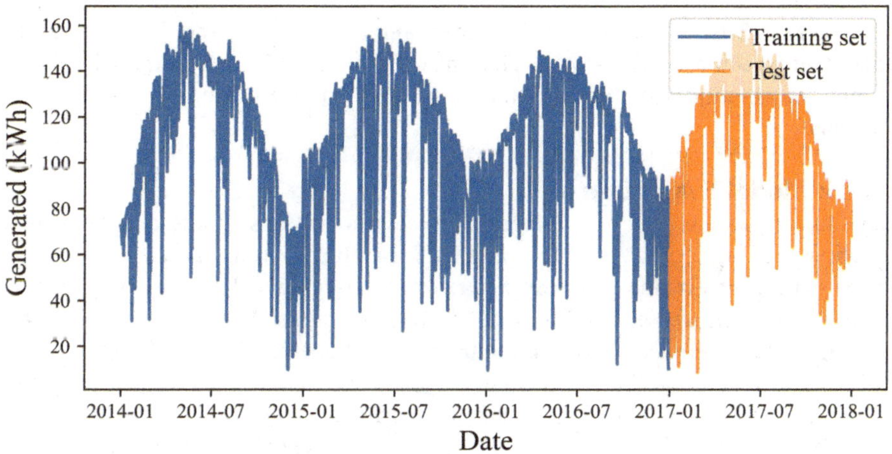

Fig. 5.12 Dataset visualization and splits

Table 5.3 Statistics of the dataset

Maximum (kWh)	Minimum (kWh)	Mean (kWh)	Standard deviation (kWh)	Kurtosis
160.820	8.712	107.915	33.549	−0.209

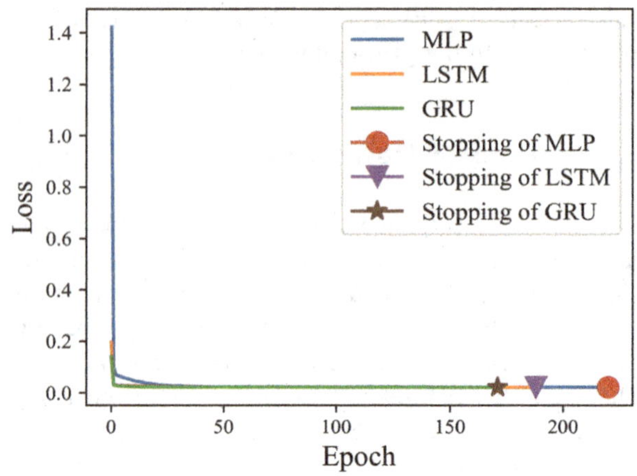

Fig. 5.13 Loss curve of the training

5.4.2 Experiments

5.4.2.1 Model Training

In the process of training the models, some of the training samples are reserved for validation to monitor the generalization ability of the models and prevent the model from over-fitting. To improve the training process, both the learning rate scheduler and early stopping strategy are again used. The training loss is visualized in Fig. 5.13.

From Fig. 5.13, it can be seen that the training loss drops rapidly in the early stage, then reach a steady status. This means that the forecasting model learned helpful information from the training sets, and no over-fitting or under-fitting happened.

5.4.2.2 Model Evaluation

The trained model is then used to perform forecasts on test samples, and the results are shown in Fig. 5.14. As shown in Fig. 5.14, the general tendency of the PV power generation in one year can be tracked satisfyingly. However, the models fail to capture the actual amplitude of the power output. This may be caused by the amplitude differences between the training and test set.

The error metrics of the forecasting are given in Table 5.4. The data in the table are the average errors of several repeated experimental results.

Table 5.4 shows that all the models demonstrate excellent performance. Two RNN models, LSTM and GRU, show slight improvements over the MLP model, and LSTM and GRU obtain the lowest error metrics. This indicates that the RNN models are better at finding the time dependence within the time series data.

5.4.2.3 Results Analysis

Daily PV power generation data is used for forecasting in this experiment, and one-step forecasting is presented. The results show that all the forecasting models

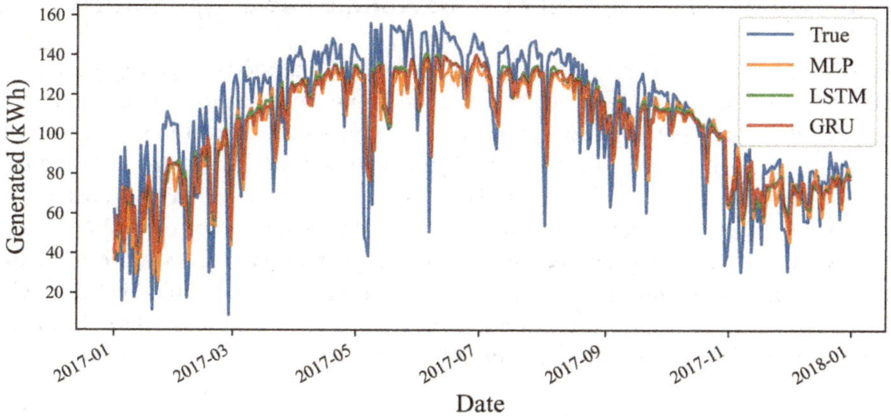

Fig. 5.14 Forecasting of the models

Table 5.4 Error metrics of the forecasting

Model	RMSE (kWh)	MAE (kWh)	MAPE (%)
MLP	23.9226	18.2675	26.0201
LSTM	**22.8760**	17.6889	**26.0134**
GRU	22.9573	**17.6311**	26.0693

Remark The value in italic represents the smallest value in each column

present comparable and relatively nice accuracy, and the RNN models again show slight improvements over the classic MLP model.

5.5 Market Economy Analysis of Photovoltaic Systems

5.5.1 Dispatch of Photovoltaic Power Integration

Places with abundant photovoltaic energy usually do not require large-scale direct consumption, so the output of the PV system usually needs to be connected to the grid for long-distance transmission and consumed by remote customers. However, the PV power generation systems are highly susceptible to environmental factors, and their output is unstable and difficult to predict. Therefore, when connected to the grid, the PV power generation systems can easily cause an impact on the grid, affecting the safety of grid operation and deteriorating the power quality of the grid. PV power forecasting technology is helpful to overcome this challenge, but the forecasting errors may also affect the power dispatch. For this reason, the power economic dispatch model that considers the impact of PV power forecasting errors is worthy of analysis and discussion.

5.5.2 Optimization Model of Photovoltaic Power Integration

Photovoltaic power generation usually needs to be integrated into the grid dominated by thermal power generation. The scheduling optimization mainly aims to optimize the total operating cost and pollution emission cost of the whole power generation system, which means that objective functions in the dispatch model have at least two parts to be minimized. Specifically, the economic cost consists of four parts, thermal power unit cost, rotating reserve unit cost, PV power generation cost, and cost caused by PV power generation forecasting errors. In addition, scheduling optimization needs to be carried out in a specific cycle, which can be one day, one week, one month, one quarter, one year, etc. The dispatch optimization model is described as follows (Liu et al.):

$$\min \begin{cases} C(t) = \sum_{t=1}^{T}\sum_{i=1}^{N_i}\left(C_{g,i}\left(P_{i,t}\right) + SU_{i,t} + SD_{i,t}\right) \\ \quad + \sum_{t=1}^{T}\sum_{j=1}^{N_j} C_{v,j}\left(P_{j,t}\right) + \sum_{t=1}^{T}\sum_{j=1}^{N_j} C_{uv,j}\left(P_{u,j,t}\right) \\ \quad + \sum_{t=1}^{T}\sum_{j=1}^{N_j} C_{ov,j}\left(P_{o,j,t}\right) + \sum_{t=1}^{T}\sum_{k=1}^{N_k} C_{r,k}\left(P_{r,k,t}\right) \\ E(t) = \sum_{t=1}^{T}\sum_{i=1}^{N_i}\left(\alpha_i + \beta_i P_{i,t} + \gamma_i\left(P_{i,t}\right)^2\right) \end{cases} \tag{5.3}$$

The meanings of each item in Eq. (5.3) are shown in Table 5.5.

The expressions of each expandable item are given as follows:

$$\begin{aligned} C_{g,i}\left(P_{i,t}\right) &= a_i + b_i P_{i,t} + c_i\left(P_{i,t}\right)^2 \\ C_{v,j}\left(P_{j,t}\right) &= K_{PV,j,t} P_{j,t} \\ C_{uv,j}\left(P_{u,j,t}\right) &= K_{PVU,j}\left(P_{j,t} - P_{f,j,t}\right) \\ C_{ov,j}\left(P_{o,j,t}\right) &= K_{PVO,j,t}\left(P_{f,j,t} - P_{j,t}\right) \\ C_{r,k}\left(P_{r,k,t}\right) &= \rho_k P_{r,k,t} \Delta t \end{aligned} \tag{5.4}$$

The meanings of each item are shown in Table 5.6.

Multiple constraints must be considered when solving the dispatch model, and these constraints are presented as follows:

5.5.2.1 Power Balance Constraint

$$\sum_{i=1}^{N_i} P_{i,t} + \sum_{j=1}^{N_j} P_{j,t} + \sum_{k=1}^{N_k} P_{k,t} = P_{L,t} \tag{5.5}$$

where $P_{L,t}$ is the load at time t.

Table 5.5 Description of items in the optimization model

Item	Meaning
t	Moment of the scheduling
T	Total time of the scheduling
i	ID of the thermal power unit
N_i	Total number of thermal power units
j	ID of the PV power station
N_j	Total number of PV power stations
k	ID of the spinning reserve unit
N_k	Total number of spinning reserve units
$C(t)$	Total economic cost at time t
$C_{g,i}$	The cost of the ith thermal power unit
$P_{i,t}$	The power of the ith thermal power unit at time t
$SU_{i,t}$	The start-up cost of the ith thermal power unit
$SD_{i,t}$	The shutdown cost of the ith thermal power unit
$C_{v,j}$	The electricity cost of the jth PV power station
$P_{j,t}$	The actual power of the jth PV power station at time t
$C_{uv,j}$	The underestimated forecasting error cost of the jth PV power station
$P_{u,j,t}$	The underestimated prediction forecasting error power of the jth PV power station at time t
$C_{ov,j}$	The overestimated forecasting error cost of the jth PV power station
$P_{o,j,t}$	The overestimated predicted forecasting error power of the jth PV power station at time t
$C_{r,k}$	Spinning reserve cost of the kth spinning reserve unit
$P_{k,t}$	The spinning reserve power of the kth spinning reserve unit at time t
Pollution cost	
$\alpha_i, \beta_i, \gamma_i$	The sulfur dioxide emission rate of the ith thermal power unit

Table 5.6 Description of items in Eq. (5.4)

Item	Meaning
a_i, b_i, c_i	Coal consumption coefficient of the ith thermal power unit
$K_{PV,j,t}$	Cost coefficient of the jth PV power station
$P_{f,j,t}$	The predicted PV output of the jth PV power station at time t
$K_{PVU,j}$	Cost coefficient of underestimated forecasting error when underestimating PV power generation
$K_{PVO,j,t}$	Cost coefficient of overestimated forecasting error when overestimating PV power generation
ρ_k	The coefficient of the kth spinning reserve unit
Δt	Time interval

5.5.2.2 Constraints on Upper and Lower Limits of Thermal Power Units

$$X_{i,t}P_i^{\min} \le P_{i,t} \le X_{i,t}P_i^{\max}, \forall t \in [1, T] \tag{5.6}$$

where $P_i^{\max}$ and $P_i^{\min}$ are the maximum and minimum output power of the ith thermal power unit. $X_{i,t}$ is the status of the ith thermal power unit at time t, 0 means the unit is stopped, and 1 means the unit is running.

5.5.2.3 Constraints on Minimum Starting and Stopping Times for Thermal Power Units

$$\begin{aligned} T_{i,t}^{st} &\ge T_i^{st} \\ T_{i,t}^{ed} &\ge T_i^{ed} \end{aligned} \tag{5.7}$$

where $T_{i,t}^{st}$ is the continuous start-up time of the ith thermal power unit at time t, T_i^{st} is the minimum start-up time of the ith thermal power unit, $T_{i,t}^{ed}$ is the continuous shutdown time of the ith thermal power unit at time t, and T_i^{ed} is the minimum shutdown time of the ith thermal power unit.

5.5.2.4 Climbing Rate Constraints for Thermal Power Units and Spinning Reserve Units

$$\begin{aligned} P_{i,t} - P_{i,(t-1)} &\le \Delta t R_i^{up} \\ P_{i,(t-1)} - P_{i,t} &\le \Delta t R_i^{dn} \\ P_{k,t} - P_{k,(t-1)} &\le \Delta t R_k^{up} \\ P_{k,(t-1)} - P_{k,t} &\le \Delta t R_k^{dn} \end{aligned} \tag{5.8}$$

where Δt is the time interval in the planned period. R_i^{up} and R_i^{dn} are the climbing rate and descending rate of the ith thermal power unit, respectively. R_k^{up} and R_k^{dn} are the climbing rate and descending rate of the kth spinning reserve unit, respectively.

5.5.2.5 Constraints for Spinning Reserve

$$\begin{aligned} \sum_{i=1}^{N_i} X_{i,t}P_i^{\max} + \sum_{k=1}^{N_k} P_{k,t} &\ge P_{L,t} + R_t, \forall t \in [1, T] \\ \sum_{i=1}^{N_i} X_{i,t}P_i^{\min} + \sum_{k=1}^{N_k} P_{k,t} &\le P_{L,t} + R_t, \forall t \in [1, T] \end{aligned} \tag{5.9}$$

where R_t is the spinning reserve of the system at time t.

5.5.2.6 Upper and Lower Limit Constraints of PV Power Generation

$$\begin{aligned} P_{j,t,e}^{-} \leq P_{j,t,e} \leq P_{j,t,e}^{+} \\ P_{j,t,e} = P_{f,j,t} - P_{j,t} \end{aligned} \tag{5.10}$$

where $P_{j,t,e}$ is the forecasting error of the PV power station at time t. $P_{j,t,e}^{-}$ and $P_{j,t,e}^{+}$ are the lower and upper limits of the forecasting error of PV power generation, respectively.

5.5.3 Single- and Multi-objective Optimization Algorithms

Once a dispatch optimization model is constructed, the most important task is to solve the model and obtain an optimal or nearly optimal solution. Meta-heuristic algorithms are powerful tools that can be used to solve the models and find one or more feasible solutions.

As shown in the last section, in the optimization of dispatch, the objective functions are generally used to describe the power plant operating costs of different generation types (such as thermal, photovoltaic, wind, hydro, etc.), pollution emission costs, and so on. The main aim is to reduce the total costs and maximize the economic revenue. Therefore, within the optimization problems, there are generally several objectives to be optimized.

Both Single-Objective Optimization (SOO) and Multi-Objective Optimization (MOO) algorithms can be adopted for solving optimization models (Liu et al., 2020). Only one objective is required to be optimized in SOO problems, while in MOO problems, two or more objectives need to be optimized simultaneously. The MOO problem can be described as follows:

$$\begin{aligned} &\text{Minimize: } y = F(\mathbf{x}) = \{f_1(\mathbf{x}), f_2(\mathbf{x}) \cdots, f_m(\mathbf{x})\} \\ &s.t. \begin{cases} \mathbf{x} = (x_1, x_2 \cdots, x_n) \in X \\ \mathbf{y} = (y_1, y_2 \cdots, y_m) \in Y \\ g_i(\mathbf{x}) \leq 0,\ i = 1, 2 \cdots h \\ h_i(\mathbf{x}) = 0,\ i = 1, 2 \cdots p \\ l_i \leq x_i \leq u_i,\ i = 1, 2 \cdots n \end{cases} \end{aligned} \tag{5.11}$$

where $\mathbf{x}$ and $\mathbf{y}$ represent the decision and objective vector, respectively. f represents the objective function. X and Y denote the decision and objective space, respectively. g and h are the inequality and equality constraints, respectively. l and u are respectively the lower and upper limits of the elements in $\mathbf{x}$.

Generally, several objectives can be integrated to form one single objective by assigning weights to each objective function. Then the converted problem can be solved with conventional SOO algorithms to simplify the solving process. The popular SOO algorithms include Genetic Algorithm (GA) (Mirjalili, 2019), Particle Swarm Optimization (PSO) (Kennedy & Eberhart, 1995), etc. This kind of method can be described as follows:

$$y = F(\mathbf{x}) = w_1 f_1(\mathbf{x}) + w_2 f_2(\mathbf{x}) + \cdots + w_m f_m(\mathbf{x}) \tag{5.12}$$

where w is the weight coefficient of the corresponding objective function.

The other method is to use the MOO algorithm, which can optimize multiple objectives simultaneously without the need for assigning weights. The MOO algorithm can generate a Pareto front that contains a lot of Pareto optimal solutions. The key in the MOO algorithm is how to guide the search process in decision space so that the true Pareto optimal solutions can be approached, and how to ensure the diversity of solutions. The popular MOO algorithms include Non-Dominated Sorting Genetic Algorithm II (NSGA-II) (Deb et al., 2002), Multi-Objective Particle Swarm Optimization (MOPSO) (Coello & Lechuga, 2002), Multi-Objective Gray Wolf Algorithm (MOGWO) (Mirjalili et al., 2016), etc. The flowchart of the NSGA-II algorithm is provided in Fig. 5.15 to illustrate the basic idea for solving a MOO problem.

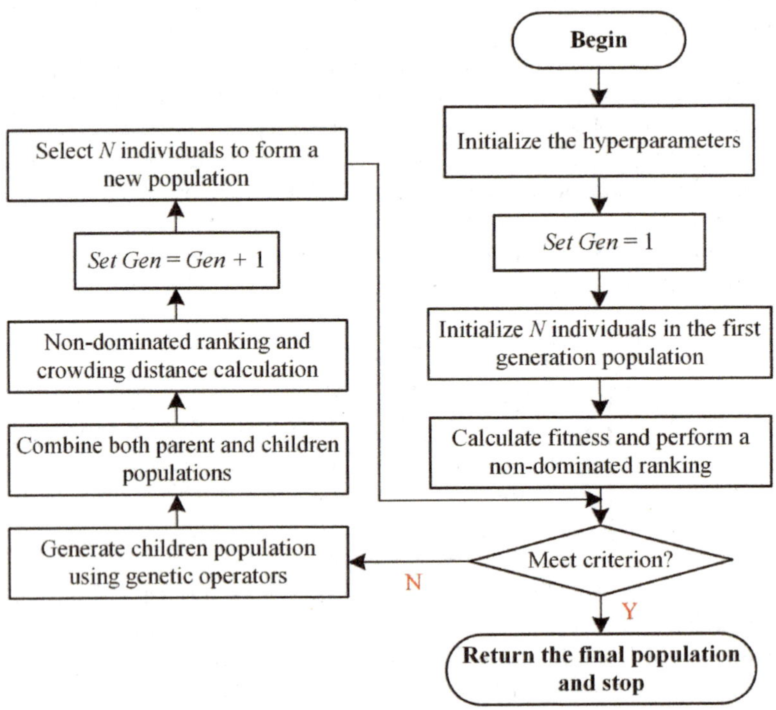

Fig. 5.15 Flowchart of NSGA-II

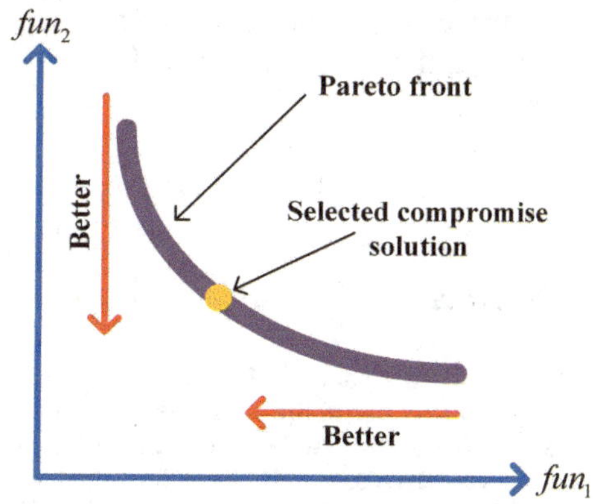

Fig. 5.16 Pareto front with two objective functions

If the problem is solvable, only one optimization solution would be given by the SOO algorithms. However, for MOO algorithms, once the optimization is finished, multiple non-dominated solutions are provided, which are distributed on the Pareto front, as shown in Fig. 5.16 (only two objectives need to be optimized). In the figure, two axes represent the value of two objective functions, respectively, and each point that falls on the curve represents a Pareto optimal solution. All of these solutions are not worse than each other.

In this way, the decision-makers must select one appropriate solution from them and apply it in practice. The selected solution is called a compromise solution, and is chosen from the Pareto optimal solution set based on some criteria, such as the Technique for Order Preference by Similarity to an Ideal Solution (TOPSIS) (Tzeng & Huang, 2011).

5.6 Conclusions

The summaries for each section in this chapter are given as follows:

(a) Section 5.1 introduces PV power generation and its impact on economic development.
(b) Section 5.2 explains big data forecasting in PV power generation. The big data analysis engine Spark is briefly introduced, followed by the commonly used forecasting strategies. Finally, the forecasting framework and model of PV power generation are summarized.
(c) Sections 5.3 and 5.4 give examples for the big data prediction of PV power generation in different scenarios. First, the dataset is organized and analyzed, and then the corresponding prediction models are built. Finally, experiments are carried out, including model training, model evaluation, and result analysis.
(d) Section 5.5 first introduces the power integration of the PV system. Then the economic scheduling model for power system involving PV power generation

is presented. Finally, intelligent meta-heuristic algorithms are discussed for solving the economic scheduling model.

References

Chen, J., Jing, H., Chang, Y., & Liu, Q. (2019). Gated recurrent unit based recurrent neural network for remaining useful life prediction of nonlinear deterioration process. *Reliability Engineering & System Safety, 185*, 372–382. https://doi.org/10.1016/j.ress.2019.01.006

Cho, K., Van Merriënboer, B., Gulcehre, C., Bahdanau, D., Bougares, F., Schwenk, H., & Bengio, Y. (2014). Learning phrase representations using RNN encoder-decoder for statistical machine translation. arXiv preprint arXiv:14061078.

Coello, C. A. C., Lechuga, M. S. (2002). MOPSO: A proposal for multiple objective particle swarm optimization. In *Proceedings of the 2002 Congress on Evolutionary Computation. CEC'02 (Cat. No. 02TH8600)* (pp. 1051–1056). IEEE.

Deb, K., Pratap, A., Agarwal, S., & Meyarivan, T. A. M. T. (2002). A fast and elitist multiobjective genetic algorithm: NSGA-II. *IEEE Transactions on Evolutionary Computation, 6*(2), 182–197.

Gardner, M. W., & Dorling, S. R. (1998). Artificial neural networks (the multilayer perceptron)—a review of applications in the atmospheric sciences. *Atmospheric Environment, 32*(14–15), 2627–2636.

Kennedy, J., Eberhart, R. (1995). Particle swarm optimization. In *Proceedings of ICNN'95-International Conference on Neural Networks* (pp. 1942–1948). IEEE.

Liu, H., Li, Y., Duan, Z., & Chen, C. (2020). A review on multi-objective optimization framework in wind energy forecasting techniques and applications. *Energy Conversion and Management, 224*, 113324. https://doi.org/10.1016/j.enconman.2020.113324

Liu, X., Geng, C., Xie, S., Tian, J., Cao, H. (2021). Day ahead thermal-photovoltaic economic dispatch considering uncertainty of photovoltaic power generation. *Journal of System Simulation*1–11.

Masson, G., Kaizuka, I. (2020). Trends in PV applications 2020– IEA-PVPS. International Energy Agency. IEA PVPS T1-36.

Mirjalili, S. (2019). Genetic algorithm. In *Evolutionary algorithms and neural networks* (pp 43–55). Springer.

Mirjalili, S., Saremi, S., Mirjalili, S. M., & Coelho, L. D. S. (2016). Multi-objective grey wolf optimizer: A novel algorithm for multi-criterion optimization. *Expert Systems with Applications, 47*, 106–119.

Shcherbakov, M. V., Brebels, A., Shcherbakova, N. L., Tyukov, A. P., Janovsky, T. A., & Kamaev, V. A. (2013). A survey of forecast error measures. *World Applied Sciences Journal, 24*(24), 171–176.

Syed, D., Refaat, S. S., Abu-Rub, H. (2020). Performance evaluation of distributed machine learning for load forecasting in smart grids. In *2020 Cybernetics & Informatics (K&I)* (pp. 1–6). IEEE.

Tzeng, G.-H., & Huang, J.-J. (2011). *Multiple attribute decision making: Methods and applications*. CRC Press.

Zhen, H., Niu, D., Wang, K., Shi, Y., Ji, Z., & Xu, X. (2021). Photovoltaic power forecasting based on GA improved Bi-LSTM in microgrid without meteorological information. *Energy, 231*, 120908.

Big Data Analysis of Power Market Energy Economics

6

6.1 Introduction

As an important energy source in the world, electric energy has promoted the prosperity of human civilization in the history of more than one hundred years. The operation mechanism of the electric power supply has been gradually improved, and a stable electric power market system has been formed (Pantoš, 2020; Zhang et al., 2014). At the same time, in the process of forming the market, the energy economy also emerges at the historic moment, and the main body members of the market economy are becoming more and more abundant. The main members of the electricity market include vertically integrated utilities, power generation companies, distribution companies, retailers, market operators, independent system operators, transmission companies, regulatory organizations, and users (Topalović et al., 2022).

With the rapid development of artificial intelligence big data in recent years, the research on energy big data has also achieved remarkable results, among which the proposal of a smart power grid has been widely recognized. Therefore, the use of new data science and engineering methods, focusing on the power market system energy economics big data research, will certainly lead to more reasonable management and utilization of electric energy, promote the combination of energy and economic research in many aspects (Giest & Mukherjee, 2018; Hafner et al., 2020; Ren et al., 2021).

The modern electric power market system consists of medium and long-term trading markets and electric power spot trading markets. As for medium and long-term electric power transactions, taking China as an example, at present, different regions have initially formed a fully functional trading mechanism, and have issued a series of relevant trading rules (Yang et al., 2019b). The spot market construction is still in its infancy, and it is planned to be fully launched at the end of 2020. To this end, it is necessary to further accelerate the exploration of the electric power spot trading mechanism, change the original market operation mode,

© Science Press 2022

H. Liu et al., *Big Data in Energy Economics*, Management for Professionals,
https://doi.org/10.1007/978-981-16-8965-9_6

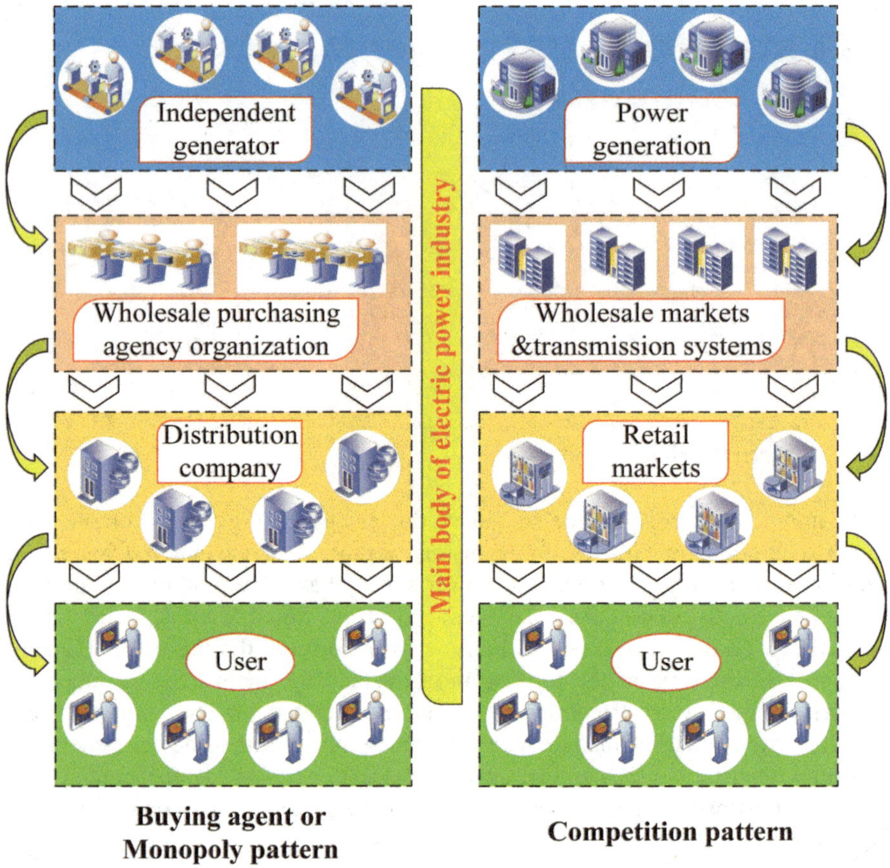

Fig. 6.1 The competitive mode of the electricity market

improve the market-based electric power quantity balance mechanism, and gradually form a medium and long-term electric power trading mode combined with the spot market. The construction of an electric power market system is the focus of current market construction (Zhang et al., 2021a). At the same time, to understand the economic problems of the electricity market, it is necessary to sort out the existing market competition relationship. According to the electric power system economics principle, the competition relationship of the electricity market is shown in Fig. 6.1 (He et al., 2021b).

6.2 Big Data Forecasting of Urban Electricity Price

Electricity price forecast and load forecast is an important part of energy management, which is mainly used for the load forecast of electric energy consumption and production in the future. At present, the short-term forecast is very important for power system production planning and safe, stable, and reasonable operation. Using the results of short-term forecasts to arrange and make production plans can reduce unnecessary spare capacity and improve economic and social benefits. Therefore, the study of electricity price and load forecasting has become an important field in the field of power engineering.

6.2.1 Electricity Price Forecasting Method Based on Empirical Mode Decomposition and Extreme Learning Machine

6.2.1.1 Electricity Price Big Data Decomposition Based on Empirical Mode Decomposition

In 1998, Huang proposed a new signal analysis method (Huang et al., 1998), Hilbert-Huang Transform (HHT). In physics, if the instantaneous frequency is meaningful, the function must be symmetric, with the local mean of zero and the same number of zero-crossing and extremum points (Yang et al., 2019a).

On this basis, Huang proposed the Intrinsic Mode Function (IMF), he believes that any signal is composed of several eigenmode functions, and a signal can contain several eigenmode functions at any time. If the eigenmode functions overlap with each other, a composite signal will be formed. The purpose of the Empirical mode decomposition (EMD) is to obtain the eigenmode function and perform the Hilbert transformation on the decomposed IMF components, thus obtaining the energy distribution spectrum (Hilbert spectrum) on the time–frequency plane.

An eigen modal function is one that satisfies the following two conditions (Yang et al., 2019a):

(a) In the whole data sequence, the number of extreme points and the number of zeros must be equal, or the difference cannot be more than one.
(b) At any time point, the average envelope value defined by the local maximum and local minimum of the signal is zero.

The EMD method assumes that any signal is composed of different intrinsic modal functions, each of which can be linear or nonlinear. The IMF component must meet the following two conditions: first, the number of extreme points and zero-crossing points is the same or at most one different; second, its upper and lower envelope is locally symmetric about the time axis. So, anyone signal can be decomposed into a finite sum of the IMFs.

The EMD decomposition process is based on the following assumptions (Bakheet & Maharatna, 2021):

(a) The signal has at least one maximum value and one minimum value.
(b) The time-domain characteristics are determined by the extremum interval.

(c) If the data sequence completely lacks extremum but only contains inflection points, it can also represent extremum points by taking one or more derivatives, and the result can be obtained by integrating these components.

The method is accomplished by a "filtering" process (Chen et al., 2018; Karan et al., 2020):

Firstly, all the maximum points of the signal $s(t)$ are found out and the envelope lines on the original data series are fitted by cubic spline function, then all the minimum points are found out and the original data are fitted by a cubic spline function.

Calculate the mean value of the upper and lower envelope, denoted as $m_1(t)$, and subtract the mean value from the original data series $s(t)$ to obtain a new data series h_1 without the low frequency:

$$s(t) - m_1(t) = h_1(t) \tag{6.1}$$

Because $h_1(t)$ is generally not an IMF component sequence, the above process needs to be repeated. Repeat k times until $h_1(t)$ meets the IMF definition: the obtained mean tends to zero. This gives the first IMF component $c_1(t)$, which represents the highest frequency component of signal $s(t)$:

$$h_{1(k-1)}(t) - m_{1k}(t) = h_{1k}(t) \tag{6.2}$$

$$c_1(t) = h_{1k}(t) \tag{6.3}$$

When $c_1(t)$ is separated from $s(t)$, that is, a difference signal $r_1(t)$ with high-frequency components removed is obtained.

$$r_1(t) = s(t) - c_1(t) \tag{6.4}$$

Take $r_1(t)$ as the original data, repeat the above steps to obtain the second IMF component $c_2(t)$, repeat n times to obtain n IMF components. So, there is:

$$\begin{cases} r_1(t) - c_2(t) = r_2(t) \\ \quad \cdots \\ r_{n-1}(t) - c_n(t) = r_n(t) \end{cases} \tag{6.5}$$

When $c_n(t)$ or $r_n(t)$ satisfies the given termination condition, the cycle ends, and the above two expressions can be obtained:

$$s(t) = \sum_{j=1}^{n} c_j(t) + r_{nn}(t) \tag{6.6}$$

where $r_n(t)$ is the residual function and represents the average trend of the signal. Each IMF component $c_1(t)$,$c_2(t)$,......$c_n(t)$ contain components of different time characteristic scales of signals, and their scales are arranged in order from small to large. Therefore, spectral analysis of each IMF component is finally carried out using Hilbert transform to obtain the instantaneous frequency of the signal:

$$s(t) = \mathrm{Re}\sum_{i=1}^{n} a_i(t)e^{j\int \omega_i(t)dt} \tag{6.7}$$

Considering the uncertainty of the long period and the information contained in the signal is mainly in the frequency division component, the residual function $r_n(t)$ is omitted, and Re represents taking the real part. The right side of the above formula is the Hilbert-type spectrum, referred to as Hilbert spectrum, which can be denoted as follows:

$$H(\omega, t) = \mathrm{Re}\sum_{i=1}^{n} a_i(t)e^{j\int \omega_i(t)dt} \tag{6.8}$$

It is the distribution of instantaneous amplitude in the frequency-time plane. The Fourier transform expansion of the same signal $s(t)$ is:

$$s(t) = \mathrm{Re}\sum_{i=1}^{\infty} a_i(t)e^{j\omega_i t} \tag{6.9}$$

where a_i and ω_i are constants. Hilbert spectrum can be further defined as follows:

$$H(\omega)=\int_{-\infty}^{\infty} H(\omega, t)dt \tag{6.10}$$

The marginal spectrum obtained by HHT here is like the Fourier spectrum. From a statistical point of view, it represents the accumulation of amplitudes overtime on the spectrum, reflecting the energy distribution at each frequency. Therefore, the local characteristics of the signal are reflected more accurately than in the Fourier spectrum.

For electricity price time-series data, the changes of data set features over time are complicated, and it is impossible to complete feature mining through simple statistical methods. Therefore, based on the principle and characteristics of the EMD, this method can be used for data mining of electricity price time series, which can screen out important characteristics of data from massive data sets and serve as data pretreatment technology for later electricity price prediction. Figure 6.2 shows an example of the EMD decomposition of electricity price data.

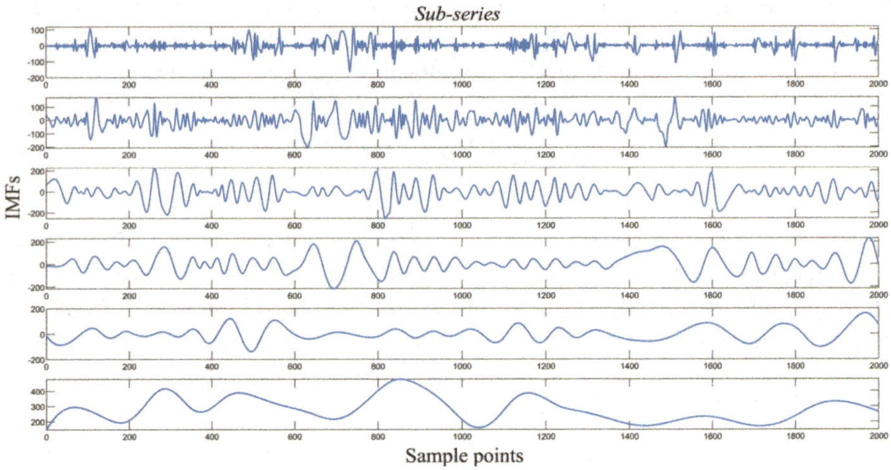

Fig. 6.2 The example of EMD feature extraction

6.2.1.2 Prediction Principle of Big Data of Power Price Based on Extreme Learning Machine

The Extreme Learning Machine (ELM) is first proposed as a learning paradigm of Single-Hidden Layer Feedforward Neural Network (SLFN) (Jamei et al., 2021; Yahia et al., 2021). In the ELM, the weights and biases of the input layer are randomly assigned, and only the output value is calculated. Therefore, the computation time of the ELM is very short. In addition, compared with the traditional Feed Forward Artificial Neural Networks (FF-ANN), ELM has a stronger generalization ability. Figure 6.3 shows a basic SLFN structure, where x_j and y_j represent network inputs and outputs, respectively (Belciug, 2020).

Fig. 6.3 Basic structure of single-hidden layer feedforward neural network

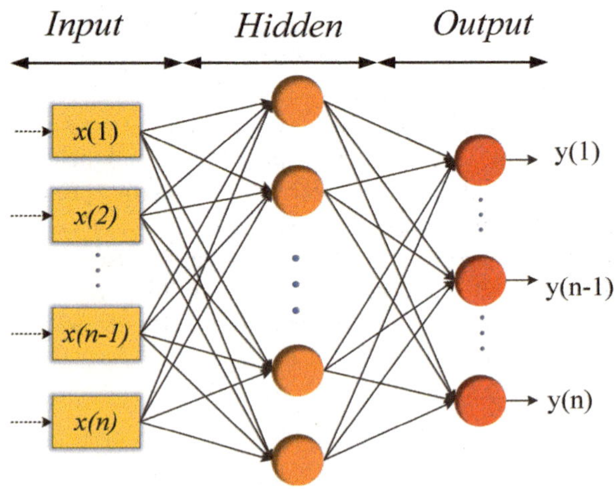

The hidden layer contains M nodes and the SLFN whose activation function is $g(x)$ can be expressed as (Cheng et al., 2021):

$$\sum_{i=1}^{M} \beta_i g(w_i \cdot x_j + b_i) = y_j \qquad j = 1, \ldots, N \tag{6.11}$$

where w is the input layer weight, b is input layer bias, β is the output layer bias, x is the input historical electricity price data, and y is the expected output electricity price of the network, which is the predicted value.

6.2.1.3 Electricity Price Forecasting Framework Based on the EMD and ELM

Power price prediction is divided into two stages. In the first stage, the original power price time-series data is decomposed into different components by using the EMD algorithm and then divided into the training set and test set. In the second stage, the sub-sequence is input as the input end of the ELM to train and predict each sub-sequence. Finally, the sub-sequence is reconstructed to obtain the final electricity price prediction result. The specific framework is shown in Fig. 6.4.

Based on the prediction model framework shown in Fig. 6.4, the electricity price data is decomposed by the EMD method first, and then each sub-sequence is input into an extreme learning machine for prediction. Finally, the prediction results are reconstructed to realize the prediction of future electricity price data. Figure 6.5 shows an example of the EMD-ELM electricity price prediction model.

6.2.2 Electricity Price Forecasting Method Based on Wavelet Packet Decomposition and Deep Belief Network

6.2.2.1 Electricity Price Big Data Decomposition Based on Wavelet Packet Decomposition

For wavelet packet analysis, the algorithm idea of Wavelet Packet (WD) analysis is basically the same as that of wavelet analysis, but the difference is that wavelet packet analysis can provide a more complex and flexible analysis method (Zhou et al., 2021). Wavelet packet analysis is a further improvement and development of wavelet analysis, which can provide higher resolution for signal data. Compared with the WD, Wavelet Packet Decomposition (WPD) technology is a more refined decomposition algorithm in essence. The WPD method can decompose not only the low-frequency part of signal information, but also the high-frequency part of signal information. Because the WPD method can effectively decompose the upper low-frequency signal information and high-frequency signal information simultaneously, WPD has a more accurate local analysis ability (Sairamya et al., 2021; He et al., 2021a).

Short-time Fourier transform of the frequency of the signal information is divided into linear intervals, such as WPD technology can carry on the effective time–frequency decomposition to the signal information, but because of the

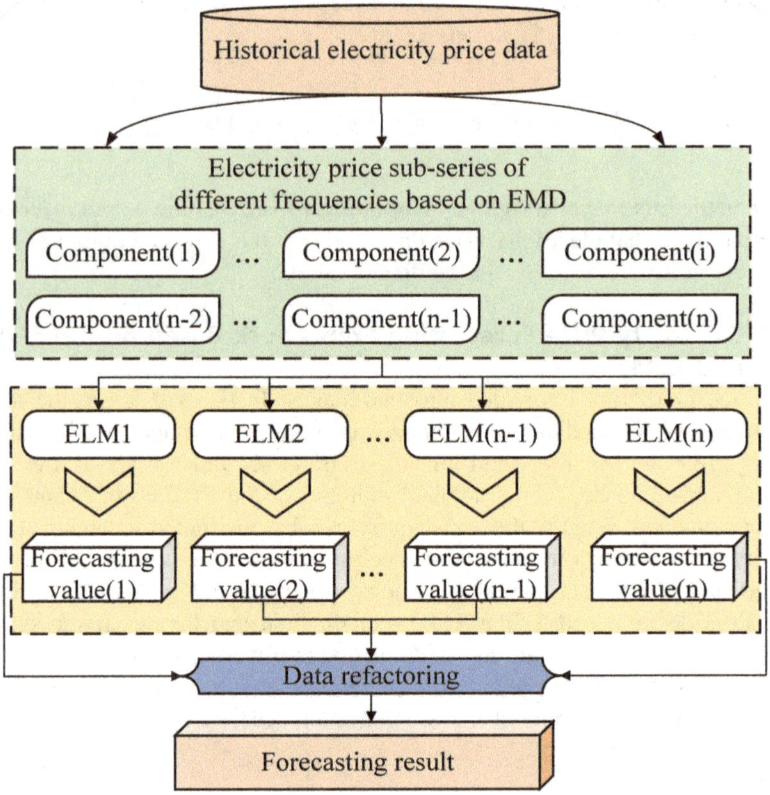

Fig. 6.4 Electricity price forecasting model framework based on the EMD and ELM

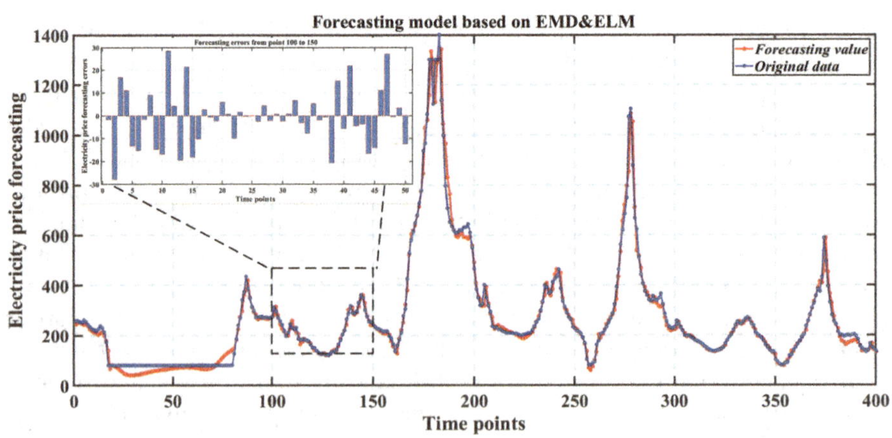

Fig. 6.5 Example of the EMD-ELM electricity price prediction model

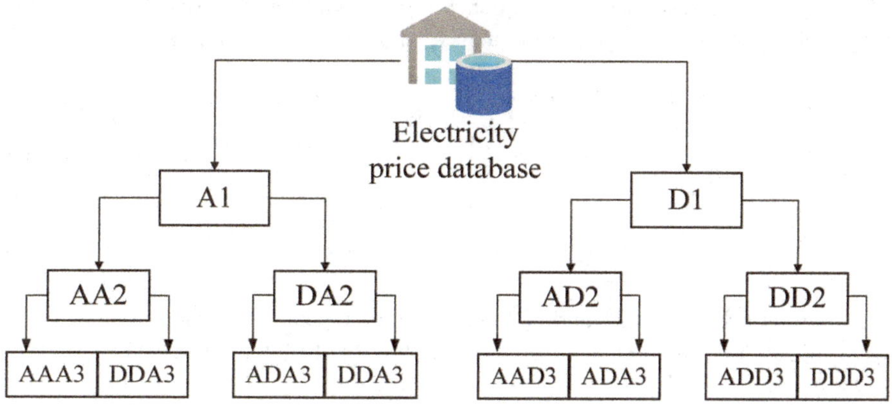

Fig. 6.6 Decomposition results of three layers WPD

scale of the WPD method is to follow the binary changes, thus in high-frequency space within the scope of the frequency resolution is poorer, and within the scope of the low-frequency space, time and frequency resolution are better. The WPD method for signal information to provide a more detailed analysis method, the frequency information of multi-level WPD will divide the, no detailed division of the high-frequency signal in the WD part for further decomposition, and can according to the characteristics of the signal information is broken down, automatically select the appropriate band, for its make band and correspond to the spectrum of the signal information, Thus, the resolution of time domain and frequency domain can be improved, which has more extensive application value. Taking three-layer WPD technology as an example, the decomposition results of three-layer WPD are shown in Fig. 6.6 (Xie et al., 2016).

As can be seen from Fig. 6.6, A is the decomposed low-frequency signal, D is the decomposed high-frequency signal, and the number in each box represents the scale number of the WPD. If the series of the WPD is larger, the scale number of wavelet packet is larger, and the spatial resolution corresponding to the wavelet packet coefficient is lower. Therefore, multi-resolution analysis and research can be carried out in different spatial resolutions. Where the signal information relationship of the WPD is (Seo & Kim, 2016):

$$S = AAA3 + DDA3 + ADA3 + DDA3 + AAD3 + ADA3 + ADD3 + DDD3 \tag{6.12}$$

The WPD method is a multi-resolution analysis method. In the multi-resolution analysis, $L^2(R) \bigoplus_{j\in Z} W_j$, indicating that the multi-resolution analysis is to decompose the Hilbert space $L^2(R)$ into the orthogonal sum of all subspaces W_j according to different scale factors j. Where W_j is the closure of the wavelet function $\psi(t)$, that is, the wavelet sub-space.

To improve the resolution of frequency signal, wavelet subspace W_j is further subdivided according to binary frequency, and the scale space V_j and wavelet subspace W_j are unified expressed $\begin{cases} U_j^0 = V_j \\ U_j^1 = W_j \end{cases}, j \in Z$ by using the new subspace U_j^n, so that the orthogonal solution $V_{j+1} = V_j \oplus W_j$ of Hilbert space can be decomposed by U as follows:

$$U_{j+1}^0 = U_j^0 \oplus U_j^1, j \in Z \tag{6.13}$$

Define the subspaces U_j^n and U_j^{2n} as the closure spaces of functions $u_n(t)$ and $u_{2n}(t)$ respectively, and let function $u_n(t)$ satisfy the following two-scale equation (Sairamya et al., 2021; He et al., 2021a):

$$\begin{cases} u_{2n}(t) = \sqrt{2} \sum\limits_{k \in Z} \xi(k) u_n(2t - k) \\ u_{2n+1}(t) = \sqrt{2} \sum\limits_{k \in Z} \varsigma(k) u_n(2t - k) \end{cases} \tag{6.14}$$

where $\varsigma(k) = (-1)^k \xi(1 - k)$, and when $n = 0$, and the calculation method can be expressed as follows (Sairamya et al., 2021):

$$\begin{cases} u_0(t) = \sum\limits_{k \in Z} \xi(k) u_0(2t - k) \\ u_1(t) = \sum\limits_{k \in Z} \varsigma(k) u_0(2t - k) \end{cases} \tag{6.15}$$

In the WPD, both $\phi(t)$ and $\psi(t)$ satisfy the two-scale equation (Xie et al., 2016):

$$\begin{cases} \phi(t) = \sum\limits_{k \in Z} \xi_k \phi(2t - k), \{\xi_k\}_{k \in Z} \in l^2 \\ \psi(t) = \sum\limits_{k \in Z} \varsigma_k \psi(2t - k), \{\varsigma_k\}_{k \in Z} \in l^2 \end{cases} \tag{6.16}$$

where $\phi(t)$ is the scaling function and $\psi(t)$ is the wavelet basis function.

In the WPD method, set $\varsigma_j^n(t) \in U_j^n$, then $\varsigma_j^n(t)$ can be expressed as follows:

$$\varsigma_j^n(t) = \sum_l d_l^{j,n} u_n(2j^t - l) \tag{6.17}$$

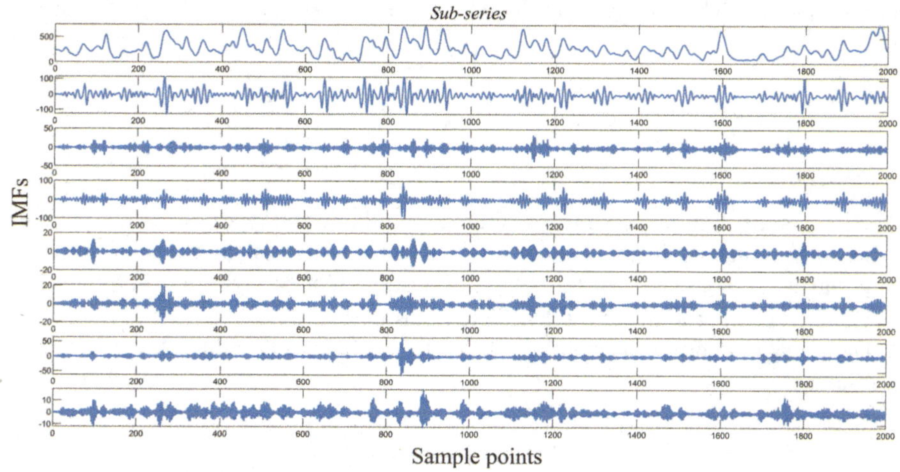

Fig. 6.7 Example of the EMD decomposition of electricity price data

The WPD algorithm is expressed as $\{d_l^{j+1,n}\}$, then $\{d_l^{j,2n}\}$ and $\{d_l^{j,2n}\}$ are obtained, namely:

$$\begin{cases} d_l^{j,2n} = \sum_k \lambda_{k-2l} d_k^{j+1,n} \\ d_l^{j,2n+1} = \sum_k \lambda_{k-2l} d^{j+1} \end{cases} \tag{6.18}$$

Wavelet packet analysis is a linear time–frequency analysis method. Compared with the WD method, the WPD can effectively decompose the high-frequency information of signal data. Due to its good time–frequency positioning characteristics and strong adaptive ability to signals, the WPD can effectively decompose the electricity price time-series data. Figure 6.7 shows an example of the EMD decomposition of electricity price data.

6.2.2.2 Prediction Principle of Big Data of Power Price Based on Deep Belief Network

The Deep Belief Network (DBN) is made up of several Restricted Boltzmann Machines (RBM) stacked on top of each other and trained in a greedy way (Scarpiniti et al., 2021). The hidden layer of the current RBM is the explicit layer of the next RBM, i.e., the output of the previous RBM is also the input of the current RBM. In the process of RBM training, the current RBM can only start training after the previous RBM training is completed and the training results are output as the input layer of the current RBM. Thus, the multi-layer RBM is trained layer by layer from the bottom layer to the top layer (Mohan et al., 2021).

Just like an ordinary Boltzmann machine, limited Boltzmann machine also has an energy model, and its joint probability distribution can be obtained from the energy function. The energy of an RBM can be expressed as follows (Chen et al., 2020):

$$E(v,h|\theta) = -\sum_{j=1}^{N_h} \chi_j v_j - \sum_{i=1}^{N_v} \varpi_i v_i - \sum_{i=1}^{N_v} \sum_{j=1}^{N_h} w_{ij} v_i h_j \tag{6.19}$$

where N_v is the number of neurons contained in the visual layer, N_h is the number of neurons contained in the hidden layer, ϖ_i is the bias of the ith neuron in the visual layer, χ_j is the bias of the jth neuron in the hidden layer, w_{ij} is the weight of visual layer neuron i and hidden layer neuron j.

The joint probability distribution of the visible layer and hidden layer is shown as follows:

$$p(v,h|\theta) = \frac{e^{-E(v,h|\theta)}}{\sum_{v,h} e^{-E(v,h|\theta)}} \tag{6.20}$$

In RBM, the probability of activation of hidden layer neuron h_j is shown in the formula below:

$$p(h_j|v) = \sigma(\chi_j + \sum_{i=1}^{N_v} w_{ij} v_i) \tag{6.21}$$

Due to the bidirectional connection between the visible layer and the hidden layer, the probability that the visual layer neuron v_i can also be activated by the hidden layer neuron is shown as follows:

$$p(v_i|h) = \sigma(\varpi_i + \sum_{j=1}^{N_h} w_{ij} h_j) \tag{6.22}$$

The electricity price prediction model constructed by DBN in this book consists of two layers of the RBM.

6.2.2.3 Electricity Price Forecasting Framework Based on WPD and DBN

Similar to the EMD-ELM electricity price prediction model, the WPD-DBN power price prediction is also divided into two stages. In the first stage, the original power price time-series data are decomposed into high frequency and low-frequency sub-sequences using the WPD algorithm and then divided into the training set and test set. In the second stage, the sub-sequence is input as the input end of the DBN, and each sub-sequence is trained and predicted. Finally, the sub-sequence is reconstructed to obtain the final electricity price prediction result. The specific framework is shown in Fig. 6.8.

Based on the prediction model framework in Fig. 6.8, electricity price data is decomposed by the WPD method first, and then each sub-sequence is input into

Fig. 6.8 Electricity price forecasting framework based on the WPD and DBN

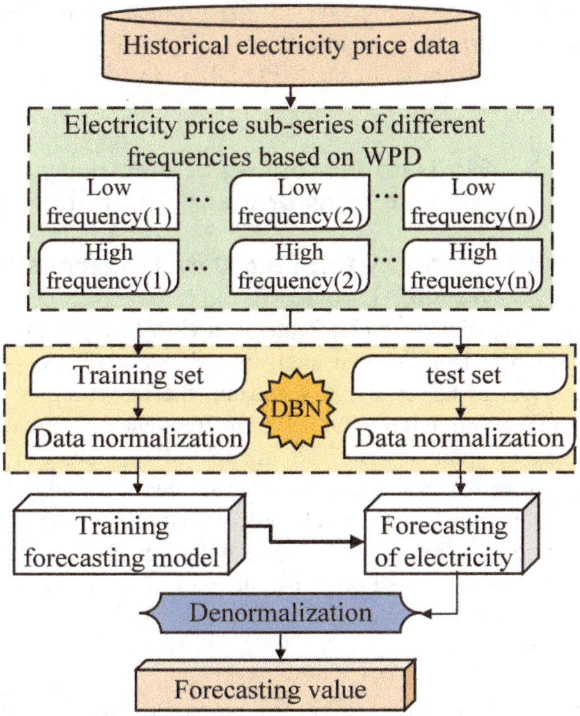

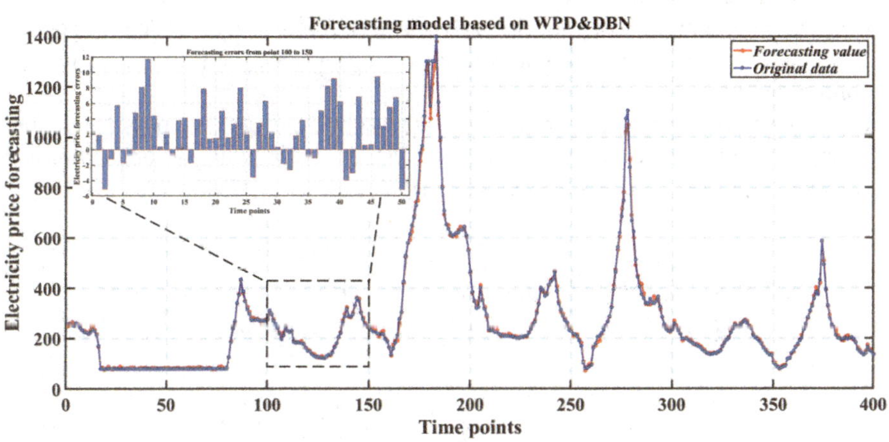

Fig. 6.9 Example of the WPD-DBN electricity price forecasting model

the DBN deep confidence network for prediction. Finally, the prediction results can be reconstructed to realize the prediction of future electricity price data. Figure 6.9 shows an example of the WPD-DBN electricity price prediction model.

6.2.3 Big Data Processing of Electricity Price Based on Empirical Wavelet Transform and Long Short-Term Memory Network

6.2.3.1 Electricity Price Big Data Decomposition Based on Empirical Wavelet Transform

The Empirical Wavelet Transform (EWT) algorithm was developed by Gilles in 2013 based on wavelet analysis (Gilles, 2013; Pearson, 2002). Through adaptive analysis of target signal spectrum, the target signal is decomposed into multiple modal components with different frequency characteristics by orthogonal wavelet filter, to achieve feature extraction of the target signal. The traditional wavelet transform cannot retain the original characteristics of the data, but the empirical wavelet transform can adaptively analyze the signal and construct its own basis function according to the data. After constructing the basis function, the wavelet transform is carried out with the help of the framework of the wavelet transform.

The EWT is an adaptive data pre-processing method, and its basic steps are as follows:

- Obtain the Fourier spectrum $F(w)$ of the signal $f(t)$, and then carry out the fast Fourier transform of the signal.
- Correct segmentation of Fourier spectrum, its boundary is determined as $\Omega = \{ \omega^k \}, k = 0, 1 \ldots k$, decomposition of the Fourier spectrum $[0, \Pi]$ into K contiguous parts.
- Littlewood-Pantry and Meyer wavelets are used to design low-pass filters and bandpass filters. Formulas (6.23) and (6.24) are scale function and empirical wavelet function respectively (Pearson, 2002).

$$\hat{\phi}_n(\omega) = \begin{cases} 1 & |\omega| \le \omega_n - \tau_n \\ \cos\left[\frac{\pi}{2}\beta\left(\frac{1}{2\tau_{n+1}}(|\omega| - \omega_n + \tau_n)\right)\right] & \omega_n - \tau_n \le |\omega| \le \omega_n + \tau_n \\ 0 & otherwise \end{cases} \tag{6.23}$$

$$\psi_n(\omega) = \begin{cases} 1 & \omega_n + \tau_n \le |\omega| \le \omega_{n+1} + \tau_{n+1} \\ \cos\left[\frac{\pi}{2}\beta\left(\frac{1}{2\tau_{n+1}}(|\omega| - \omega_n + \tau_n)\right)\right] & \omega_{n+1} - \tau_{n+1} \le |\omega| \le \omega_{n+1} + \tau_{n+1} \\ \sin\left[\frac{\pi}{2}\beta\left(\frac{1}{2\tau_n}(|\omega| - \omega_n + \tau_n)\right)\right] & \omega_n - \tau_n \le |\omega| \le \omega_n + \tau_n \\ 0 & otherwise \end{cases} \tag{6.24}$$

where w represents the frequency, w_n represents the frequency of the nth boundary, and the values of τ_n and $\beta(x)$ are shown in Formulas (6.25) and

(6.26).

$$\tau_n = \gamma\omega_n(0 < \gamma < 1) \tag{6.25}$$

$$\beta(r) = \begin{cases} 1 & r \geq 1 \\ 0 & r \leq 0 \\ \beta(r) + \beta(1-r) = 1 & otherwise \end{cases} \tag{6.26}$$

- The scale coefficient is obtained according to the inner product of the scale function and the original signal, and the empirical wavelet coefficient is obtained by the inner product of the empirical wavelet function and the original signal. The calculation methods of the scale coefficient $W(0, t)$ and the empirical wavelet coefficient $W(k, t)$ are presented as follows (Mohan et al., 2021):

$$W(0, t) = \int f(\tau)\cdot\overline{\varphi_1(\tau - t)}d\tau = F^{-1}[f(\omega)\hat{\varphi}_1(\omega)] \tag{6.27}$$

$$W(k, t) = \int f(\tau)\cdot\overline{\psi_k(\tau - t)}d\tau = F^{-1}[f(\omega)\hat{\psi}_1(\omega)] \tag{6.28}$$

where $\hat{\varphi}_1(\omega)$ and $\hat{\psi}_k(\omega)$ are the Fourier transforms respectively.
- After the original signal is reconstructed.

$$f(t) = W(0, t) * \varphi_1(t) + \sum_{k=1}^{n} W(k, t) * \psi_k(t) \tag{6.29}$$

The empirical modal function can be obtained as follows:

$$fi(t) = \begin{cases} W(0, t) * \varphi_1(t), & i = 1 \\ W(i, t) * \psi_1(t), & i > 1 \end{cases} \tag{6.30}$$

In fact, Empirical Wavelet Transform (EWT) is a data analysis method, which combines the adaptive decomposition idea of EMD with the theoretical framework of Wavelet Transform (WT). Firstly, the Fourier spectrum of the original signal is divided adaptively, and then a wavelet filter bank is constructed to extract the tightly supported modal components. The EWT is suitable for smoothing non-stationary and nonlinear time series or signals. Therefore, the EWT can also well identify the time series characteristics of electricity price data, as shown in Fig. 6.10 shows an example of the EWT decomposition of electricity price data.

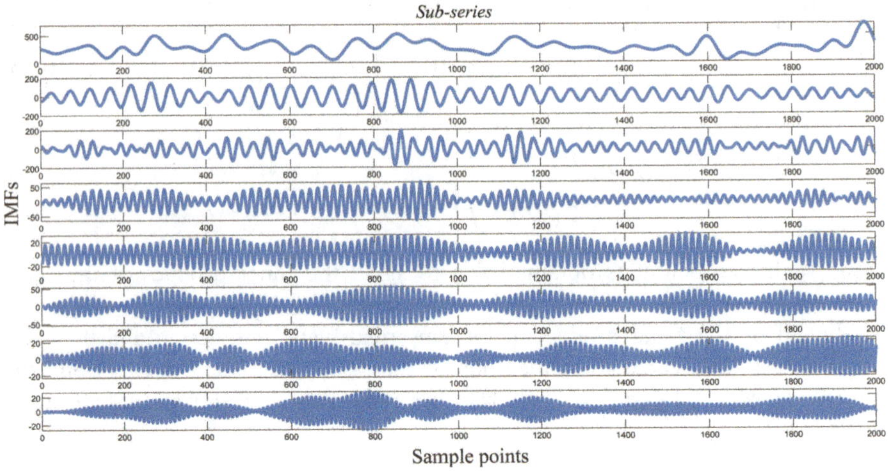

Fig. 6.10 Example of EWT decomposition of electricity price data

6.2.3.2 Prediction Principle of Big Data of Power Price Based on Long Short-Term Memory Network

The LSTM defines and maintains an internal memory unit state, namely cell state c, and updates, maintains, or deletes information in cell state through gate structures of input gate j, forgetting gate f, and output gate o_t (Solgi et al., 2021). Its memory unit structure is shown in Fig. 6.11.

In Fig. 6.11, x_t represents the input, h_{t-1} represents the hidden layer state at $t-1$, c_{t-1} represents the long-term memory state at $t-1$, f_t represents the output signal of the forgetting gate, j_t represents the output signal of the input gate, o_t represents the output signal of the output gate, k_t represents the preparatory information to be input into long-term memory, l_t represents the preparatory information to be entered into the hidden layer state h_t. The input and output from

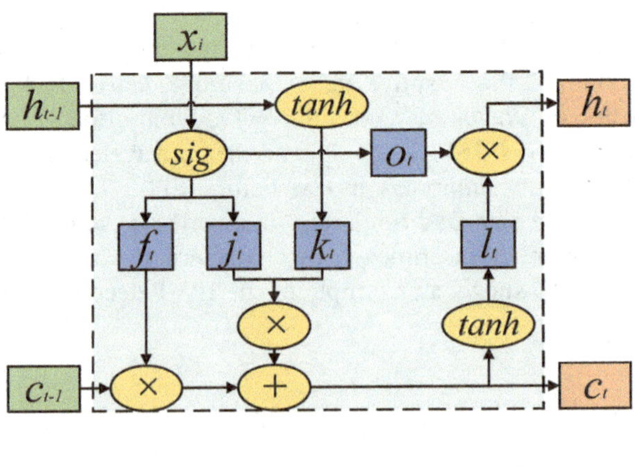

Fig. 6.11 LSTM memory unit structure diagram

the memory unit are calculated as follows (Solgi et al., 2021):

$$\begin{cases} f_t = \text{sigmoid}(W_{fx}x_t + W_{fh}h_{t-1} + b_f) \\ j_t = \text{sigmoid}(W_{jx}x_t + W_{jh}h_{t-1} + b_j) \\ k_t = \text{sigmoid}(W_{kx}x_t + W_{kh}h_{t-1} + b_k) \\ o_t = \text{sigmoid}(W_{ox}x_t + W_{oh}h_{t-1} + b_o) \end{cases} \tag{6.31}$$

where f_t, j_t and o_t are all real numbers ranging from 0 to 1. f_t determines the forgetting proportion of long-term memory, j_t determines the proportion of current information input to long-term memory c, and o_t determines the proportion of long-term memory c output to the current state h. W and b are parameter matrices. After the filtering of the input gate, k_t can be multiplied by j_t to obtain part of the information in long-term memory c:

$$l_t = \tanh(c_t) \tag{6.32}$$

After screening by the output gate, l_t and o_t are multiplied to obtain the information in the hidden layer state h_t, which are shown as follows:

$$h_t = o_t l_t \tag{6.33}$$

After the screening of the forgetting gate and input gate, the long-term memory value c_t at a time t can be obtained, which are shown as follows:

$$c_t = f_t c_{t-1} + j_t k_t \tag{6.34}$$

6.2.3.3 Electricity Price Forecasting Framework Based on the EWT and LSTM

Similar to the two electricity price prediction models proposed above, the EWT-LSTM power price prediction is also divided into two stages. In the first stage, the original power price time-series data are decomposed into different empirical mode components using the EWT algorithm and then divided into the training set and test set. In the second stage, each empirical mode component is input as the input end of the LSTM, and each sub-sequence is trained and predicted. Finally, the sub-sequence is reconstructed to obtain the final electricity price prediction result. The specific framework is shown in Fig. 6.12.

Based on the prediction model framework in Fig. 6.12, the electricity price data is decomposed by the EWT method at first, and then each sub-sequence is input into the LSTM for prediction. Finally, the prediction results can be reconstructed to realize the prediction of future electricity price data. Figure 6.13 shows an example of the EWT-LSTM electricity price prediction model.

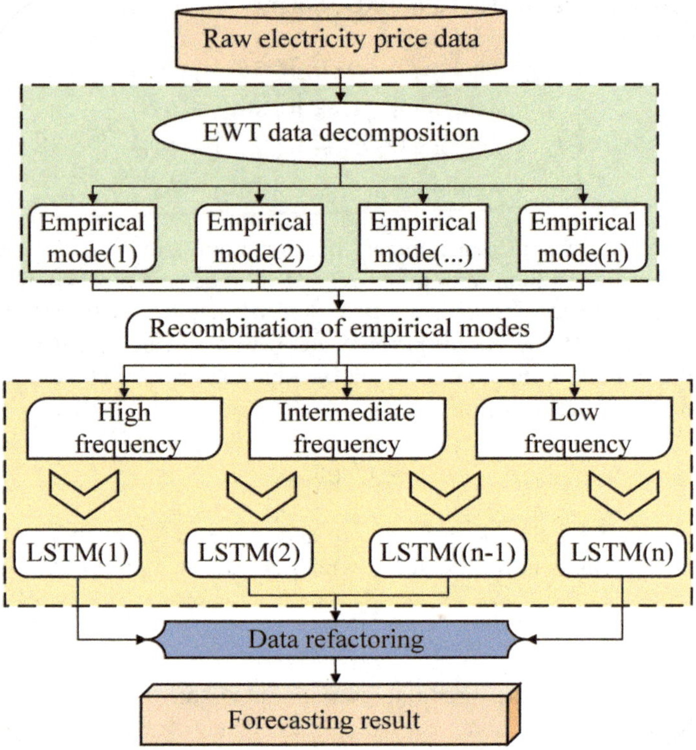

Fig. 6.12 Electricity price forecasting model framework based on the EWT and LSTM

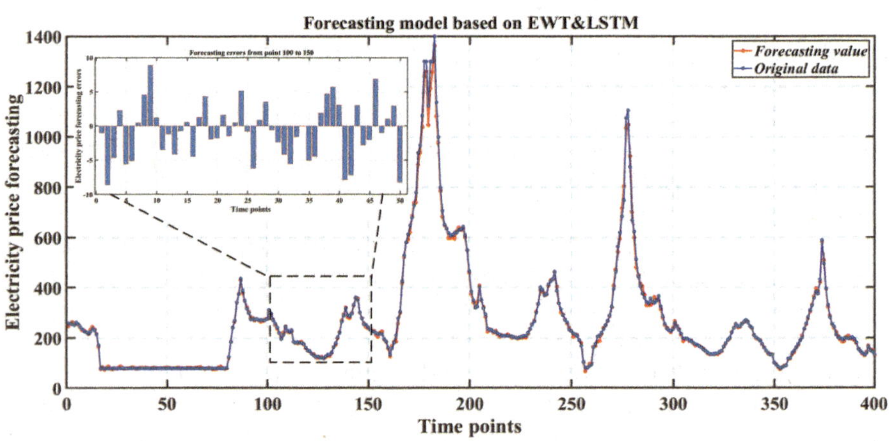

Fig. 6.13 Example of the EWT-LSTM electricity price forecasting model

6.3 Correlation Analysis of Urban Energy Consumption and Economic Growth

Energy is an important material basis for economic development and social progress and plays an important role in the sustainable economic growth of a country. But energy consumption and environmental protection influence and restrict each other. Energy is an important material basis for economic development and social progress, but when its development cannot meet the requirements of national economic growth, it will become one of the factors that hinder rapid economic growth. Obviously, there is a certain correlation between energy consumption and economic growth. This paper will take China as an example to explore the correlation analysis between urban energy consumption and economic growth.

6.3.1 Grey Correlation Model in the Energy Economy

Grey System Theory (GST) was founded in 1982. Grey correlation analysis has significant advantages in studying time series with small samples and poor information (Fan et al., 2021). It can effectively make up for the defects of traditional statistical analysis methods, that is, there is no need to consider the sample size and the rule of sample change, let alone the inconsistency between quantitative results and qualitative results. This analysis method makes up for the deficiency of systematic analysis of mathematical statistics. The idea of grey correlation analysis is to judge the degree of correlation between parent series and child series according to the similarity of geometric shape between parent series and child series curves of the reaction system. Now the grey correlation analysis method is adopted, and the process of building the grey correlation model is as follows (Kumar & Soota, 2020):

Step 1: Select the dependent variable reference sequence $Y = Y(k)|k = 1, 2...n$ from the original sequence, and the self-variable ratio of the sequence is $X_i = X_i(k)|k = 1, 2...n, i = 1, 2...m$.

Step 2: The reference sequence and comparison sequence are dimensionless processed by means method, and the formula is given as follows:

$$\begin{cases} \overline{x_i} = \dfrac{1}{m}\sum\limits_{i=1}^{m} x_i \quad , i = 1, 2...m \\ x_i(k) = \dfrac{x_i(k)}{\overline{x_i}} \quad , k = 1, 2...n \end{cases} \tag{6.35}$$

Step 3: The grey correlation coefficient of the reference sequence and comparison sequence is calculated as follows:

$$\zeta_i(k) = \frac{\min\limits_i \min\limits_k |y(k) - x_i(k)| + \rho \max\limits_i \max\limits_k |y(k) - x_i(k)|}{|y(k) - x_i(k)| + \rho \max\limits_i \max\limits_k |y(k) - x_i(k)|} \tag{6.36}$$

In which, ρ is the number of discriminating systems, the general value is 0.5, $\rho \in (0, 1)$.

Step 4: The grey correlation degree R is calculated, and the R value is the mean value of multiple grey correlation values in the comparison sequence, and the formula is as follows:

$$R_i = \frac{1}{n}\sum_{k=1}^{n} \zeta_i(k) \quad , k = 1, 2 \ldots n \tag{6.37}$$

6.3.2 Grey Correlation Analysis of Economic Growth and Energy Consumption Varieties

Step 1: Data sources of energy consumption varieties.

The statistical data used in this paper are from "China Energy Statistical Yearbook 2018", and the growth rate of China's Gross Domestic Product (GDP) and various energy consumption data (tons of standard coal) from 1980 to 2017 are selected.

Step 2: Dimensionless processing of original data.

China's GDP growth rate data as the dependent variable reference sequence $Y(k)$, all kinds of energy consumption data sequence to be compared to the independent variables $X_i(k)$, according to the above formula, after taking no dimension of the initial data, to eliminate the dimension of various factors, strengthen the proximity between the various factors, increase the comparability.

Step 3: Calculate the difference sequence.

The difference between the GDP growth rate series and energy consumption series are calculated, $\Delta_i(k) = |y(k) - x_i(k)|$.

Step 4: Calculate the correlation coefficient.

Based on the above difference sequence, the discriminant number $\rho = 0.5$ is substituted into Formula (6.36) to calculate the correlation coefficient. Figure 6.14 shows the grey correlation between different energy types and GDP growth rate.

Step 5: Calculate the correlation coefficient.

According to Formula (6.37), the grey correlation degree is calculated and sorted, and the relationship between economic growth and energy consumption variety structure is further studied. The results are shown in Table 6.1.

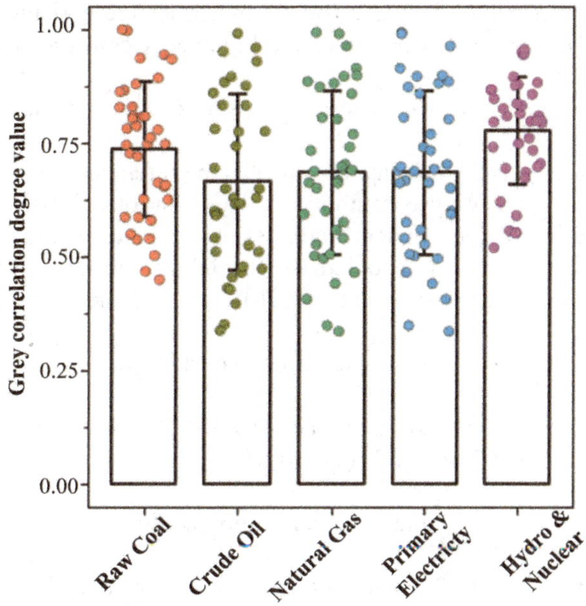

Fig. 6.14 Grey correlation between different energy types and GDP growth rate

Table 6.1 The rank of gray correlation between energy consumption type and GDP growth rate

Energy consumption type	Raw coal	Crude oil	Natural gas	Primary electricity	Hydro and nuclear
Grey correlation degree	0.844	0.777	0.738	0.685	0.665
Rank	1	2	3	4	5

From Table 6.1, raw coal has the largest grey correlation, followed by crude oil and natural gas, followed by primary electricity. Hydropower and nuclear power have the lowest correlation. The closer the correlation value is to 1, the greater the impact of the correlation between energy consumption and economic development.

6.3.3 Grey Correlation Analysis of Economic Growth and Energy Consumption Industrial Structure

Step 1: Data sources of the industrial structure of energy consumption.

The statistical data used in this paper are from "China Energy Statistical Yearbook 2018", and the data of China's GDP and energy consumption of various industries (tons of standard coal) from 2009 to 2017 are selected. The GDP value is taken as the reference sequence, and the energy consumption of six industries is taken as the comparison sequence. The six industries are agriculture, industry, manufacturing, transportation, wholesale, and life energy consumption.

Step 2: Dimensionless processing of original data.

Taking The GDP data of China as the reference series $Y(k)$ of dependent variables and the energy consumption data of various industries as the comparative series $X_i(k)$ of independent variables, the dimensionless reduction of the dimensions of all factors is adopted for the initial data according to the above formula to strengthen the proximity of all factors and increase their comparability.

Step 3: Calculate the difference sequence.

Calculate the difference between GDP series and energy consumption series of various industries, $\Delta_i(k) = |y(k) - x_i(k)|$.

Step 4: Calculate the correlation coefficient.

Based on the above difference sequence, the discriminant number $\rho = 0.5$ is substituted into Formula (6.36) to calculate the correlation coefficient. Figure 6.15 shows the grey correlation between energy consumption and GDP in different industries.

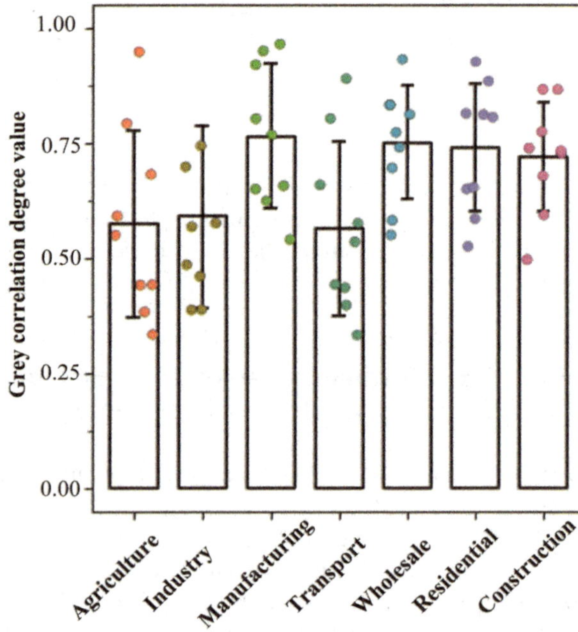

Fig. 6.15 Grey correlation between energy consumption and GDP in different industries

Step 5: Calculate the correlation.

According to the formula, the grey correlation degree is calculated and sorted, and the correlation between economic growth and energy consumption industrial structure is further analyzed. The results are shown in Table 6.2.

6.4 Metering Charge Adjustment Analysis of City Electricity Prices

6.4.1 Background of the K-means Algorithm for Characteristic Analysis of Electricity Price

The K-means algorithm is a clustering algorithm based on the partition (Shrifan et al., 2021). Suppose that data set D contains n data objects and divides these data objects into k clusters $C_1, .., C_k$ according to certain rules, such that for $l \leq i, j \leq k$, there is $C_i \subseteq D$ and $C_i \cap C_j = \emptyset$.

The basic description of the algorithm is given as follows (Zhang & Peng, 2022):

Step 1: Select K objects from the dataset as the initial cluster center of K clusters, or divide all objects into k initial classes and take the center of gravity (mean) of K clusters as the initial cluster center.
Step 2: For each remaining object, the distance between it and the center of each cluster is calculated, and it is assigned to the nearest (or most similar) cluster by comparison one by one.
Step 3: The average value of objects in each cluster is calculated as the new cluster center.
Step 4: Judge whether the clustering results converge according to the set threshold. If convergence conditions are not met, go back to step 2 and continue.
Step 5: Output clustering results.

The distance relationship between the data object $x \in C_i$ and the clustering center c_i of the cluster is represented by the distance function. The distance function used in cluster analysis has various definition forms. In general, the distance function is defined as the Euclidean distance between objects, which is expressed as follows (Zhang et al., 2021b):

$$dist(x_i, x_j) = \sqrt{\sum_{m=1}^{m} |x_{im} - x_{jm}|^2} \tag{6.38}$$

where m represents the data dimension of objects x_i and x_j.

Usually, Sum of Squared Error (SSE) is used as the threshold target to measure clustering completion, that is, for all objects contained in the data set, the sum of

Table 6.2 Rank of gray correlation between different industries and GDP

Different industries	Agriculture	Industry	Manufacturing	Transport	Wholesale	Residential	Construction
Grey correlation degree	0.590	0.564	0.575	0.740	0.720	0.751	0.765
Rank	5	7	6	3	4	2	1

squares of the distance from the object x to the cluster center c_i corresponding to its cluster C_i is calculated. The SSE is defined as follows:

$$SSE = \sum_{i=1}^{k} \sum_{x \in C_i} |x - c_i|^2 = \sum_{i=1}^{k} \sum_{x \in C_i} dist(x, c_i)^2 \tag{6.39}$$

This criterion tries to make k clusters as compact as possible within clusters and independent as possible between clusters.

6.4.2 Analysis of User Electricity Price Consumption Characteristics Based on the K-means Algorithm

The K-means algorithm is applied to cluster the residential users according to the historical electricity consumption data, and the users with similar electricity consumption levels are divided into a cluster, finally forming multiple clusters. In a cycle, a similar level of electricity consumption can largely reflect the similar behavior characteristics of electricity consumption, and then indirectly reflect the various influencing factors behind it. Figure 6.16 shows the clustering results of each user's electricity load under ideal conditions.

As can be seen from Fig. 6.16, the K-means algorithm can better classify data sets, to realize the classification of users with the same power consumption characteristics, thus laying data foundation for the optimization design of residents' stepped electricity price.

It can be considered that the K-means algorithm can avoid the error risk caused by improper consideration of influencing factors, thus skipping the defects

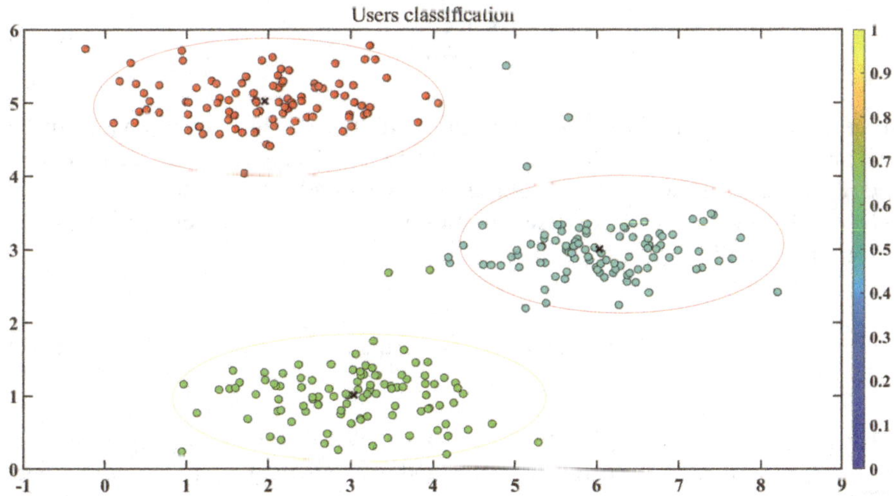

Fig. 6.16 The ideal clustering result of users' electricity load

of traditional modeling and directly dealing with the problem from the actual phenomenon.

6.4.3 Optimization Design of Residential Stepped Electricity Price

The formation mechanism of unit electricity price is a complex process, and its formulation needs to consider the influence of many factors, such as power generation costs of power generation enterprises, power supply costs of power grid companies and depreciation of fixed assets, national energy strategic planning and regional resource scheduling, international energy market changes, geopolitical factors, etc. Considering the actual needs of both power supply enterprises and consumers, it can make appropriate adjustments and optimization on the basis of the current residential electricity price scheme, which can better reflect the value of the stepped electricity price system (Jia et al., 2020; Sovacool et al., 2021).

- Objective function.

Clustering results in the formation of multiple clusters, which can be used as the natural basis for the division of step electricity. The number of clusters corresponds to the number of steps of sections, and the boundary quantity between clusters corresponds to the limit value of the electric quantity of each step.

For data set D containing n household electricity consumption data samples, k electricity steps are formed after clustering according to the method in ***Sect. 6.4.2***. Then, the electricity sales income P of power supply enterprises to residential users under the optimized scheme is (Jia et al., 2020):

$$P = \sum_{r=1}^{n} \sum_{i=1}^{k} z_i q_{r,i} \tag{6.40}$$

where z_i represents the unit electricity price of step i, which is a quantity to be determined, $q_{r,i}$ represents the electricity quantity of class i in the accumulated electricity consumption of the rth household user in a billing cycle.

Similarly, the income P' of residential users of power supply enterprises under the original plan is:

$$P' = \sum_{\mathrm{r}=1}^{\mathrm{n}} \sum_{\mathrm{j}=1}^{\mathrm{k}'} \mathrm{z}'_{\mathrm{j}} \mathrm{q}'_{\mathrm{r,j}} \tag{6.41}$$

where k' represents the ladder division of the original scheme, z'_{j} represents the unit electricity price of the original *j-class*, and is a known quantity. $q'_{\mathrm{r,j}}$ represents the electricity value of class j in the accumulated electricity consumption of the r household user in a billing cycle, $j = 1, 2, \ldots, k'$.

Based on the premise that electricity sales revenue of power supply enterprises remains basically unchanged before and after the adjustment of the scheme and only adjusts the grading level and unit electricity price of residential ladder electricity price, the 2 norms of the difference between electricity sales revenue of power supply enterprises before and after the adjustment of the scheme is minimized as the optimization objective, and the objective function is established as follows:

$$F = \min \frac{1}{2} \| P - P' \|_2^2 \tag{6.42}$$

- Constraints.

The value of solving quantity z_i is limited as follows:

(a) Step constraint of electricity price.

The scheme adopts the incremental stepwise electricity price model, requiring that the unit price of high-order electricity should be higher than that of low-order electricity by a certain percentage, namely:

$$z_i = 0.95 z_{i+1} \tag{6.43}$$

(b) Price increase constraint.

Unit stepped-price should be set according to the appropriate increasing rate to realize the differentiated attribute function of the stepped-price system and guide users to use electricity reasonably. Based on the unit stepped price z_i corresponding to the first stage, the price increase rate constraints are as follows:

$$r_i z_1 \leq z_i \tag{6.44}$$

In the formula, r_i represents the increasing rate of unit step electricity price z_i relative to z_1 of class i.

(c) Constraints on upper and lower limits of electricity price.

From the perspective of market law, the z_{low} should be at least higher than the initial step electricity price and lower than the upper limit z_{high}, that is:

$$z_{low} \leq z_i \leq z_{high} \tag{6.45}$$

- Optimization model.

To sum up, the optimization model of unit stepped electricity price is obtained as follows:

$$\min \frac{1}{2}\left\| \sum_{r=1}^{n}\sum_{i=1}^{k} z_i q_{r,i} - \sum_{r=1}^{n}\sum_{j=1}^{k'} z'_j q'_{r,j} \right\|_2^2$$
$$\text{s.t.}\begin{cases} z_i = 0.95 z_{i+1} \\ r_i z_1 \le z_i \\ z_{low} \le z_i \le z_{high} \end{cases} \tag{6.46}$$

According to the established optimization model of residential ladder electricity price, combined with specific historical data of residential electricity consumption, using cluster analysis method and optimization function solving tool, the adjusted residential ladder electricity price scheme can be obtained.

- Example calculation.

According to the current residential ladder electricity price method of a city in China, the classification of the ladder electricity price under the current plan of the city and the distribution of sample users are sorted out, as shown in Table 6.3.

Based on the above conditions, the electricity price of each step unit of the adjustment plan can be obtained by substituting it into the electricity price optimization model for the solution. The specific results are shown in Table 6.4.

By comparing the data in Tables 6.3 and 6.4, some similarities and differences between the adjustment plan and the current plan can be obtained:

(a) The lowest and highest unit electricity price basically maintains the original level, inheriting the functions of user subsidies and adjustment limits of the current scheme. The middle three order electricity price has a large step separation, which can play a guiding and regulating role in users' electricity consumption behavior as well as the electric quantity classification.
(b) Comparing the corresponding electricity charges of the two schemes in each sample, about 87% of users will realize the effect of reducing electricity

Table 6.3 Tiered level and consumer distribution under the current scheme

Classes	1	2	3
Electricity consumption lower limit (kWh/year)	0	2161	4201
Electricity consumption upper limit (kWh/year)	2160	4200	-
Unit price (CNY/kWh)	0.4983	0.5483	0.7983
User coverage	47.94	50.69	1.37
Proportion of power consumption of each class (%)	30.65	65.7	3.64
Proportion of electricity cost of each class (%)	30.02	65.73	4.25

Table 6.4 Tiered electricity pricing and electricity bills distribution under the adjusted scheme

Classes	1	2	3	4	5
Electricity consumption lower limit (kWh/year)	0	1561	2401	3241	4801
Electricity consumption upper limit (kWh/year)	1560	2400	3240	4800	-
Unit price (CNY/kWh)	0.50	0.52	0.55	0.62	0.89
User coverage	26.66	26.86	35.67	10.07	0.74
Proportion of power consumption of each class (%)	11.54	24.97	44.14	17.02	2.34
Proportion of electricity cost of each class (%)	11.13	24.38	43.89	17.59	3.00

charges, with significant social benefits. A small number of medium—and high-power consumers have seen their electricity bills rise, making up for the shortfall in fee cuts. Under the dynamic condition, the increase and decrease of electricity fee can stimulate the increase of electricity consumption of the users who reduce the fee and lead to the decrease of electricity consumption of the users who increase the fee, to achieve the dynamic balance of electricity income of electric power enterprises.

(c) Under the adjustment scheme, the proportion of electricity consumption of users in each file is basically the same as the proportion of electricity consumption of users in each file, which is like the law of the current scheme.

6.5 Conclusions

From the perspective of the energy economy, this chapter mainly expounds on the big data method and application of urban electricity price prediction, analyzes the correlation between urban energy consumption and economic growth, and puts forward the adjustment plan of urban electricity price metering and charging. More importantly, experimental verification is carried out based on the collected data set, and the experimental results are shown as follows:

(a) Compared with the EMD and WPD, the EWT decomposition algorithm can better identify and extract the features of complex electricity price time-series data, which lays an important foundation for the prediction model. The LSTM model has outstanding performance in the application of electricity price prediction, and its performance is superior to the DBN and ELM models. The combined model of the EWT and LSTM is the preferred model for electricity price prediction, which has high prediction accuracy and good robustness.

(b) Grey correlation analysis has achieved important results in the correlation analysis of energy consumption and the national economy. By establishing the grey correlation model, based on the data set of China's energy consumption and the GDP growth in decades, the experiment shows that different industries and different types of energy have different degrees of correlation with

China's economic growth. In terms of energy industry classification, the rise of the construction industry is closely related to China's economic growth, which is also in line with the actual situation of China. In terms of energy classification, raw coal has the highest correlation with the growth rate of China's GDP, which also indicates that China's primary energy comes from raw coal. Therefore, it is necessary to adjust the energy structure to stimulate economic growth.

(c) In terms of urban electricity price metering and charging, to balance users' electricity demand and capacity allocation and ensure efficient energy use, it is necessary to carry out new planning for some existing electricity consumption structures. This chapter conducts clustering analysis based on the K-means algorithm, divides different electricity consumption groups, and conducts optimization design of residential ladder electricity price. The adjustment plan has a more finely differentiated guiding role in stimulating the sustainable development of the energy economy.

References

Bakheet, D., & Maharatna, K. (2021). Linear and nonlinear analysis of intrinsic mode function after facial stimuli presentation in children with autism spectrum disorder. *Computers in Biology and Medicine, 133*, 104376. https://doi.org/10.1016/j.compbiomed.2021.104376

Belciug, S. (2020). Logistic regression paradigm for training a single-hidden layer feedforward neural network Application to gene expression datasets for cancer research. *Journal of Biomedical Informatics, 102*, 103373. https://doi.org/10.1016/j.jbi.2019.103373

Chen, D., Lin, J., & Li, Y. (2018). Modified complementary ensemble empirical mode decomposition and intrinsic mode functions evaluation index for high-speed train gearbox fault diagnosis. *Journal of Sound and Vibration, 424*, 192–207. https://doi.org/10.1016/j.jsv.2018.03.018

Chen, Y., Sohel, F., Ali, S. S. A., & Ding, S. (2020). Deep Boltzmann machine for corrosion classification using eddy current pulsed thermography. *Optik, 219*, 164828. https://doi.org/10.1016/j.ijleo.2020.164828

Cheng, S., Wu, Y., Li, Y., Yao, F., & Min, F. (2021). TWD-SFNN: Three-way decisions with a single hidden layer feedforward neural network. *Information Sciences, 579*, 15–32. https://doi.org/10.1016/j.ins.2021.07.091

Fan, Y., Liu, C., & Wang, J. (2021). Prediction algorithm for springback of frame-rib parts in rubber forming process by incorporating Sobol within improved grey relation analysis. *Journal of Materials Research and Technology, 13*, 1955–1966. https://doi.org/10.1016/j.jmrt.2021.05.102

Giest, S., & Mukherjee, I. (2018). Behavioral instruments in renewable energy and the role of big data: A policy perspective. *Energy Policy, 123*, 360–366. https://doi.org/10.1016/j.enpol.2018.09.006

Gilles, J. (2013). Empirical wavelet transform. *IEEE Transactions on Signal Processing, 61*(16), 3999–4010. https://doi.org/10.1109/TSP.2013.2265222

Hafner, S., Anger-Kraavi, A., Monasterolo, I., & Jones, A. (2020). Emergence of new economics energy transition models: A review. *Ecological Economics, 177*, 106779. https://doi.org/10.1016/j.ecolecon.2020.106779

He, H., Chen, Y., & Lan, B. (2021). Damage assessment for structure subjected to earthquake using wavelet packet decomposition and time-varying frequency. *Structures, 34*, 449–461. https://doi.org/10.1016/j.istruc.2021.07.087

He, W., King, M., Luo, X., Dooner, M., Li, D., & Wang, J. (2021). Technologies and economics of electric energy storages in power systems: Review and perspective. *Advances in Applied Energy, 4*, 100060. https://doi.org/10.1016/j.adapen.2021.100060

Huang, N. E., Shen, Z., Lon, S. R. (1998). The empirical mode decomposition and the Hilbert spectrum for nonlinear and non-stationary time series analysis. In *The empirical mode decomposition and the hilbert spectrum for nonlinear and non-stationary time series analysis* (pp. 903–995).

Jamei, M., Hasanipanah, M., Karbasi, M., Ahmadianfar, I., & Taherifar, S. (2021). Prediction of flyrock induced by mine blasting using a novel kernel-based extreme learning machine. *Journal of Rock Mechanics and Geotechnical Engineering*. https://doi.org/10.1016/j.jrmge.2021.07.007

Jia, X., Du, H., Zou, H., & He, G. (2020). Assessing the effectiveness of China's net-metering subsidies for household distributed photovoltaic systems. *Journal of Cleaner Production, 262*, 121161. https://doi.org/10.1016/j.jclepro.2020.121161

Karan, B., Sahu, S. S., & Mahto, K. (2020). Parkinson disease prediction using intrinsic mode function based features from speech signal. *Biocybernetics and Biomedical Engineering, 40*(1), 249–264. https://doi.org/10.1016/j.bbe.2019.05.005

Kumar, J., & Soota, T. (2020). Multi-response optimization of machining parameter for Zircaloy by response surface methodology and grey relation analysis. *Materials Today: Proceedings, 21*, 1544–1550. https://doi.org/10.1016/j.matpr.2019.11.084

Mohan, A., Nakano, A., & Ferrara, E. (2021). Graph signal recovery using restricted Boltzmann machines. *Expert Systems with Applications, 185*, 115635. https://doi.org/10.1016/j.eswa.2021.115635

Pantoš, M. (2020). Market-based congestion management in electric power systems with exploitation of aggregators. *International Journal of Electrical Power & Energy Systems, 121*, 106101. https://doi.org/10.1016/j.ijepes.2020.106101

Pearson, R. K. (2002). Outliers in process modeling and identification. *IEEE Transactions on Control Systems Technology, 10*(1), 55–63. https://doi.org/10.1109/87.974338

Ren, D., Guo, X., & Li, C. (2021). Research on big data analysis model of multi energy power generation considering pollutant emission—Empirical analysis from Shanxi Province. *Journal of Cleaner Production, 316*, 128154. https://doi.org/10.1016/j.jclepro.2021.128154

Sairamya, N. J., Subathra, M. S. P., Suviseshamuthu, E. S., & Thomas, G. S. (2021). A new approach for automatic detection of focal EEG signals using wavelet packet decomposition and quad binary pattern method. *Biomedical Signal Processing and Control, 63*, 102096. https://doi.org/10.1016/j.bspc.2020.102096

Scarpiniti, M., Colasante, F., Di, T. S., Ciancia, M., Lee, Y.-C., & Uncini, A. (2021). Deep belief network based audio classification for construction sites monitoring. *Expert Systems with Applications, 177*, 114839. https://doi.org/10.1016/j.eswa.2021.114839

Seo, Y., & Kim, S. (2016). River stage forecasting using wavelet packet decomposition and data-driven models. *Procedia Engineering, 154*, 1225–1230. https://doi.org/10.1016/j.proeng.2016.07.439

Shrifan, N. H. M. M., Akbar, M. F., & Isa, N. A. M. (2021). An adaptive outlier removal aided k-means clustering algorithm. *Journal of King Saud University—Computer and Information Sciences*. https://doi.org/10.1016/j.jksuci.2021.07.003

Solgi, R., Loáiciga, H. A., & Kram, M. (2021). Long short-term memory neural network (LSTM-NN) for aquifer level time series forecasting using in-situ piezometric observations. *Journal of Hydrology, 601*, 126800. https://doi.org/10.1016/j.jhydrol.2021.126800

Sovacool, B. K., Hook, A., Sareen, S., & Geels, F. W. (2021). Global sustainability, innovation and governance dynamics of national smart electricity meter transitions. *Global Environmental Change, 68*, 102272. https://doi.org/10.1016/j.gloenvcha.2021.102272

Topalović, Z., Haas, R., Ajanović, A., & Hiesl, A. (2022). Economics of electric energy storage. *The Case of Western Balkans. Energy, 238*, 121669. https://doi.org/10.1016/j.energy.2021.121669

Xie, J., Towsey, M., Zhang, J., & Roe, P. (2016). Adaptive frequency scaled wavelet packet decomposition for frog call classification. *Ecological Informatics, 32*, 134–144. https://doi.org/10.1016/j.ecoinf.2016.01.007

Yahia, S., Said, S., & Zaied, M. (2021). Wavelet extreme learning machine and deep learning for data classification. *Neurocomputing*. https://doi.org/10.1016/j.neucom.2020.04.158

Yang, C., Ling, B. W. K., Kuang, W., & Gu, J. (2019). Decimations of intrinsic mode functions via semi-infinite programming based optimal adaptive nonuniform filter bank design approach. *Signal Processing, 159*, 53–71. https://doi.org/10.1016/j.sigpro.2019.01.022

Yang, Z., Gao, C., & Zhao, M. (2019). Utilizing big data to explore the running opportunity of power-to-gas in energy system. *Energy Procedia, 158*, 2341–2347. https://doi.org/10.1016/j.egypro.2019.01.281

Zhang, C.-K., He, Y., Jiang, L., Wu, M., & Wu, Q. H. (2014). Stability analysis of sampled-data systems considering time delays and its application to electric power markets. *Journal of the Franklin Institute, 351*(9), 4457–4478. https://doi.org/10.1016/j.jfranklin.2014.05.014

Zhang, H., & Peng, Q. (2022). PSO and K-means-based semantic segmentation toward agricultural products. *Future Generation Computer Systems, 126*, 82–87. https://doi.org/10.1016/j.future.2021.06.059

Zhang, Y., Xu, S., Zhang, L., & Yang, M. (2021). Big data and human resource management research: An integrative review and new directions for future research. *Journal of Business Research, 133*, 34–50. https://doi.org/10.1016/j.jbusres.2021.04.019

Zhang, Z., Feng, Q., Huang, J., Guo, Y., Xu, J., & Wang, J. (2021). A local search algorithm for k-means with outliers. *Neurocomputing, 450*, 230–241. https://doi.org/10.1016/j.neucom.2021.04.028

Zhou, K., Lei, D., He, J., Zhang, P., Bai, P., & Zhu, F. (2021). Real-time localization of micro-damage in concrete beams using DIC technology and wavelet packet analysis. *Cement and Concrete Composites, 123*, 104198. https://doi.org/10.1016/j.cemconcomp.2021.104198

Big Data Management of Smart City Energy Conservation and Emission Reduction

7

7.1 Introduction

7.1.1 Background and Introduction

The smart city is the inevitable direction of urban development, while energy conservation and emission reduction are closely related to the sustainable development of the urban economy. Electric energy consumption is the main source of energy consumption in a city, including electricity consumption in urban buildings and transportation systems (Yu et al., 2021). With the widespread use of new energy vehicles, environmental problems such as carbon emissions caused by traffic are gradually becoming reasonable planning problems of electric energy. The electricity consumption information and the rational planning of buildings are conducive to the construction of the smart power grid (Lin & Tsai, 2015), the reduction of urban energy economic expenditure, and the promotion of benign cycle of the urban economy.

With the development of the power industry, distributed power supply network with high complexity and variability is involved on a large scale. The high coupling and spatial-temporal mixing characteristics of different types of power terminal equipment will complicate the overall structure of the power system. With the popularity of smart grid, traditional statistical analysis methods will encounter many challenges in power engineering system scheduling, overall planning, and transaction rules. Artificial intelligence, which combines current advances in sensors and electronic information, will play an increasingly important role in the power industry (Zhang & Zhou, 2014). The application of artificial intelligence in the electric power system can promote the transformation of power supply network. Besides, it is also conducive to building a new power user data network and ecological system, so as to further guarantee the safety and stability of the electric power system (Raza & Khosravi, 2015; Rigas et al., 2014).

© Science Press 2022

H. Liu et al., *Big Data in Energy Economics*, Management for Professionals,
https://doi.org/10.1007/978-981-16-8965-9_7

Load identification technology is one of the core technical difficulties to be solved in realizing the overall layout planning of the smart power grid (Himeur et al., 2021). Statistics show that with the message provided by the Non-intrusive Load Monitoring (NILM) system, the energy costs can be cut down by 10 to 15% (Darby, 2006). In the international scope, the current load identification technology mainly adopts the traditional template matching method to establish the feature database and uses algorithms such as hidden Markov chain (Makonin et al., 2016) and K-nearest neighbor (Zhang & Zhou, 2007) to match the feature database, or adopts the machine learning algorithm to recognize the pattern.

In general, the identification accuracy is greatly affected by load types, leading to weak generalization ability and low identification efficiency. The research of load identification by using a neural network or deep learning has been widely concerned by various universities and research institutions and has a high application prospect and academic value. At present, the electrical load identification in buildings mainly includes signal decomposition, steady-state separation, and electrical switching event identification (Jian et al., 2010). This chapter introduces the load identification methods mainly around these three methods.

Smart grid construction is a necessary means of urban power consumption behavior management and energy conservation and emission reduction implementation. It is a general trend to apply the smart grid to smart city planning, which is conducive to the power distribution network and more guarantee of urban power supply. In practical application, during the operation of an urban distribution network, it is often restricted by various elements, and power failures occur frequently. The implementation of smart grid technology helps to improve the overall operational efficiency of the distribution network, achieve the effect of all-around monitoring, and improve the quality of voltage acquisition.

In the construction of the urban smart grid, it is necessary to consciously stand on the professional perspective and lock the planning scheme steadily. Continuously improving the scientific degree of urban distribution network planning is also important. This chapter gives feasible suggestions from two aspects of urban power consumption behavior planning, energy saving (Liao et al., 2021), and emission reduction efficiency (Li et al., 2017) to promote urban smart grid construction.

7.1.2 Dataset Description

7.1.2.1 WHITED Dataset

The load identification experiment in this study is carried out on WHITED data set (Kahl et al., 2016). The WHITED dataset contains the current and voltage of all devices operating independently. The data were measured at 44.1kHz. Examples of these appliances were collected from eight regions that contained two power supply standards. The dataset contains 1100 samples of 110 electrical appliances, which can be divided into 47 different types. Each device has 10 running instances, each recording 5 seconds of data. In this study, 22 commonly used appliances were selected, including some with high energy consumption and some with the largest

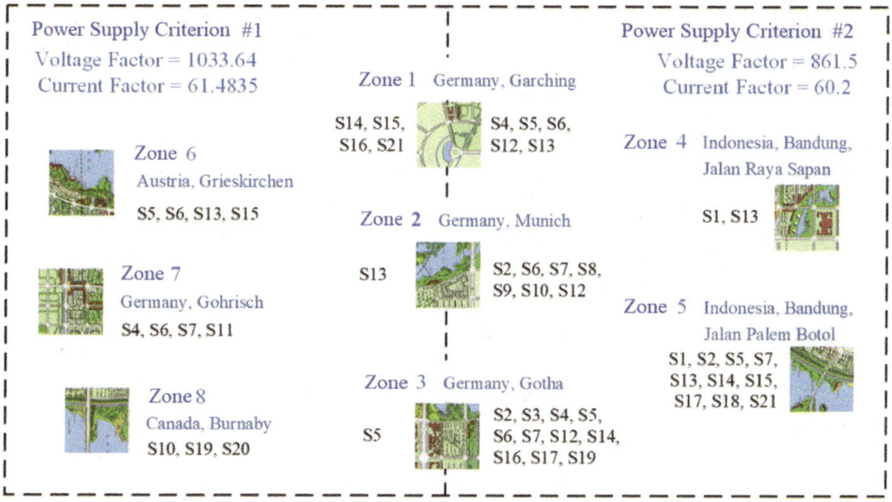

Fig. 7.1 The power supply information and area information of the appliances

sample size. Fig. 7.1 shows the power supply and collection area distribution of the device.

In this chapter, six kinds of electrical data in WHITED dataset are selected as processing objects. The six categories of appliances are LED lights, washing machines, fan heaters, fluorescent lights, air conditioners, and hair dryers. Table 7.1 lists the label values and codes corresponding to these six appliances.

7.1.2.2 REDD Dataset

REDD data sets are the most classic and commonly used data sets for non-invasive load identification (Kolter & Johnson, 2011). The data set contains total voltage, current, and power data collected from five households, as well as power data for each household appliance. Each piece of data contains a timestamp and a data value. The total voltage and current data were sampled at a frequency of 15kHz, and the power data of each electrical appliance were sampled at a frequency of 1/3Hz. Only family 3 and family 5 collected the high-frequency total current and voltage data, and stored the data in the way of saving only one cycle wave with similar waveform. The specific data is shown in Table 7.2.

7.2 Non-intrusive Load Identification of Electrical Equipment

Non-intrusive load decomposition can analyze the operation state of all kinds of electric equipment in the system. At the same time, it can make the electric power consumption acquirement more accurate (Gopinath et al., 2020). The purpose of non-invasive load decomposition is to develop a monitoring tool that has no impact on the monitoring object and has as little impact as possible. This monitoring tool

Table 7.1 The information of selected appliances types in the WHITED dataset

Appliance	Abbreviation	Number of samples	Number of appliances
LED light	S1	90	9
Charger	S2	50	5
Fan heater	S3	10	1
Light bulb	S4	60	6
Hairdryer	S5	60	6
Kettle	S6	60	6
Toaster	S7	40	4
Game console	S8	40	4
Washing machine	S9	10	1
TV	S10	20	2
Stove	S11	10	1
Mixer	S12	40	4
Fan	S13	60	6
Vacuum cleaner	S14	40	4
Water heater	S15	40	4
Drilling machine	S16	20	2
Iron	S17	30	3
Massager	S18	30	3
Microwave	S19	20	2
Refrigerator	S20	10	1
Air conditioning	S21	10	1
Water pump	S22	10	1

can provide the electric energy consumption of various electrical equipment for the power enterprises so as to promote energy conservation and sustainable economic development. At present, the main methods include signal decomposition, multi-label classification, multi-classification of electrical switching events, etc. (Wang et al., 2018).

7.2.1 Nonintrusive Load Identification Based on Signal Decomposition

Nonintrusive load identification based on signal decomposition, also known as nonintrusive load disaggregation based on steady current or power decomposition, is very popular in this domain. The main idea of steady-state decomposition

Table 7.2 Appliances in the REDD dataset

House number	Sampling rate	Facility types
1	1 Hz for mains 1/3 Hz for devices	Dishwasher, furnace, washer dryer, smoke disposal, alarms, bathroom gfi, kitchen outlets, lighting, refrigerator, microwave, electronics
2	1 Hz for mains 1/3 Hz for devices	Washer dryer, bathroom GFI, dishwasher, kitchen outlets, microwave, oven, electric heat, refrigerator, stove, lighting
3	15 kHz for mains 1/3 Hz for devices	Disposal, dishwasher, washer dryer, furnace, bathroom gfi, refrigerator, lighting, kitchen outlets, electric heat, microwave, outdoor outlets, electronics
4	1 Hz for mains 1/3 Hz for devices	Smoke alarms, bathroom gfi, kitchen outlets, washer dryer, stove, furnace, disposal, dishwasher, air conditioning, lighting
5	15 kHz for mains 1/3 Hz for devices	Dishwasher, washer dryer, kitchen outlets, disposal, microwave, refrigerator, stove, lighting

is that the load decomposition system should be able to accurately identify the type, quantity, and energy consumption of the load in operation. The main methods include evolutionary algorithm (Coello et al., 2004; Michael & Jiirgen, 2004; Chang et al., 2013; Piga et al., 2016), hidden Markov (Mengistu et al., 2018), quadratic programming (Panten et al., 2016), and so on.

7.2.1.1 The Theoretical Basis of the Models

Hidden Markov Model

Hidden Markov Model (HMM) originated in the 1960s, and *Baum et al.* published their research results through a series of articles, which attracted the attention of researchers from all walks of life, and finally developed into an important research hotspot today (Makonin et al., 2016). The most obvious feature of hidden Markov model is that it can accurately model the feature series of events with probability. In various fields such as fault diagnosis, deep learning, speech and text recognition, hidden Markov chain has been widely used. In addition, its strong robustness, high accuracy, and solid mathematical foundation enable it to perform good simulation of dynamic event data in various situations, so it is often used for time sequence information processing (Johnson & Willsky, 2013).

HMM can be summarized as a double random process, it is composed of two key random process sequences. One is its own hidden state transition process sequence, which corresponds to the Markov process of the discrete-time state, and the states are related to each other by state transition probability matrix. The other is the sequence of output states that can be directly observed, which can be directly described by the probability of observed values. The two stochastic processes represent each other and together constitute the intrinsic characteristic of the original signal.

The basic structure of the hidden Markov model is detailed explained in the part. Each hidden state corresponds to an output value. When HMM is applied to load identification, its power characteristic parameters become the corresponding output values, and the running state of load is its implied state quantity.

The general hidden Markov model is mainly composed of five state parameters, including state-space set and probability transition distribution, which are expressed as $\gamma=(A, B, S, \pi, O)$ (Johnson & Willsky, 2013).

S represents the implicit state set, where $S = [S_1, S_2, ..., S_N]$ and N represents the number of Markov models contained therein. These states cannot be directly observed, and Q_t is generally used to represent the state at time t. The probability distribution of mutual transition between states is the correlation between internal states.

O represents the sequence set that can be directly observed, where $O = [O_1, O_2, ..., O_k]$ and k represents the number of possible observed values in the output of each state. The k value and N value need not be the same, and the number of directly observed states is not necessarily the same as the number of implicit state sets.

π is the initial state distribution matrix, which is generally used to represent the probability distribution form of the state Q_1 in the hidden Markov model at the initial moment of the direct observation sequence, that is, the form satisfying the equations below. The uniform distribution is usually used as the probability density function of the initial state.

$$\pi_j = P\left(Q_0 = S_j\right) \quad j = 1, 2, \cdots, N \tag{7.1}$$

$$\sum_{j=1}^{N} \pi_j = 1 \tag{7.2}$$

A represents the probability distribution matrix of the hidden state transition, representing the probability of each state transferring to each other in the hidden Markov model, which is shown as below:

$$A = \begin{bmatrix} a_{11}, a_{12}, \cdots, a_{1M} \\ a_{21}, a_{22}, \cdots, a_{2M} \\ \vdots \quad \vdots \quad \vdots \\ a_{N1}, a_{N2}, \cdots, a_{NM} \end{bmatrix} \tag{7.3}$$

where $a_{ii} = P(Q_t = S_i | Q_{t-1} = S_i)$ represents the size of transition probability distribution from initial state I to final state j. This shows that the state of the random process sequence at Q_t moment is only related to the state at the last moment, and has nothing to do with the state at any other moment. At the same time, in non-special cases, the transition between states cannot take the form of step, that is, the transition between states cannot jump. This characteristic is in line with the practical constraints of non-invasive load identification system design, and is of great significance.

B represents the probability distribution matrix of observation state transition. This is the probability distribution of the corresponding observation state O_k at time t under the premise that the implied state is S_N. The Gaussian distribution is usually used as a directly observed probability density function.

To simplify the description, the expression form of triples, namely, $\gamma = (A, B, \pi)$, is generally used to characterize the parameter set equation of a hidden Markov model.

Particle Swarm Optimization

Particle swarm optimization (PSO) is an evolutionary algorithm invented according to the predation behavior of birds (Kennedy, 1995). The nature of particle swarm optimization (PSO) is a random search algorithm. PSO mimics the predation behavior of birds randomly searching for food. Assume that area only a piece of food in search of food, the flock of birds all do not know where the food is, so the bird exists between exchange information behavior, measured by its fitness value to analysis of the current location how far is from food, and search for the nearest food near the area of the bird is simple and efficient way to find food, The colony is optimized through group cooperation between birds.

PSO takes inspiration from this model and is used to deal with optimization problems. In PSO, the possible solution to any optimization problem can be assumed to be a bird in the search area, known as a "particle". PSO is initialized into a group of random particles, and then the optimal solution is obtained by iterative method. During each iteration, each particle updates its position by following two "extreme values", one of which is the optimal solution that can be found by the particle itself, called individual extreme value. The other is the optimal solution currently found for the whole population, which is called the global extremum of the population. These two optimal variables move particles in these directions to a certain extent, and each particle keeps updating its velocity and position with these two extremes until it finds the optimal solution.

In the search for these two optimizations, the particle updates its velocity and position according to the following formula (Yang et al., 2009):

$$v_j = wv_j + \alpha r_1(p_j - x) + \beta r_2(p_{global} - x_j) \tag{7.4}$$

$$x_j = x_j + \lambda v_j \tag{7.5}$$

where v_j is the velocity of the particle j, x_j is the position, α and β are learning rates, r_1 and r_2 are random numbers within a range from 0 to 1.

In the non-intrusive load decomposition problem, the objective function of the optimization algorithm is given as follows:

If the power data is utilized, then:

$$f_{opt} = \min\left|\sum P_a - P_m\right| \tag{7.6}$$

where P_m is the power of mains and P_a is the power of an appliance.

If the current data is utilized, then:

$$f_{opt} = \sum_{k=1}^{n} (I_k \cdot \cos\theta_k) - I_m \cdot \cos\theta_m \tag{7.7}$$

where I_m is the current value of mains and θ_m is the phase angle of the current of mains; I_k is the current of appliance k and θ_k is the relevant phase angle.

7.2.1.2 Performance of the Signal Decomposition Algorithms

In this section, some data from WHITED dataset are used to conduct comparative experiments on the above algorithms and compare the performance of the two algorithms. As shown in Fig. 7.2, the total current signals of the two sections both contain the superposition of the opening state of three appliances. The above two methods realize the decomposition of the total signal by combining and optimizing the steady-state current per second in the characteristic database. Besides, the information of the two current signals is listed in Table 7.3.

As shown in Figs. 7.3 and 7.4, The classification recall rate of hidden Markov model and particle swarm optimization algorithm is compared in two total current decomposition experiments, which reflects its steady-state current decomposition performance. It can be found that the decomposition accuracy of hidden Markov model is slightly worse than that of PSO, and the decomposition effect of PSO method is more stable under the condition of large or small difference of electric current value.

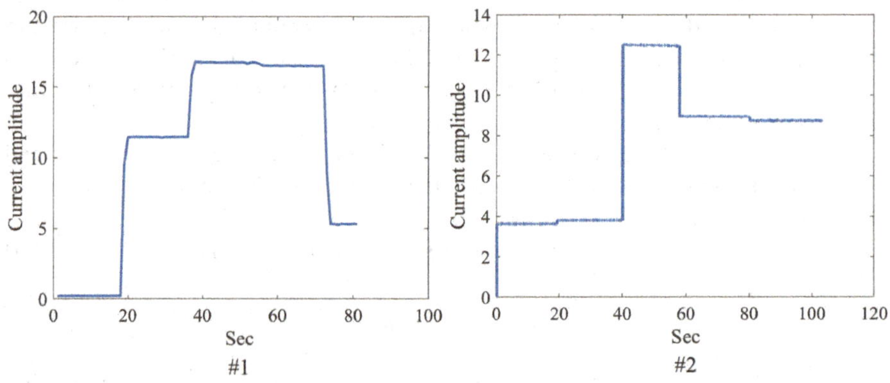

Fig. 7.2 The current amplitudes

Table 7.3 Information of the two current signals in comparison experiments

Number of the current	Appliances in the current	Running time of each appliance (s)	Total time (s)
#1	S4+S11+S7	54.7+54.4+45.9	81.9
#2	S6+S4+S11	40.8+44.6+59.5	87.4

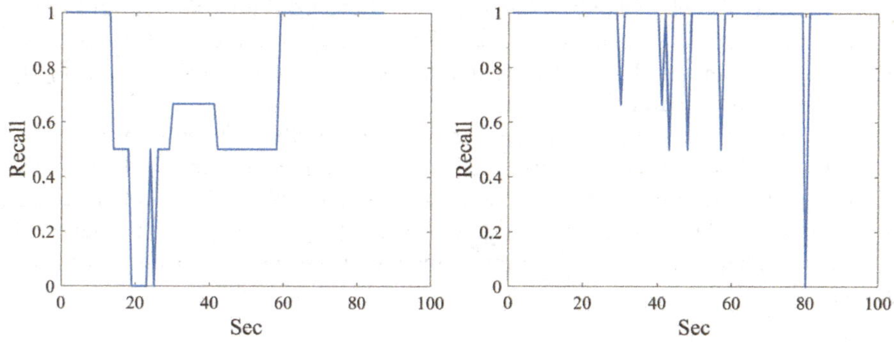

(a) Hidden Markov (b) PSO

Fig. 7.3 The recall value of the two algorithm of Current #1

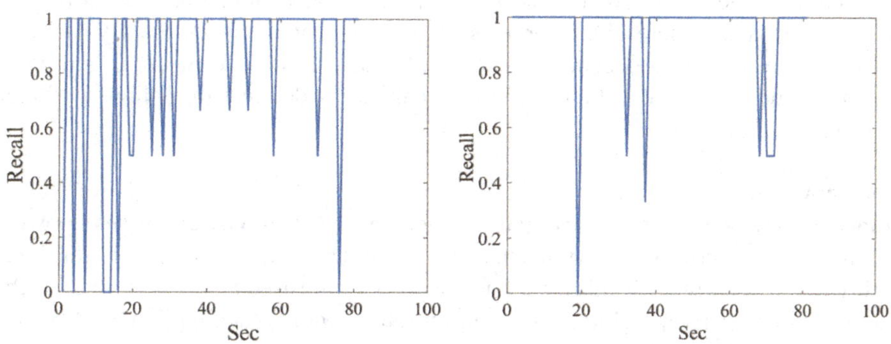

(a) Hidden Markov (b) PSO

Fig. 7.4 The recall value of the two algorithm of Current #2

7.2.2 Non-intrusive Load Identification Based on Electrical Switching Event Classification

Nonintrusive load recognition based on switch event refers to positioning electric load of opening and closing events, extracting transient characteristics, combining the corresponding load identification algorithm. It can determine the on-off states of electrical appliances. Finally, match the various electrical events and achieve the goal of nonintrusive monitoring of load.

The transient characteristic of the electrical switch refers to the change characteristic of current from on-off to steady state. Resistive load directly into the steady state; When the nonlinear load current is started, the pulse current will appear. Therefore, different electrical equipment has different characteristics when it is on or off. Switching time, the peak value of starting pulse, and current step can be used as the characteristic value of switching transient. The current envelope is also a basic current characteristic. The stability of the current envelope of different loads can also be used as a characteristic. Transient features are highly differentiated and accurately identified.

This section mainly introduces the identification method of the transient characteristics of electrical switches in order to stimulate readers' enthusiasm for the research of this technical idea. In fact, the difficulty of this research idea lies in the final switching event matching, which is still in the research stage with low accuracy. Therefore, switching event matching will not be introduced in this section.

7.2.2.1 Transient Features

Due to the limitation of steady-state feature in load identification, transient feature extraction technology has stronger applicability and can distinguish loads with different functions. In addition, the duration of transient process of load is shorter than that of steady-state process, so the overlap of transient characteristics is less than that of steady-state characteristics. Transient characteristics mainly include instantaneous power, transient current peak, transient waveform area, duration, response time, etc. Transient characteristics can be divided into the following three categories.

Transient Power

Transient power characteristics mainly include transient power curves, power differences, and power spectrum characteristics. The characteristics of transient power class have the advantages of no need for high-frequency sampling, high feature discrimination, and high recognition rate of multi-state appliances.

Current Features

Current characteristics mainly include transient current peak, waveform area, duration, transient waveform, response time, current harmonic peak. The current waveform can distinguish different switching states of electrical appliances, and it is better to distinguish between switching appliances and multi-state electrical appliances.

Voltage Noise

Transient voltage noise feature can distinguish multi-state electrical appliances, and it is a feature that will be used in special cases, because it requires expensive measuring equipment and complex calculation process.

7.2.2.2 The Basic Theory of Classification Models

Convolutional Neural Network

The main difference between Convolutional Neural Network (CNN) and general neural networks lies in the introduction of convolution operation. The nature of CNN is a kind of feedforward neural network with deep operational structure.

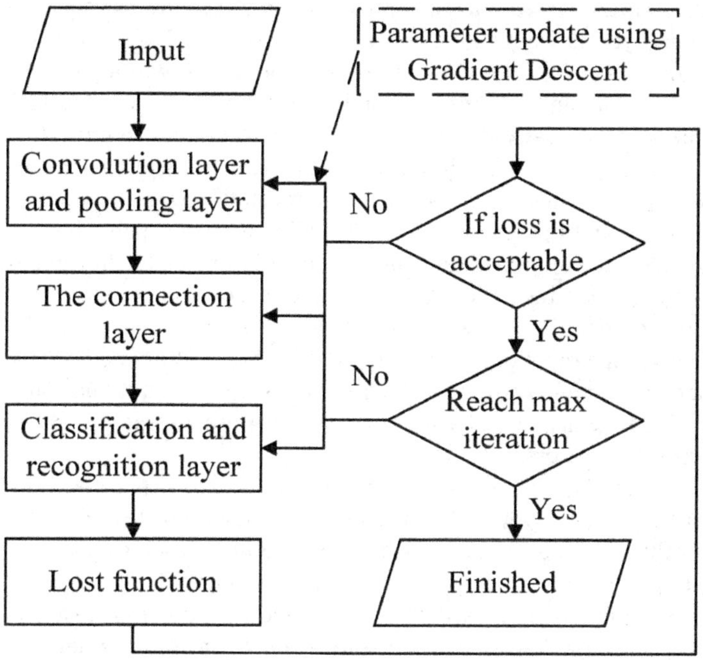

Fig. 7.5 Structure of CNN

CNN's image feature extraction ability is mainly realized through the local perception ability of its convolution kernel. After the image data passes through the multi-layer superimposed convolution kernel, the deep features of the image are extracted, and subsequent neural networks or computing models with different structures can complete regression or classification tasks according to these features. Due to the special structure of CNN, it has excellent representation learning ability and can carry out geometric invariant classification of neural network input information according to different hierarchical structures, especially excellent feature extraction ability for deep features in two-dimensional grid-like data such as time series and images. At present, CNN has been widely used in many fields of recognition and regression.

CNN is mainly accumulated by network structures such as convolutional layer, pooling layer, and full connection layer. Meanwhile, some researchers have proposed batch normalization layer, random inactivation layer, and global pooling layer in order to improve the capability of SOME specific aspects of CNN, which have been widely used in CNN. The convolutional layer and pooling layer are two unique network structures of CNN. During CNN's training phase, the network weights and bias in the way of regulating parameters are consistent with the traditional ANN. The training process of CNN network can be shown in Fig. 7.5.

The convolutional layer is not only the core structure of CNN but also an indispensable and extremely important part of the whole neural network model. The main function of the convolutional layer is to realize feature extraction of data. A convolutional layer may contain one or more convolutional kernels. Each convolution kernel in the convolution layer is directly connected with a part of the region of the adjacent convolution kernel through neurons, which is called the receptive field, and its size generally depends on the size of the convolution kernel. Each convolution kernel in the convolution layer carries out a sliding scan on the data with given step size and carries out matrix convolution operation on the relevant areas in the data, and the result obtained is called the feature graph.

Just like the convolution kernel in the convolution layer, the pooling kernel also exists in the pooling layer. The convolution kernel completes the convolution operation by studying the area weighted sum of input data nodes. The pooled kernel calculates the input data according to different pooling modes without training parameters. Therefore, the introduction of aggregation layer not only does not increase the complexity of CNN network, but also greatly reduces the number of CNN network parameters.

At present, there are two common pooling methods in academia. (1) Maximum pooling: in the feature graph obtained by the convolution kernel, the maximum eigenvalue of the rectangular region where the pooling kernel is located is extracted as the pooling value of this region. Maximum pooling can effectively reduce the network recognition bias caused by the parameters and transmission errors of the convolutional layer, so that the texture features in the original input can be preserved more. (2) Average pooling: in the feature graph obtained from the convolution kernel, the average eigenvalue of the rectangular region where the pooling kernel is located is extracted as the pooling value of this region. Average pooling can effectively reduce the variance of network recognition caused by parameter training and transmission error at the convolutional layer, so as to retain more background information in the original input.

As the full connection layer plays the role of classifier and discriminator in the whole CNN model, it is usually added to the last several layers of the CNN network. The full connection layer can effectively fuse the discrete features extracted from the convolutional layer and the pooling layer in CNN into the mapping sample tag. The main use of the full connection layer in the CNN model lies in the case that the $M \times N$ feature graph generated by the convolutional layer or pooling layer can be flattened into a one-dimensional feature vector containing $M \times N$ data. In addition, if the layer before the full connection layer in the CNN network is also the full connection layer, the main function of the full connection layer is to transform the dimension of the output feature vector of the upper layer.

In order to further simplify the computational complexity of the CNN network and prevent it from over-fitting, the global pooling layer is widely used in CNN as a neural network structure. Unlike the pooling layer, which only pools part of feature maps in CNN, the global pooling layer pools complete feature maps in CNN. Similarly, the global pooling layer can be divided into global average pooling and global maximum pooling according to the pooling mode. The introduction of the global pool shell lies in its improvement of the neural network can effectively identify the target. Due to the fact that the global pool shell itself has no need to train parameters, the global connection layer can effectively reduce the number of parameters in the network. The model calculation complexity will be cut down and the fitting phenomenon will be improved. At the same time, the global pooling layer is more sensitive to global spatial information.

Long Short-Term Memory

Traditional neural networks lack the ability to model historical events in sequential data. When dealing with Natural Language Processing (NLP), Machine Translation (MT), Speech Recognition (SR), and other related scenarios, it is difficult to predict effectively. However, due to the existence of cyclic structure, Recurrent Neural Network (RNN) can operate the historical information in time series data many times, ensuring its long-life cycle, so it can very well solve the limitation that traditional neural network does not have the ability to retain historical information.

A represents a series of typical neural networks, x_t is the input data of the network, h_t is the output of the network. In the loop structure of RNN, the historical information of timing data can be transferred from the last training state to the next training. The loop structure of RNN is similar to duplicating the same neural network many times, and the internal state information of the neural network is one-way transmitted between adjacent neural networks from first to last. Therefore, RNN has been widely used in NLP, MT, VR, and other fields requiring timing modeling.

Although RNN has achieved good results in the process of timing sequence modeling, the long-term dependency problem is still not effectively solved. Therefore, Long Short-Term Memory (LSTM) network as a special RNN has been proposed and developed rapidly. *Sepp Hochreiter* and *Jurgen Schmidhuber* first proposed the concept of LSTM in 1997 (Hochreiter & Schmidhuber, 1997). In the subsequent development, researchers found that the current version of LSTM may appear gradient explosion phenomenon during training, which limited its retention of historical information. To solve this problem, *Jurgen Schmidhuber* developed the LSTM structure through continuous improvement.

As an RNN network with the best performance, LSTM can obtain any desired output results according to the input data when its network scale is large enough. Meanwhile, its ability to retain historical information is very strong compared with other RNNs. Therefore, other RNN networks, HMM, and other timing models are gradually replaced by LSTM. Since 2016, many top technology companies in China and other countries have incorporated research related to LSTM networks into their R&D strategies. When the LSTM network processes the information in the current stage, it can either use the previous historical information or obtain the long-term dependence information from the current learning stage.

LSTM network is mainly controlled by forgetting gate, input gate, and output gate to control the state of historical information and transfer the training information inside the network. The main function of the forgetting gate is to analyze the internal state of LSTM and distinguish the state information that needs to be discarded. It is mainly implemented based on the sigmoid network channel. The value range of its network output vector is 0 to 1, and the vector size is consistent with $C\_(t-1)$. The main function of input gate is to select the information that can be retained by memory unit in LSTM, which is realized by the sigmoid network channel and TANH network channel.

The updating process of memory unit state information in LSTM can be expressed as follows: LSTM actively discards part of previously trained state data under the action of forgetting gate; Selectively transmit the training data of the current training phase to LSTM; Through the function of the input gate, the network state and control information of the training can be updated. The main function of the output gate is to determine the output of the state information of the LSTM internal memory unit according to the training information. The completion of the gate mechanism is mainly based on the sigmoid network channel and tanh network channel, and combined with the dot product of its output vector, the output gate is formed.

7.2.2.3 Performance of the Models

WHITED data set is used to compare the above characteristics and models. All samples of nine kinds of electrical appliances with many samples were selected to form the experimental data set, and 80% samples were randomly selected as the training set and 20% samples as the test set. Experimental data information is shown in Table 7.4.

It can be seen from Figs. 7.6 and 7.7 that both CNN and LSTM have good electrical recognition effects, which is attributed to the strong learning ability of deep neural network, which is able to mine deep features of data. However, due to the lack of training data, it is easy to overfit the deep network. Therefore, it can be found that the recognition effect of the two-deep networks for some electrical appliances is slightly worse.

Table 7.4 Information of the samples used in experiments

Appliance	Abbreviation	Number of samples	Number of appliances
LED light	S1	90	9
Charger	S2	50	5
Light bulb	S4	60	6
Hairdryer	S5	60	6
Kettle	S6	60	6
Toaster	S7	40	4
Mixer	S12	40	4
Fan	S13	60	6
Vacuum cleaner	S14	40	4
Water heater	S15	40	4

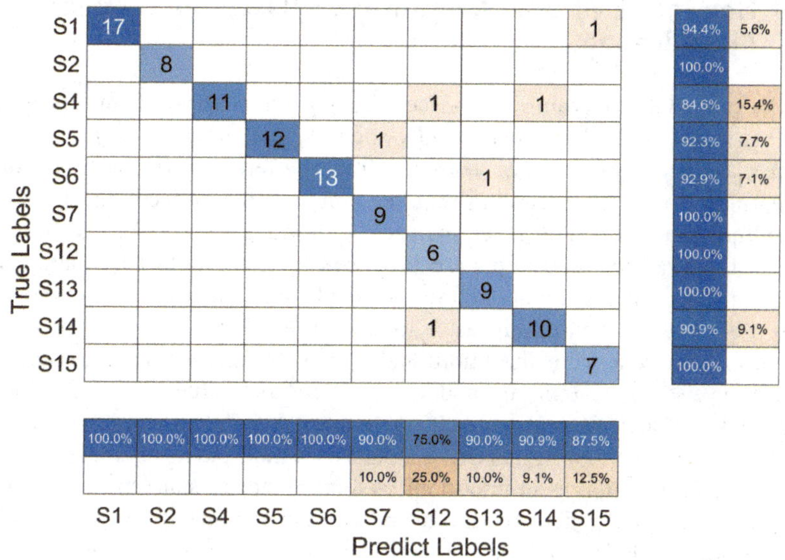

Fig. 7.6 Confusion chart of the results predicted by CNN

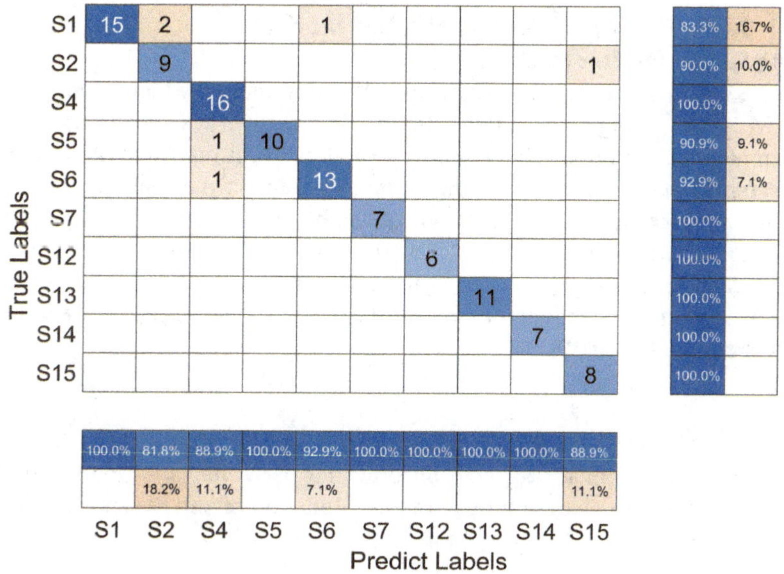

Fig. 7.7 Confusion chart of the results predicted by LSTM

7.2.3 Non-intrusive Load Identification Based on Multi-label Classification

Recently, multi-label learning has attracted extensive attention from researchers, and it has been applied in many fields, such as text, image, video, music and emotion classification (Gibaja & Ventura, 2014). Different from traditional single-label learning, multi-label learning requires simultaneous processing of instances with multiple class tags, that is, an instance with multiple tags (Yeh & Li, 2019). Based on this, various algorithms are proposed to learn multi-label data. Multi-label classification is mainly applied to the steady-state signal decomposition idea proposed in Sect. 7.2.1 in non-invasive load identification.

However, different from the traditional signal decomposition, the multi-label classification uses a pre-training model to predict label values for load identification. It does not need the independent operation data of electrical appliances as a template, but only the total signal data marked with multiple labels for pre-training. Multi-label classification is an important method of steady-state current decomposition (Luo et al., 2017). This section introduces the application of two multi-label classification models in non-invasive load identification, namely Ranking Support Vector Machines (SVMs) (Feng et al., 2019) and Rank Extreme Learning Machine (RELM) (Su et al., 2021).

7.2.3.1 Basic Theory of the Multi-label Models

Ranking Support Vector Machine

Ranking SVM is a popular multi-label classification model, which is derived from the SVM model (Feng et al., 2019). It was first used in text analysis to achieve text summary. In the multi-label classification domain, it ranks the multiple labels of all samples and then classifies the samples based on the ranking results. The basic theory of Ranking SVM is introduced as below:

Assuming that the input is X, with n features. $Y = \left[\mathrm{y}_1, y_2, \ldots, y_m\right]$ is the label group of each samples, in which k is the number of labels. Then, there exists a set of sorting functions that can represent the sorting relationship between tags as follows (Feng et al., 2019):

$$x_i \succ x_j \Leftrightarrow f(x_i) > f\left(x_j\right) \tag{7.8}$$

The sorting problem for a given unsorted tag is to find the optimal function from the set of functions to minimize the loss function of sorted samples. Suppose the function satisfies the following condition:

$$f_w(x) = \langle w, x \rangle \tag{7.9}$$

where w represents the weight vector, thus,

$$x_i \succ x_j \Leftrightarrow \left(w, x_i - x_j\right) > 0 \tag{7.10}$$

For any two samples in the feature space, the new vector $x_i - x_j$ and the corresponding new label can be used to represent the ordered relationship between pairs. Then, a new data set S_n can be formed by the labeled samples in the known training set S. Therefore, the sorting problem becomes a classification problem, and the relationship of tags can be obtained by constructing SVM to solve the sorting problem of tags. Achieving multi-label classification with the ranking information of samples is the core of Ranking SVM. The L1 cost function is proposed to transform ranking problems into classification issues:

$$C(s) = \mu \sum \max\left(1 - r^T\left(s_i - s_j\right), 0\right) + \frac{\|r\|^2}{2} \tag{7.11}$$

where μ is a parameter. The function utilizes $\left(s_i - s_j\right)$ to replace the independent variable s in the ranking function, and then the sorting problem becomes a classification problem that SVM can deal with.

Multi-label Rank Extreme Learning Machine

Multi-Label Radial Basis Function (ML-RBF) neural network is an effective multi-label learning algorithm which introduces radial basis neural network into multi-label learning and has good global approximation ability. RELM is an improved ELM that adds regularization parameters to solve the problems existing in traditional ELM. It also has good generalization ability and fast learning ability. The combination of ML-RBF and RELM makes the model have good generalization ability and learning ability while retaining good global approximation ability.

The number of nodes at the output layer of RELM corresponds to the total number of tags in multi-label learning. To use RELM for multi-label learning, you need to change the output form of RELM at the output layer. Since RELM's output nodes correspond to labels one by one, it is possible to determine which class samples the samples contain according to the values output by RELM's output layer. The predicted label value of RELM can be set as follows: If the output label value is greater than 0, it indicates that the sample is likely to contain this label and set the label value to 1, it indicates that the sample contains this label. Otherwise, it indicates that the sample is less likely to contain the label, and the label value is set to -1, indicating that the sample does not contain the label value.

$$TestL = \begin{cases} 1, & OutputL > 0 \\ -1, & OutputL < 0 \end{cases} \tag{7.12}$$

RELM can learn multi-label data directly by simply setting the output judgment form of labels. The original multi-label RBF neural network determines the center of the hidden layer RBF function and the number of hidden layer nodes in the multi-label RBF neural network by means of K-means clustering of the samples contained in each label. For the K-means clustering of samples contained in each label, the K value is selected based on a fixed proportion of the samples contained in each label. The selected K value cannot well reflect the internal information of the original sample. For example, a particular tag may contain many samples, but these samples may not contain much useful training data.

However, in the process of K-means clustering, if the data with many samples are selected according to a fixed proportion of the number of samples, the clustering centers will be set more. The number of samples does not reflect the number of categories of samples, and this selection method cannot well reflect the internal characteristics of samples. Increasing or decreasing the number of hidden nodes will also affect the classification accuracy of the model. The nearest neighbor propagation clustering algorithm, which can complete clustering without any prior knowledge, is used to determine the K value of each type of sample. In this way, the selected K value can better represent the internal characteristics of the sample, and meanwhile, the parameter adjustment of sample proportion can be avoided.

7.2.3.2 Performance of the Introduced Models

A REDD data set is used to conduct a comparative experiment on the above models. The total current data of house 3 in the REDD data set is taken as the training data, and the switching state of each electrical appliance is judged by the current of each electrical appliance and the switching state label of each electrical appliance is marked for the total current data. 80% of the samples are trained and the remaining samples are tested. The results are shown in Table 7.5. It can be seen from the table that RELM's recognition effect is obviously better than that of Ranking SVM. The recognition effect of both models is poor for the electric lamp. This is because of the small number of lamp samples and the large imbalance between lamp samples and other electrical appliances.

The calculation method of indicators in the table is shown as follows (Liu et al., 2020):

Table 7.5 Experiments of the introduced models

Devices	Ranking SVM			RELM		
	Recall	Precision	F1score	Recall	Precision	F1score
Disposal	0.916	0.922	0.919	0.998	0.916	0.955
Dishwasher	0.817	0.859	0.838	0.989	0.904	0.945
Washer dryer	0.745	0.756	0.751	0.958	0.911	0.934
Furnace	0.887	0.583	0.704	0.873	1.000	0.932
Bathroom GFI	0.439	0.222	0.295	0.671	0.846	0.748
Refrigerator	0.621	1.000	0.766	0.932	0.974	0.952
Lighting	0.038	0.667	0.072	0.481	0.806	0.603
Electric heat	0.289	0.295	0.292	0.644	0.935	0.763
Microwave	0.702	0.379	0.492	0.733	0.887	0.803
Total indexes	0.865	0.867	0.866	0.956	0.962	0.959

Table 7.6 Confusion matrix of classification indicators

Real situation	Predicted results	
	Positive	Negative
Positive	TP (true positive)	FN (false negative)
Negative	FP (false positive)	TN (true negative)

Precision effectively reflects the probability that all predicted sample values are true. It can be calculated by the following formula:

$$Precision = \frac{TP}{TP + FP} \tag{7.13}$$

Recall is the probability that a sample that is actually positive is predicted to be positive. The calculation formula of recall rate is given as follows:

$$Recall = \frac{TP}{TP + FN} \tag{7.14}$$

The $F1score$ represents the weighted average of accuracy and recall. When the numerical difference between accuracy and recall rate is large, $F1score$ value can effectively combine the two indicators. The formula is given as follows:

$$F1score = \frac{2 \times Precision \times Recall}{Precision + Recall} \tag{7.15}$$

It is worth mentioning that the confusion matrix of the classification indicators is given in Table 7.6.

7.3 Guide to Smart City Electricity Behavior

7.3.1 Smart Grid Planning of a City

Smart power grid is a new type of power grid that integrates modern sensing measurement, communication information, and control technology based on physical power grid. It is so flexible, intelligent, and efficient that can adapt to large-scale renewable energy. Meanwhile, due to its safe, reliable, and self-adaptive, it has great advantages in distributed power supply access and user-side demand response. Since the development of smart grid construction in China, power grid companies, research institutions, and government officers have put forward smart grid planning and implementation plans from all levels.

In recent years, China's economic and social development has put forward new requirements for the smart grid. On the one hand, China's economy is shifting from a stage of high-speed growth to a stage of high-quality growth, and new trends are emerging in economic and social development. A clean, low-carbon, safe and efficient energy system is gradually being built. The power grid is needed

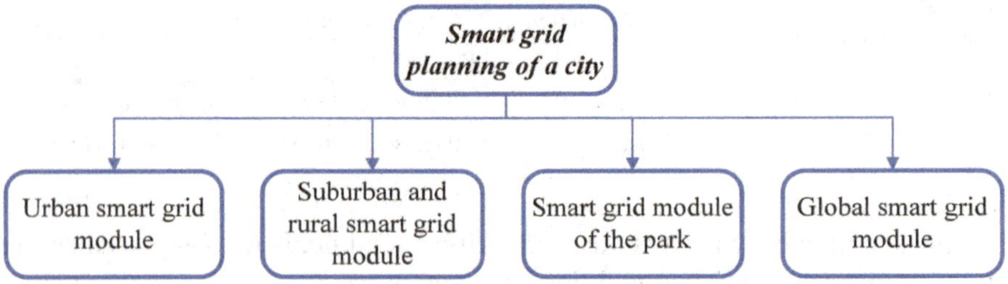

Fig. 7.8 The framework of city smart grid planning

to better support economic and social development. On the other hand, the reform of the electric power system has been further deepened, with the marketization of electric power transactions and the liberalization of incremental distribution and sale of electricity to social capital to promote the optimal allocation of resources.

In addition, energy storage, big data, artificial intelligence, and other new technologies and materials are gradually being applied to the power grid on a pilot basis. This plays an important role in improving the efficiency of power grid operation and ensuring the safe and reliable operation of the power grid. In this context, how to make full use of advanced technology to build an urban smart grid is an important issue facing smart grid planning. This section proposes a smart grid planning system that comprehensively covers urban, rural, and garden city levels. Smart power grid construction is reasonably planned according to the power supply demand of different regions, so as to improve the power grid operation efficiency and service, and to adapt to the power demand of urban economic and social development. As shown in Fig. 7.8, the framework of city smart grid planning is introduced

7.3.1.1 The Urban Smart Grid Module

Urban smart grid planning focuses on high power supply reliability and smart power consumption. The improvement of power supply and peak regulation capacity can be adapted to controllable loads such as electric vehicles. Only in this way can we meet the needs of current urban economic and social development. Combined with the objectives of urban smart grid planning, the planning content can be put forward as follows.

Ensure High Reliability of Power Supply

Improve the level of distribution network frame and promote the automatic upgrade of distribution. According to the urban development plan and high-quality electricity demand, it will plan and build the distribution network with a high starting point and high standard to ensure sufficient power supply capacity. Using demand-side management and virtual power plant technology to improve peak load management capability. We need to make full use of various controllable resources, and carry out group control and demand side management of the Group in the way of

"Internet + intelligence + market means". Adjusting power consumption behavior and optimizing power flow distribution can effectively reduce peak load. That's accelerate the energy distribution. In areas where power grid construction conditions are not available, energy storage, demand side management, virtual power plant, and other technologies can be fully utilized to change local power flow distribution and improve power supply capacity. This method delays the upgrading of the distribution network and provides a new way for the improvement of the power network.

Meet Smart Electricity Demand

Promote intelligent business halls to improve customer service capabilities. Through the Internet of things, artificial intelligence, mobile applications, and other new technologies can realize the intelligent upgrade of business, equipment, service content, service environment; Customer experience diversified service channels can be improved; Customers can be guided to change their business processing habits; Remote service channel transformation can be promoted. It is a good way to improve the efficiency of traditional business volume processing and introduce diversified value-added power services such as fine billing, electricity consumption analysis, and energy-saving services in time.

We will develop comprehensive energy services and improve energy services. Deploy distributed energy management systems and collect information in residential, commercial, or public services. The information can be used for distributed energy and deep information mining. It can be used to analyze the user's electricity consumption, identify the user's behavior and draw the user's electricity consumption portrait. Further, it facilitates regional power generation and enables the analysis and display of information, ensuring the accuracy of energy-saving services.

7.3.1.2 The Suburban and Rural Smart Grid Module

To upgrade rural power grids should be moved faster. Under the current trend of urbanization and rural revitalization, measures are supposed to be adjusted to local conditions. To develop intelligent area construction of comprehensive monitoring system, realize the distribution area comprehensive display, auxiliary decision-making, low-pressure area monitoring, and other functions, comprehensive control area running state, using the operational mobile terminals, realize the intelligent single, fault repair, low-pressure area monitoring, the repair order processing process monitoring functions, promote grass-roots power substation fast active operations support ability.

7.3.1.3 Smart Grid Module of the Park

The construction of smart grid in the park should focus on reducing the energy cost of enterprises. Based on differentiated power supply services, energy management in the park should be promoted to provide diversified energy supply services, so as to help enterprises "travel lightly" and improve their competitiveness. According to the development needs of smart grid in the park, the following planning contents are proposed.

Differentiated power supply services are provided to improve power supply reliability in the park. Optimize distribution network architecture: customize power services for users according to power supply location, load distribution, and industry characteristics in the park; At the same time, it is necessary to consider the different needs of users to plan the corresponding distribution network architecture, wiring form, and equipment selection. Strengthen the construction of distribution automation, popularize the centralized distribution network automation terminal mode, and gradually realize the full coverage of "three remote" terminals. Promote the flexibility of distribution network. In areas where Direct Current (DC) resources are concentrated, such as distributed power sources, energy storage and electric vehicles, the DC transformation of low-voltage power grids will be promoted, and flexible DC converters will be built in parks to form low-voltage DC grids, which will operate as independent islands or ac and DC parallel. The construction of medium and low voltage flexible Alternating Current (AC)-DC hybrid distribution networks will be carried out to efficiently accept DC power supplies and loads, provide dynamic reactive power compensation, reduce system losses, and improve power quality.

7.3.1.4 Global Smart Grid Module

The global smart grid is based on the construction of smart grids in cities, rural areas and industrial parks, and is fully integrated and regulated by advanced technologies such as cloud computing, artificial intelligence, big data, and the Internet. On the one hand, regional information is collected, analyzed, and integrated for power grid planning; On the other hand, the utilization efficiency of the power grid should be improved internally, the management and service capacity of the power grid should be optimized, and the construction of the smart city and smart transportation should be supported externally. Based on urban, rural, and park smart grid planning and urban smart grid development needs, the following points are proposed as the main contents of the global smart grid.

Power energy big data system construction. It is important to integrate grid operation and equipment monitoring, marketing, and management data for electricity production and substation use and hybrid dispatching, production, and marketing. It can provide data support for smart city construction; It can also promote the construction of urban public charging networks to meet the charging demand of electric vehicles. According to the transition from economically developed areas to less developed areas, the distribution density of public charging infrastructure should be gradually increased. This can reasonably support urban areas and road parking areas and build a charging infrastructure network combining speed and slow.

7.3.2 Urban Public Electricity Behavior Research

On the basis of in-depth analysis of smart city residents' energy use behavior and formulation of relevant policies and measures, the behavioral energy conservation of smart city residents is comprehensively considered. In terms of home behavior energy conservation persuasion, it is necessary to consider that each household often uses a variety of energy consumption equipment, and different types of home equipment, and different habits. Each household should be treated as an independent entity in the design process. By analyzing relevant data of usage behavior, abnormal usage behavior can be accurately identified and appropriate energy-saving measures can be matched for it. In the formulation of energy conservation and emission reduction policies and measures, although the target is family groups rather than individual individuals, its main purpose is similar to household behavior energy-saving persuasion, effectively promoting households to implement energy conservation behavior, and maximize the realization of energy conservation and emission reduction.

7.3.2.1 Household Electricity Behavior Data Collection

Collecting information and data of household energy use behavior is the basis of designing household behavior energy saving strategy and developing corresponding products. The data collected mainly include the following three categories: The first category is the influencing factors of household energy use behavior, such as physical environment factors (such as indoor temperature and humidity), socioeconomic factors (such as education level and income level) and subjective psychological factors (such as energy conservation and environmental awareness). Second, household energy use behavior data, including direct energy use behavior, indoor energy use behavior, and adaptive behavior. The third category is household energy equipment data, such as equipment brand, model, power, etc.

It is worth emphasizing that since the core process is based on the application of the NILM model to judge the motivation and capability of home energy saving, various motivation and capability elements in this model should be included in the design of the data acquisition system. In addition, considering that the same energy using equipment usually has different energy-saving measures from different angles, it is necessary to establish an information database including energy-saving measures from different angles of various energy-using equipment for selection.

7.3.2.2 Targeted Policies and Measures to Conserve Energy

After the completion of data collection and the establishment of an information base, it is necessary to formulate targeted policies and measures. The government can analyze the energy-saving potential of abnormal users from a macro perspective. Considering that different policies and measures often involve and affect different energy-using behaviors, the policies and measures with high correlation should be selected from the established information base to ensure that the formulated policies and measures have strong operability.

On this basis, the government should further combine the regional characteristics of each factor and its own conditions for regulation, and constantly change the specific content of alternative policies and measures. At the same time, tenants should also be combined with the motivation and ability of key target groups to judge the possible impact range of energy-saving, to achieve different alternative policies and measures after each change of supervision and its energy-saving effect evaluation combination forecast accurately. Through the above continuous adjustment and evaluation, government departments can finally choose and implement some policies and measures with the best energy-saving effect.

7.3.2.3 Interactive Influence and Mutual Promotion

In order to realize the sustainable development of residential building behavioral energy saving, it is necessary to give full play to the positive influence of the two to realize the mutual promotion. For example, household energy use behavior may change with the implementation of energy conservation and emission reduction policies and measures. Government departments can track and evaluate the actual application effects of policies and measures through macro analysis of household energy use behavior data, and further adjust and improve the specific content according to the evaluation results.

At the same time, the policy measures of energy saving and emission reduction will have a certain impact on the energy-saving ability and enthusiasm of households and the existing energy-saving measures. In the process of household behavior energy-saving persuasion, relevant data of household energy consumption behavior and information database of energy-saving measures should be updated in time in combination with the formulation and implementation of policies and measures. On this basis, more suitable energy-saving measures are provided for families and corresponding trigger methods are selected to help families achieve long-term efficient behavior energy saving.

7.4 Analysis of Energy Conservation and Emission Reduction of Smart Cities

This section mainly takes Jiangxi Province of China as an example to analyze the effect of energy conservation and emission reduction. Specifically, the efficiency of energy saving and emission reduction in Jiangxi province will be analyzed from four aspects: energy consumption index, carbon emission intensity per unit Gross Domestic Product (GDP), energy processing and conversion efficiency, and energy consumption per unit product.

According to the analysis in Fig. 7.9, energy consumption per unit GDP and industrial output value in Jiangxi province decreased significantly from 2006 to 2015, while power consumption per unit GDP showed a trend of fluctuating decline. From 2006 to 2015, energy consumption per unit GDP of the province decreased from 0.81 tons of standard coal per 10,000 yuan in 2006 to 0.54 tons of standard coal per 10,000 yuan, and electricity consumption per unit GDP of the

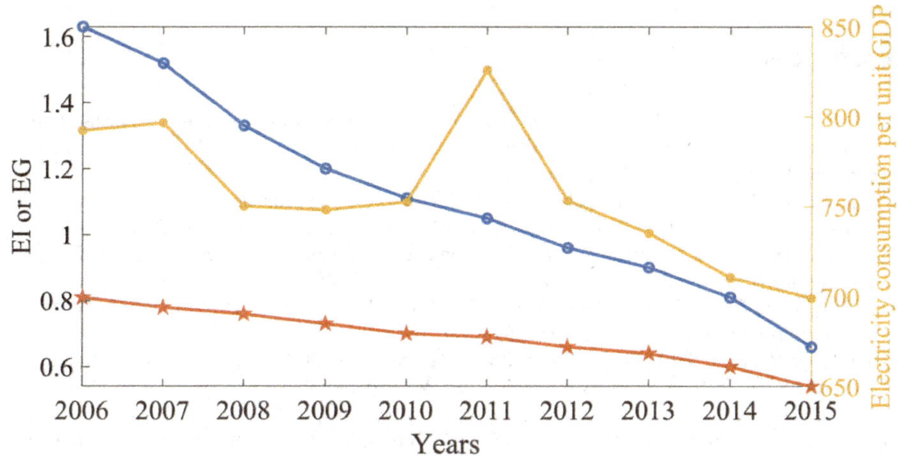

Fig. 7.9 Analysis of the effect of energy conservation and emission reduction of Jiangxi Province in China

province decreased from 792.4 kWh per 10,000 yuan to 699.1 kWh per 10,000 yuan. Energy consumption per unit of industrial output decreased from 1.63 tons of standard coal per ten thousand yuan to 0.66 tons of standard coal per ten thousand yuan, which is more obvious than energy consumption per unit of GDP, indicating that industrial energy saving plays a leading role in energy saving work in Jiangxi Province.

7.5 Conclusions

Smart city construction is the main direction of urban development at present. Energy management, urban intelligent grid planning, individual energy-saving behavior planning, and reasonable effectiveness analysis are all quite important. Energy and economy, as two key elements of urban development, must be considered together. It is an important way for the sustainable development of the urban economy to realize the rational planning of energy management level and user level through non-invasive load identification and realize energy conservation through the information interaction of the smart grid. On the one hand, the energy-saving goal can be achieved by applying NILM. The energy-consuming information of a building or a house does benefit the owners to make better energy management. The information also benefits the government to achieve better grid planning and promotes the sustainable development of a city's economy. On the other hand, the guideline of energy-saving and healthy economic development is of great importance for a city. The city managers and citizens should obey the energy-saving guideline proposed by the government to build a healthy and beautiful living environment.

References

Chang, H. H., Lin, L. S., Chen, N., & Lee, W. J. (2013). Particle-swarm-optimization-based non-intrusive demand monitoring and load identification in smart meters. *IEEE Transactions on Industry Applications, 49*(5), 2229–2236.

Coello, C. A. C., Pulido, G. T., & Lechuga, M. S. (2004). Handling multiple objectives with particle swarm optimization. *IEEE Transactions on Evolutionary Computation, 8*(3), 256–279.

Darby, S. (2006). The effectiveness of feedback on energy consumption. *A Review for DEFRA of the Literature on Metering, Billing direct Displays, 486*(2006), 26.

Feng, P., Qin, D., Ji, P., Ma, J., & Yang, S. (2019). Multi-label learning algorithm with SVM based association. *High Technology Letters, 25*(1), 97–104.

Gibaja, E., & Ventura, S. (2014). Multilabel learning: A review of the state of the art and ongoing research. *Wiley Interdisciplinary Reviews: Data Mining and Knowledge Discovery, 4*(6), 411–444.

Gopinath, R., Mukesh, Kumar, Joshua, C. P. C., & Srinivas, K. (2020). Energy management using non-intrusive load monitoring techniques—state-of-the-art and future research directions. *Sustainable Cities and Society, 62*, 102411. https://doi.org/10.1016/j.scs.2020.102411

Himeur, Y., Ghanem, K., Alsalemi, A., Bensaali, F., & Amira, A. (2021). Artificial intelligence based anomaly detection of energy consumption in buildings: A review, current trends and new perspectives. *Applied Energy, 287*, 116601. https://doi.org/10.1016/j.apenergy.2021.116601

Hochreiter, S., & Schmidhuber, J. (1997). Long short-term memory. *Neural Computation, 9*(8), 1735–1780.

Jian, L., Ng, S. K. K., Kendall, G., & Cheng, J. W. M. (2010). Load signature study—Part II: Disaggregation framework, simulation, and applications. *IEEE Transactions on Power Delivery, 25*(2), 561–569.

Johnson, M. J., & Willsky, A. S. (2013). Bayesian nonparametric hidden semi-markov models. *Journal of Machine Learning Research, 14*(1), 673–701.

Kahl, M., Haq, A. U., Kriechbaumer, T., & Jacobsen, H. A. (2016). WHITED—a worldwide household and industry transient energy data set. *International Workshop on Non-intrusive Load Monitoring, 2016*, 1–5.

Kennedy, J. (1995). Particle swarm optimization. In C. Sammut, G. I. Webb (Eds.), *Encyclopedia of machine learning* (pp. 760–766). Springer US. https://doi.org/10.1007/978-0-387-30164-8_630.

Kolter, J. Z., Johnson, M. J. (2011). REDD: A public data set for energy disaggregation research. In *Paper presented at the Proceedings of the SustKDD Workshop on Data Mining Applications in Sustainability.*

Li, H., Tang, B., Che, L., Liu, J., & Su, X. (2017). Energy saving potential of air conditioning system of equipment stations in metro systems. *Procedia Engineering, 205*, 3519–3524. https://doi.org/10.1016/j.proeng.2017.09.919

Liao, J., Zhang, F., Zhang, S., Yang, G., & Gong, C. (2021). Energy-saving optimization strategy of multi-train metro timetable based on dual decision variables: A case study of Shanghai Metro line one. *Journal of Rail Transport Planning & Management, 17*, 100234. https://doi.org/10.1016/j.jrtpm.2021.100234

Lin, Y.-H., & Tsai, M.-S. (2015). The integration of a genetic programming-based feature optimizer with fisher criterion and pattern recognition techniques to non-intrusive load monitoring for load identification. *International journal of green energy, 12*(3), 279–290.

Liu, H., Yu, C., Wu, H., Chen, C., & Wang, Z. (2020). An improved non-intrusive load disaggregation algorithm and its application. *Sustainable Cities and Society, 53*, 101918. https://doi.org/10.1016/j.scs.2019.101918

Luo, F., Guo, W., Yu, Y., & Chen, G. (2017). A multi-label classification algorithm based on kernel extreme learning machine. *Neurocomputing, 260*, 313–320. https://doi.org/10.1016/j.neucom.2017.04.052

Makonin, S., Popowich, F., Bajić, I. V., Gill, B., & Bartram, L. (2016). Exploiting HMM sparsity to perform online real-time nonintrusive load monitoring (NILM). *IEEE Transactions on Smart Grid, 7*(6), 2575–2585.

Mengistu, M. A., Girmay, A. A., Camarda, C., Acquaviva, A., & Patti, E. (2018). A cloud-based on-line disaggregation algorithm for home appliance loads. *IEEE Transactions on Smart Grid, 99*, 1–1.

Michael, B., & Jiirgen, V. (2004). Genetic algorithm for pattern detection in nialm systems. *IEEE SMC, 4*, 3462.

Panten, N., Abele, E., & Schweig, S. (2016). A power disaggregation approach for fine-grained machine energy monitoring by system identification ☆. *Procedia Cirp, 48*, 325–330.

Piga, D., Cominola, A., Giuliani, M., Castelletti, A., & Rizzoli, A. E. (2016). Sparse optimization for automated energy end use disaggregation. *IEEE Transactions on Control Systems Technology, 24*(3), 1044–1051.

Raza, M. Q., & Khosravi, A. (2015). A review on artificial intelligence based load demand forecasting techniques for smart grid and buildings. *Renewable and Sustainable Energy Reviews, 50*, 1352–1372.

Rigas, E. S., Ramchurn, S. D., & Bassiliades, N. (2014). Managing electric vehicles in the smart grid using artificial intelligence: A survey. *IEEE Transactions on Intelligent Transportation Systems, 16*(4), 1619–1635.

Su, H., Xiang, L., Hu, A., Gao, B., & Yang, X. (2021). A novel hybrid method based on KELM with SAPSO for fault diagnosis of rolling bearing under variable operating conditions. *Measurement, 177*, 109276. https://doi.org/10.1016/j.measurement.2021.109276

Wang, Y., Chen, Q., Hong, T., & Kang, C. (2018). Review of smart meter data analytics: Applications, methodologies, and challenges. *IEEE Transactions on Smart Grid, 99*, 1–1.

Yang, J., Zhou, J., Liu, L., & Li, Y. (2009). A novel strategy of pareto-optimal solution searching in multi-objective particle swarm optimization (MOPSO). *Computers and Mathematics with Applications, 57*(11), 1995–2000.

Yeh, M.-C., & Li, Y.-N. (2019). Multilabel deep visual-semantic embedding. *IEEE Transactions on Pattern Analysis Machine Intelligence, 99*, 1–1.

Yu, Y., You, S., Zhang, H., Ye, T., Wang, Y., & Wei, S. (2021). A review on available energy saving strategies for heating, ventilation and air conditioning in underground metro stations. *Renewable and Sustainable Energy Reviews, 141*, 110788. https://doi.org/10.1016/j.rser.2021.110788

Zhang, M.-L., & Zhou, Z.-H. (2007). ML-KNN: A lazy learning approach to multi-label learning. *Pattern Recognition, 40*(7), 2038–2048. https://doi.org/10.1016/j.patcog.2006.12.019

Zhang, M. L., & Zhou, Z. H. (2014). A review on multi-label learning algorithms. *IEEE Transactions on Knowledge and Data Engineering, 26*(8), 1819–1837.

Optimization Analysis of Clean Energy Transformation

8

8.1 Introduction

The traditional energy structure dominated by fossil energy has brought negative environmental impact. The transformation of the global energy structure has become an urgent task (Fouquet, 2012). The low-carbon revolution, which aims at improving energy efficiency, optimizing consumption patterns, and reducing energy consumption emissions, has become an inevitable choice for countries around the world to promote sustainable economic development. The structural adjustment trend with clean energy as the main body has initially taken shape. Clean energy generally refers to various environment-friendly and renewable energy sources such as nuclear, wind, and solar energy. The clean energy industry is an industrial form mainly aimed at the improvement of various traditional energy-efficient development and utilization technologies and the development of various environment-friendly and renewable energy utilization technology systems. A low-carbon energy economy has become an important strategy for developed countries to cultivate new economic growth points and seize innovation highland in the new round of technological revolution (Fang et al., 2010). It is of great practical significance to draw lessons from the characteristics of the transition and development of the clean energy industry in developed countries and formulate our strategic path in combination with the existing problems in the clean energy industry in China.

8.1.1 Global Status of Clean Energy Development

Figures 8.1, 8.2, and 8.3 respectively give the global annual change in energy production, total primary energy supply, and world electricity generation mix from 1971 to 2019. Data is from the International Energy Administration's statistic report World Energy Balances: Overview (updated to August 2021) (IEA).

© Science Press 2022

H. Liu et al., *Big Data in Energy Economics*, Management for Professionals,
https://doi.org/10.1007/978-981-16-8965-9_8

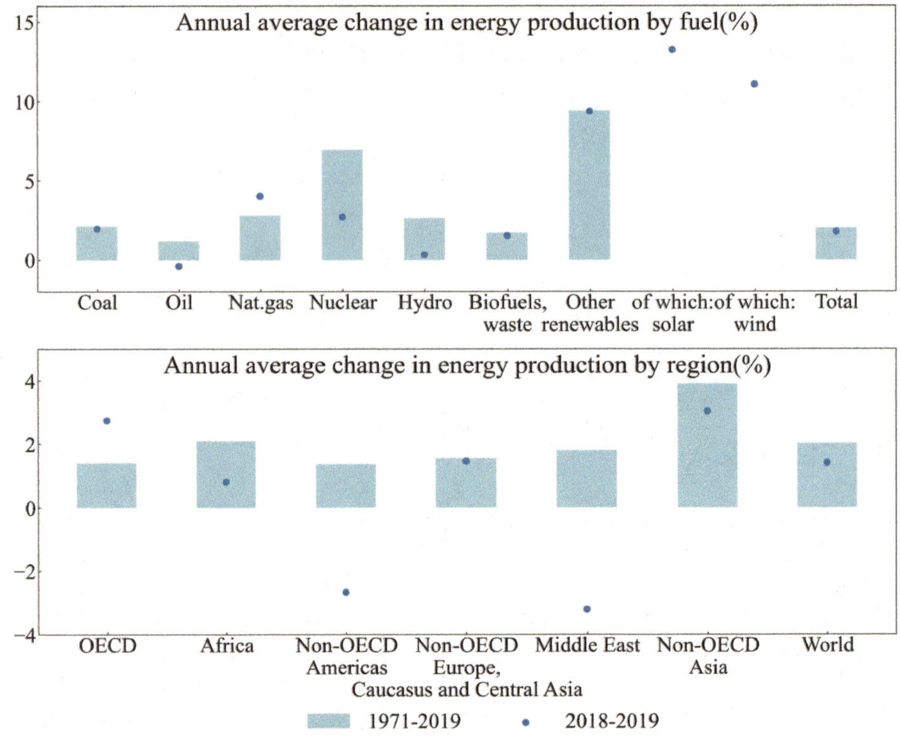

Fig. 8.1 Annual average change in global energy production, 1971–2019

As can be seen from Fig. 8.1, in terms of fuels, in 2019, the global energy output reached 617EJ, a growth of 2% over 2018. It is mainly due to the growth of natural gas (+ 4%) and coal (+ 2%), although some renewables saw larger relative increases, such as solar (+14%) and wind (+12%). Regarding nuclear, however, the figure has gradually slowed down in recent years, from 6.95% in 1971–2019 to 2.98% in 2018–2019. Hydropower production remained at 15 EJ, with a growth rate of almost 0. Despite the slight decrease in production in 2019, oil is still the most productive energy form, with an output of 190EJ. In terms of regions, the growth rate of energy production in non-Organization for Economic Co-operation and Development (OECD) Asia was the highest in both the long run (+3.91%) and the short run (+3.81%). The energy production of non-OECD Americas and the Middle East in 2019 showed a negative growth trend compared with that in 2018. Regarding total energy production, OECD member countries accounted for the most, reaching 195EJ, with a year-on-year increase of 3.5%. Non-Asia came second, reaching 179EJ. Non-OECD Europe, Caucasus, and Central Asia were the third (85 EJ, + 2%).

As can be seen from Fig. 8.2, the world's Total Energy Supply (TES) rose from 230 EJ in 1971 to 606 EJ in 2019, an increase of 2.6 times. In terms of fuels, the

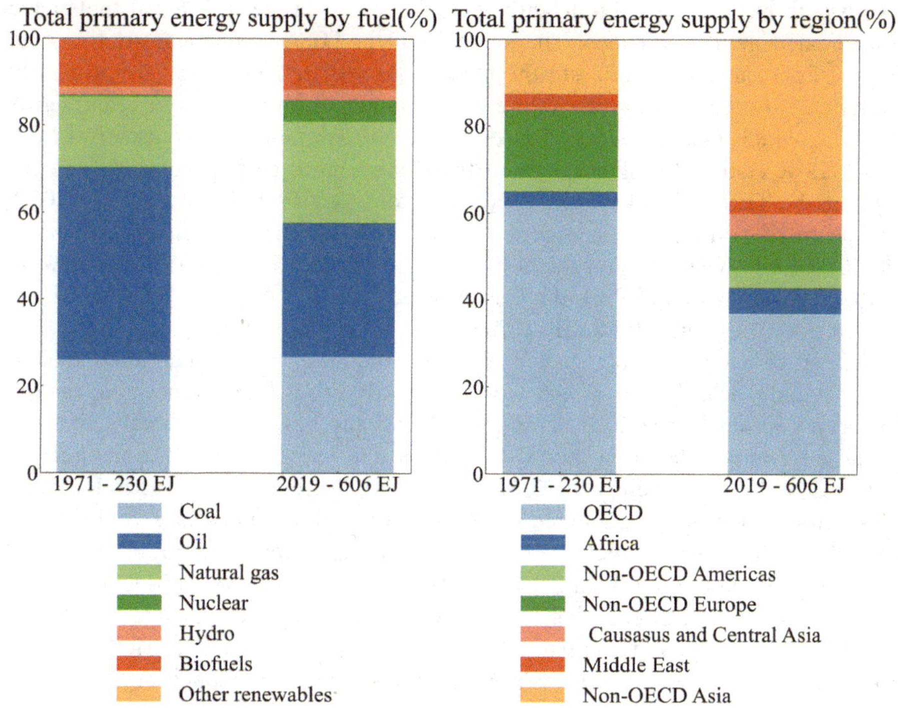

Fig. 8.2 Total primary energy supply shares, 1971–2019

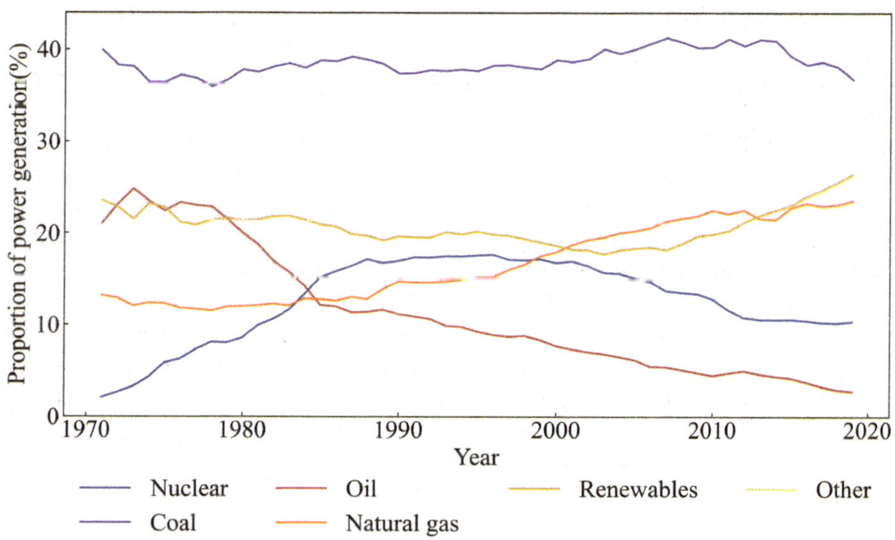

Fig. 8.3 World power generation structure by fuel, 1971–2019

energy mix has changed significantly over the years. In general, fossil fuels have consolidated their dominance. The share of oil in TES has fallen from 44.3% to 30.9%. Coal ranked second, and the natural gas's share rose from 16% to nearly a quarter. However, it is worth noting that clean energy is gradually expanding its share, in which nuclear energy has experienced a significant increase (from 0.5 to 5%). The status of traditional fossil fuels is bound to be impacted. In terms of regions, the TES share of OECD decreased from 61.7% in 1971 to 37% in 2019. The share of non-OECD Asia tripled, from 12.5 to 37.1%, equating to that of OECD. Non-OECD Europe ranked third, falling from 15.5% in 1971 to 8.1% in 2019, ahead of Africa (5.9%), the Caucasus and Central Asia (5.1%), non-OECD Americas (3.9%), and Middle East (2.9%).

As can be seen from Fig. 8.3, coal is still the main power generation fuel so far, providing nearly 37% of the world's electricity in 2019. Renewable energy has been booming since 2007. This growth not only has squeezed coal's share of global electricity generation, but also allowed renewables to overtake natural gas as the second largest fuel by 2013, with the gap widening. In 2019, the electricity provided by renewables accounted for 26.5% of the global electricity. In the 1970s and 1980s, the share of nuclear energy of global electricity generation expanded at a robust pace, surging from 2.1% in 1971 to 17.2% in 1988. Subsequently, the development of nuclear energy slowed down, and its share plateaued at about 10% after 2012. After years of continuous decline, oil's share in total electricity in 2019 was less than 3%, even though its share in energy supply remained high.

8.1.2 International Experience in the Transformation of Clean Energy Industry

Most developed countries went through the period of coal as the main energy source before the petroleum age. However, the development and utilization of coal, on the one hand, has brought about the emission of many local pollutants such as SO_2 and NO_x, and on the other hand, has aggravated the scarcity of natural resources such as water resources. Therefore, since the 1950s, developed countries have realized "reducing coal, increasing oil, and increasing gas" with the help of energy internationalization, greatly improved energy efficiency, vigorously promoted pollutant control technologies such as desulfurization and denitration, and intensified research and development of non-fossil energy such as nuclear energy and renewable energy (Zheng & Ligang, 2012). Coal's share of global primary energy consumption (excluding China) fell from 70% in 1925 to 36% in 1965, and 19% in 2013. Traditional pollutants and greenhouse gas emissions have been effectively controlled. Developed countries can achieve the optimization of energy structure, attributed to the following reasons:

8.1.2.1 Establish the Development Concept of Low-Carbon Transformation Early

In 1975, Sweden came up with the idea of meeting its energy needs 100% from renewable sources. In 1980, the United States proposed a strategic vision of 100% reliance on renewable energy by 2100. Subsequently, some Western European countries also put forward such goals, with Denmark and Germany taking the lead in developing renewable energy (Luis et al., 2019). Denmark established the Energy Agency in 1976. In the following 20 years, the energy structure has been continuously optimized, with oil and coal consumption both falling by about 36% and natural gas consumption rising to 20%. It has set the goal of 50% wind energy, 15% solar energy, and 35% other renewable energy by 2030 and strives to achieve the grand goal of 100% renewable energy supply by 2050. At present, Denmark has established a relatively perfect distributed energy system. In addition to wind power, more than 50% of electricity comes from cogeneration, and biomass also accounts for 18% of primary energy. Denmark's leading position in the process of the energy transition is largely attributed to its green growth and development concept, which has been formed and perfected since the 1990s (Yan et al., 2015).

8.1.2.2 Make Room for Renewable Energy Through Concrete Actions

Taking Germany as an example, in the 1990s, it established the development goal of taking nuclear energy as the transitional energy and solar, wind, and other clean energy as the main energy in the future (Jacobsson & Lauber, 2006). In 2012, it was proposed that the share of renewable energy in the German power supply should reach 35, 50, 65, and 80% in 2020, 2030, 2040, and 2050, respectively, which defined the dominant position of renewable energy in the future energy structure. These are all substantial preparations and support for the development of the clean energy industry.

8.1.2.3 Encourage Competition in the Electricity Market

Since the 1990s, the EU has promoted three rounds of power market-oriented reform and accelerated the integration of the power industry within the EU. In 2015, the EU proposed that the installed power generation capacity of cross-border power grid operation should reach 10% of the total installed capacity of Member States by 2020 and strive to increase this proportion to 15% by 2030. In the competitive power market, renewable energy power, as a green product, is introduced to the market and supported by relevant on-grid tariffs and subsidies, which had high market competitiveness (Castagneto-Gissey & Giorgio, 2014).

8.1.2.4 Complement Energy Advantages Globally

Japan, which is short of fossil energy resources, guarantees its continuous overseas energy supply through active energy diplomacy. Europe, long dependent on Russian oil and gas supplies, is shifting its focus as the crisis in Ukraine escalates. In addition, Europe is committed to building strategic energy partnerships with other regions of the world, including central Asia, the Middle East, North Africa, the United States, Canada, and other energy exporters (Hewitt et al., 2017). At the same time, countries have made great efforts to develop clean and renewable energy to reduce energy dependence and improve energy supply and efficiency (Quitzow et al., 2019).

8.2 Efficiency Analysis of Energy Utilization Under Diversified Development

8.2.1 Evaluation Indexes and Methods of Energy Efficiency

Energy efficiency is an important subject in the field of energy economics, and there is no unified definition so far. It can be measured by a variety of quantitative indicators. It is generally believed that energy efficiency refers to producing as many services or useful outputs as possible with as little energy input as possible. The indicators to measure a country's energy efficiency can be summarized into two categories: energy economic efficiency and energy technical efficiency (Diaconescu et al., 2019). Figure 8.4 shows several evaluation indexes and methods of energy efficiency.

Economic indicators of energy efficiency measure energy input and output in monetary terms. It can be further divided into single-factor productivity and total factor productivity according to the number of production factors investigated. The single factor productivity, also called energy intensity, only considers the energy input and output in production, and does not consider other production factors. The most representative one is the energy consumption per Gross Domestic Product (GDP), that is, the ratio of total energy consumption to GDP. The single factor productivity method is simple, clear, and understandable, but it ignores the contribution of other factors to output and the substitution effect between various factors. Total factor productivity comprehensively considers the interaction of multiple input factors such as capital, labor, energy, and technical conditions in the process of social production (Zhang, 2009). The measurement methods mainly include the multiple regression method, simultaneous equations method, factor decomposition method (Sharma et al., 2007), Data Envelopment Analysis (DEA) method (Hossain et al., 2012), etc.

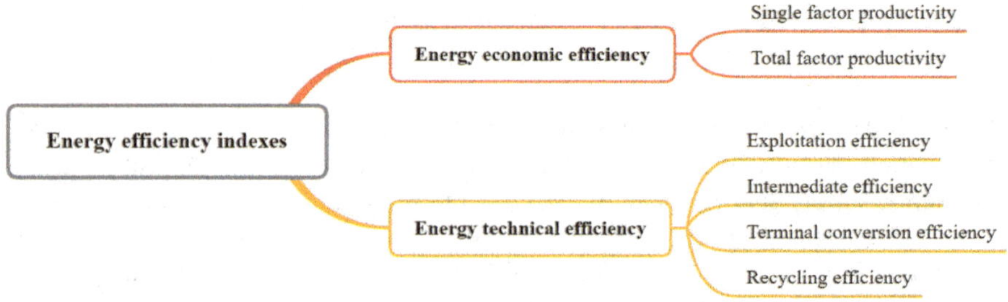

Fig. 8.4 Energy efficiency indexes

Energy technical efficiency refers to the ratio between the initial input energy after exploitation, processing, conversion, storage and transportation, terminal utilization, and the final effective utilization of energy (Shen & Lin, 2017). It can be divided into exploitation efficiency, intermediate efficiency, terminal conversion efficiency, and recycling efficiency. Energy technical efficiency can also be expressed as the product of the above four indicators. Since the energy technical efficiency often involves complex data, it is rarely used in the studies. Most of them take energy economic efficiency indicators instead.

8.2.2 Analysis of Influencing Factors and Mechanism of Energy Efficiency

The factors affecting energy efficiency are diverse, which can be divided into direct factors and indirect factors. Direct factors refer to technology and management level; Indirect factors refer to factors other than technology, which are further divided into economic factors, institutional factors, regional and natural factors (Yang et al., 2011). Figure 8.5 shows the influencing factors and mechanism of energy efficiency.

8.2.2.1 Technological Progress

First, the most direct impact of technological progress on energy efficiency lies in the improvement of energy technical efficiency. The invention, improvement, and popularization of energy technology can promote the progress of energy application technology in exploitation, production, processing, terminal conversion, recycling, and other links, and then affect the supply–demand relationship and structural change of energy input.

Second, technological progress can improve basic production conditions such as equipment and optimize production elements other than energy input. The saving

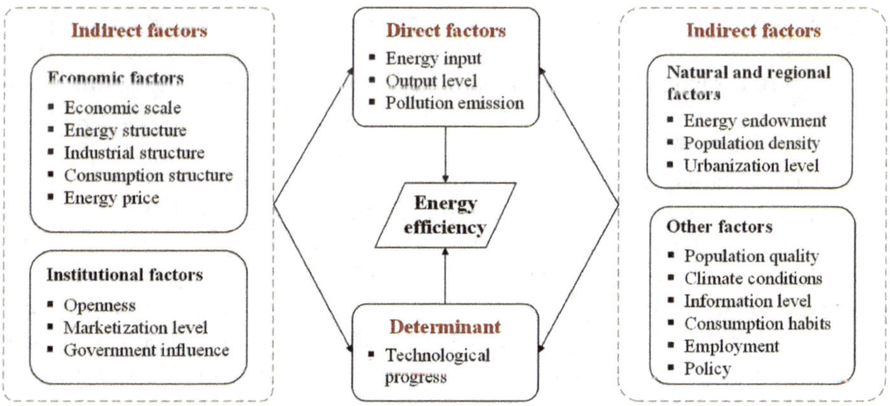

Fig. 8.5 Influencing factors and mechanism of energy efficiency

of energy input, the increase of economic output, and the control of pollution emissions have been realized to varying degrees, to improve energy efficiency.

Third, the "soft" technological progress such as management level can realize the macro control of the whole production activities, reduce the intermediate transaction cost, and promote the rational optimization of resource allocation (Li & Lin, 2018).

8.2.2.2 Economic Factors

Economic Scale

According to the theory of scale economy, when the production activities reach a certain economic scale, the organic combination of various production factors can reduce the average cost of production and produce scale effect. The growth rate of economic output will be greater than that of energy consumption, which will reduce the intensity of energy consumption and improve energy efficiency (Song et al., 2015).

Economic Structure

- Industrial structure.

  The impact of industrial structure on energy efficiency depends on the energy efficiency level of each industry and its proportion. The greater the proportion of industries with low energy efficiency levels, the lower the overall energy efficiency level. When the energy efficiency level among industries is relatively unchanged, energy flows from industries (or departments) with low-efficiency levels to industries (or departments) with high-efficiency levels. The allocation of energy among different industries (or departments) can be optimized, thus improving the energy efficiency of the whole economic system (Xiong et al., 2019).

  The secondary industry consumes the most energy and has more serious pollution emissions than other industries (He, 2011). The tertiary industry is mainly a service industry with high added value, low energy consumption, clean and environmental protection. It has low energy consumption intensity, less pollution emission, and high energy efficiency. Therefore, increasing the proportion of the tertiary industry in the national economy and realizing the optimization and upgrading of industrial structure will play a positive role in improving the overall energy efficiency.
- Industrial internal structure.

  Although the industrial sector only accounts for about 40% of China's national economy, its energy consumption in recent years has exceeded 70% of the country's total energy consumption. Therefore, the change of the internal structure of the industry will have a significant impact on the overall energy consumption and efficiency level of China. The higher the proportion of low energy efficiency in industrial sectors is, the more unfavorable it is to improve the energy efficiency level of the whole economic system.

Heavy industry has always been a large consumer of energy, and its economic growth model is extensive. It is a typical industry with high energy consumption and high pollution. The amount of energy consumed, and pollution discharged by each unit of added value is much higher than that of the light industry. Therefore, the higher the proportion of heavy industry within the industry, the greater the energy consumption and pollution emissions, the lower the energy efficiency (Han et al., 2019).

- Energy consumption structure.

  Energy consumption structure refers to the combination of energy consumed by production activities under the restriction of a certain energy efficiency level and energy endowment. Different energy sources have different characteristics, and their utilization efficiency is also different. Obviously, at the same output level, efficient and clean energy can save more energy input and relatively fewer pollution emissions than low-quality energy. Natural gas and electric power are efficient and clean energy, and the increase of their proportion will be conducive to the improvement of overall energy efficiency (Tao & Li, 2018). For coal with low utilization efficiency, the increase of its proportion will not be conducive to the improvement of overall energy efficiency.

Energy Price

Energy prices fully reflect the market supply and demand relationship of energy, the current level of energy technology, the substitution effect between different energy varieties, and the substitution technology between energy and other factors (Hang & Tu, 2007). Its impact mechanism on energy efficiency is mainly realized through the following ways:

First, the income effect. Rising energy prices will directly lead to higher production costs and lower output levels. In the next stage, the demand for energy and energy consumption intensity will decrease.

Second, the substitution effect. Rising energy prices will force enterprises to take energy-saving measures to improve energy efficiency, reduce energy consumption and save costs as much as possible. However, using energy-saving technology to improve energy efficiency will increase the cost of enterprises. There is a game relationship between energy consumption and increased costs. Finally, the improvement of energy efficiency will be taken as the equilibrium point of the whole game.

8.2.2.3 Institutional Factors

Different from the obvious influence of technological progress and economic factors on energy efficiency, institutional factors, as non-economic factors, also play a potential role in energy efficiency.

Openness

First, the international division of labor and specialization brought by opening to the outside world can make full use of various elements and obtain more output.

It is conducive to improving the input–output efficiency of elements, including energy efficiency.

Second, the trade structure of goods has a significant impact on energy efficiency. From the perspective of export, the proportion of low energy consumption and low pollution products in export commodities is higher to promote the development of the efficient clean industry. The energy demand of economic development is relatively reduced, thus improving the overall energy efficiency. From the perspective of import, the import of high energy consumption products and the introduction of advanced equipment is conducive to the country to save energy (Sebri & Ben-Salha, 2014).

Third, the technology spillover effect brought by foreign direct investment. With the deepening of the degree of opening to the outside world, the spillover effect of technology becomes more and more significant. The higher the level of actual utilization of foreign direct investment, the more likely it is to use foreign technology to improve the energy efficiency of the region.

Government Influence

The blindness, irrationality, and one-sided profit-seeking of the market economy led to energy waste and environmental problems. As the main department of macro-control, the government shoulders the responsibility of standardizing the behavior of market subjects and overcoming the defects of the market economy.

Porter Analysis advocates that appropriate environmental regulation will stimulate technological innovation, reduce costs, improve product quality, and possibly improve industrial productivity. Then, (Lanoie et al., 2008) studied the impact of environmental regulation on enterprise productivity based on the Porter Analysis. It confirms the Porter Analysis that environmental regulation promotes the productivity of enterprises.

8.2.2.4 Regional and Natural Factors

Energy Endowment

Energy endowment refers to the comprehensive evaluation of the quantity and quality of energy by a country (region) to measure the contribution of energy factors to economic development. The Resource Curse theory holds that abundant resources are not a favorable condition for economic growth, but a restriction, and there is a negative correlation between resource abundance and economic growth (Wu et al., 2021). The expansion of resource-based industries in resource-rich areas has squeezed out the accumulation effect of human capital and induced the opportunistic behavior of resource use and rent-seeking activities. It causes a waste of resources and predatory exploitation, which is not conducive to economic growth and efficiency improvement.

Population Density and Urbanization Level

When other factors remain unchanged, the energy consumption of a region is usually directly proportional to the population. The higher the population density, the greater the energy consumption and the heavier the environmental pollution.

Therefore, it can be predicted that population density is inversely proportional to energy efficiency.

The level of urbanization has a dual impact on energy efficiency. On the one hand, the improvement of urbanization level is conducive to economic development and the improvement of income level, resulting in income effect. The increasing energy demand is not conducive to the improvement of energy efficiency. On the other hand, the improvement of urbanization level is conducive to optimize resource allocation, improve energy technology, and then improve energy efficiency (Li et al., 2018). Hence, it is difficult to judge the impact direction of urbanization level on energy efficiency in theory.

8.2.2.5 Other Factors

In addition to the above factors, many other factors have an indirect impact on energy efficiency, such as population quality, climate conditions (Lu et al., 2010), information level, consumption habits, employment, policy factors, etc. These factors can explain the differences in energy efficiency levels in various regions to a certain extent. Due to space constraints, this chapter will not discuss it.

Through the above theoretical analysis, it is easy to know that the change in energy efficiency is the result of the joint action of many factors. Firstly, statistically speaking, energy efficiency is determined by energy input, output level, and pollution emission. Secondly, technological progress is the most direct and fundamental factor to promote energy efficiency. Thirdly, economic factors, institutional factors, regional and natural factors, and other factors indirectly affect energy efficiency by acting on energy input, output level, pollution emission, and technical level.

8.2.3 International Comparative Analysis of Energy Efficiency

8.2.3.1 Energy Economic Efficiency

Energy economic efficiency, also known as energy intensity, refers to the amount of energy consumed per economic output. The lower the energy intensity, the more efficient the energy economy (Cornillie & Fankhauser, 2004). At present, there are mainly two ways to calculate energy economic efficiency, one is calculated at the exchange rate, the other is calculated at Purchasing Power Parity (PPP) (Phylipsen et al., 1997).

Table 8.1 and Fig. 8.6 give the comparison of energy use per GDP for different countries and regions from 1990 to 2015, which is calculated by PPP at constant prices in 2017. The data is from the database World Development Indicators (WDI) from the World Bank (WB).

From Table 8.1 and Fig. 8.6, it can be found that:

From a horizontal perspective, in 2014, China's energy intensity was 187.7koe/$1,000, 1.54 times that of the United States, 2.98 times that of the United Kingdom, 2.12 times that of Japan, 1.92 times that of OECD members, and 1.55 times that of the world average. Compared with other BRICS countries, China's

Table 8.1 Energy use (kg of oil equivalent) per $1,000 GDP (constant 2017 PPP) by region, 1990–2015

	1990	2000	2005	2010	2012	2014	2015
Brazil	89.4	92.4	92.1	91.3	91.3	95.0	–
Canada	–	222.2	200.3	173.6	167.7	166.0	160.6
China	538.7	260.4	261.3	220.0	207.3	187.7	–
Germany	120.7	95.5	93.1	85.2	77.9	74.4	74.7
India	193.1	161.8	138.6	132.6	130.7	124.4	–
Japan	110.4	114.7	108.4	103.6	92.5	88.4	86.2
Russian Federation	276.6	289.9	226.6	201.1	199.2	186.7	–
South Africa	240.0	240.4	234.2	222.3	206.9	209.2	–
United States	189.8	160.7	144.4	131.8	123.6	121.7	116.2
United Kingdom	118.1	98.9	86.5	76.8	71.4	62.9	61.7
OECD	137.3	125.2	116.6	107.9	101.7	97.7	95.8
World	*165.4*	*143.9*	*138.0*	*130.9*	*126.5*	*121.0*	–

Source The World Bank

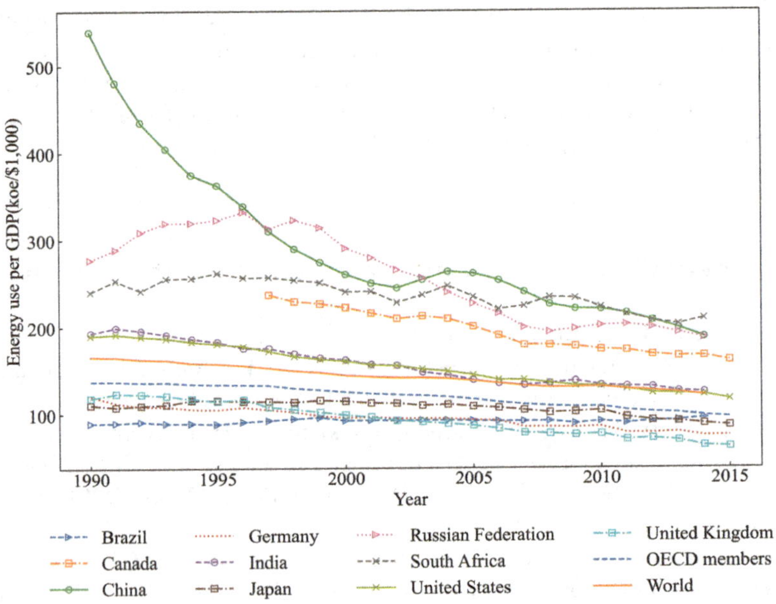

Fig. 8.6 Energy use per $1,000 GDP

energy intensity was higher than that of all but South Africa, 1.98 times that of Brazil, 1.51 times that of India, and 1.01 times that of Russia.

From a vertical perspective, China's energy intensity declined steadily year by year except for a small increase from 2002 to 2004. Compared with the world average, China's energy intensity decreased from 3.26 times that of the world in 1990 to 1.89 times that of the world in 2005, and then to 1.55 times that of the world in 2014. However, China's energy intensity is still far from that of developed countries. Compared to the United States, the figure dropped from 1.62 in 2000 to 1.54 in 2014. Compared to OECD members, the figure fell from 2.08 in 2000 to 1.92 in 2009. The energy intensity of developed countries has been maintained between 100 and 150koe/$1,000 for years.

To sum up, although China's energy intensity is approaching the world average, it is still far away from developed countries. The problem of excessive energy intensity remains.

8.2.3.2 Energy Technical Efficiency

Energy technical efficiency, also known as energy system efficiency, refers to the ratio of the effective energy obtained in the process of using energy activities to the actual energy input. The total efficiency of the energy system consists of three parts: exploitation efficiency, intermediate efficiency, and terminal utilization efficiency.

Table 8.2 and Fig. 8.7 show the physical energy efficiency of China from 2000 to 2018 and its comparison with the estimation of the Economic Commission of Europe (ECE) in the early 1990s. In Table 8.2, η_1, η_2, η_3, η and ξ denote exploitation efficiency, intermediate efficiency, terminal utilization efficiency, energy efficiency, and energy system efficiency respectively. Intermediate

Table 8.2 Comparison of physical energy efficiency

	2000	2005	2010	2014	2016	2018	Estimation of ECE in the early 1990s
η_1	33.0	33.3	35.9	36.2	36.3	36.6	59–71
η_2	68.5	70.8	70.9	68.7	68.7	70.8	67–75
η_3	–	–	–	–	–	–	–
Agriculture	32.0	33.0	34.0	36.2	36.6	36.8	33–36
Industry	46.0	47.3	50.5	53.8	54.8	55.2	65.0
Transportation	28.9	29.2	29.1	33.1	34.5	34.7	25–30
Construction	66.0	68.4	74.2	74.2	74.8	76.2	50–65
Total	46.7	48.3	51.0	53.5	55.2	55.6	51–55
η	32.0	34.2	36.0	36.8	37.9	39.4	34–41
ξ	10.6	11.4	12.9	13.3	14.1	14.4	20–30

Source China's 2020 Energy Data; Energy policy research, No. 6, 2003

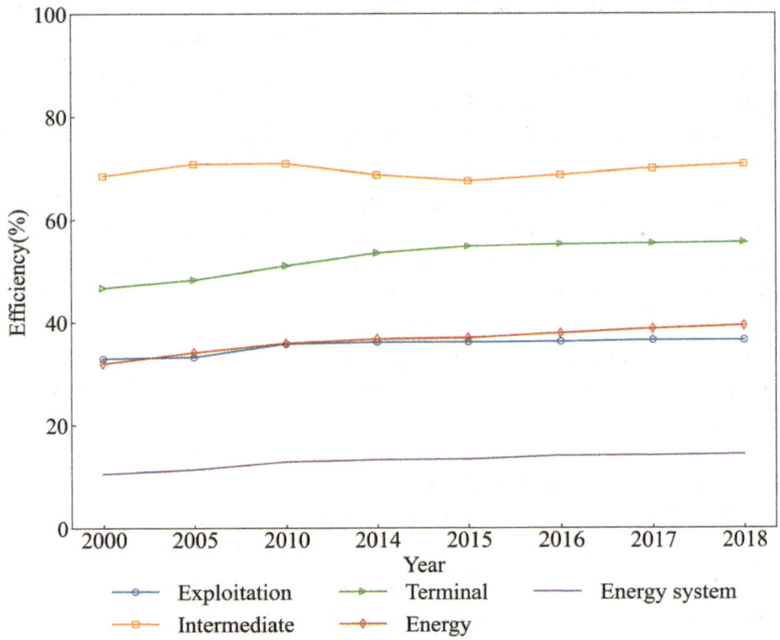

Fig. 8.7 Physical energy efficiency, 2000–2018

links include energy processing, transformation, storage, and transportation. Terminal utilization efficiency is the ratio of the energy obtained by the end-user to the input energy. Energy efficiency is the product of intermediate efficiency and terminal utilization efficiency, i.e., $\eta = \eta_2 \times \eta_3$. Energy system efficiency is the product of exploitation efficiency and energy efficiency, i.e.,$\xi = \eta_1 \times \eta$.

From Table 8.2 and Fig. 8.7, it can be found that:

From a horizontal perspective, among the three parts of the total efficiency of the energy system, the intermediate link has the highest efficiency, followed by the terminal utilization efficiency, and the exploitation efficiency is the lowest. The total efficiency of the energy system was only 14.4% 85.6% of the energy was seriously lost and wasted in the process of exploitation, processing, conversion, storage, and terminal utilization. Only 14.4% of recoverable energy reserves were converted to useful energy.

From the vertical point of view, the efficiency of each link shows a rising trend. China's energy efficiency was 39.4 percent in 2018, with intermediate efficiency at 70.8 percent and terminal utilization efficiency at 55.6 percent. With the development of technology, China's energy efficiency has gradually caught up with that of Europe in the early 1990s. However, the overall efficiency of the energy system is still low. There is still a certain gap in exploitation, industrial utilization, and other links compared with developed countries.

8.3 Analysis of Reasonable Energy Consumption Patterns

The transformation of clean energy development mode should not only consider the links of energy development, transformation, transmission, and allocation but also attach great importance to the end consumption of energy. The ideal state of clean energy development is full consumption. However, clean energy consumption must have supporting policies, facilities, and markets. Clean energy consumption is affected by many factors, such as local load, grid-connected lines, transmission channels, system peak shaving, power quality, operation characteristics, and so on.

8.3.1 Challenges Facing Energy Consumption

8.3.1.1 Inverse Distribution of Resources and Demands

Wind and photovoltaic resources are mostly distributed in Northeast, North, and Northwest China. Hydropower resources are mainly distributed in Southwest China. However, the power load is mainly distributed in the central, eastern, and southern regions. The resulting cross-provincial transmission pressure is large (Li et al., 2021).

8.3.1.2 Unbalance Development of Clean Energy, Load, and Power Grid

The mismatch between supply and demand has caused great consumption pressure. Insufficient analysis of grid conditions leads to the abandonment of power after the completion of incremental clean energy power plants. In the later stage, the power grid construction lags, resulting in insufficient local consumption and delivery capacity.

8.3.1.3 Great Fluctuation of Wind and Photovoltaic Power Generation Output

Due to the natural conditions, the intermittent and inverse peak shaving characteristics of clean energy challenge the peak shaving ability and real-time power balance of the power grid. From the perspective of annual balance, the load is heavy during peak summer, but the wind power is weak. In winter, the load is light, and heating is required, but wind power is strong. From the perspective of real-time balance, the load at night is light but the wind power is strong, and the load at noon is light but the photovoltaic power is strong. There is a natural conflict between the development of clean energy and load characteristics, and sometimes measures such as low valley shutdown and deep peak shaving must be taken (Chang et al., 2001).

8.3.1.4 Undesirable Voltage, Frequency, and Power Response Characteristics of Generator Set

Clean energy has become a weak link in the power grid since it was connected to the grid. To ensure the safety of system operation, power grid enterprises must increase investment in the construction of the backbone grid and security control measures (Zhi et al., 2016). Voltage fluctuation caused by clean energy may cause large-area disconnection of wind power and photovoltaic power and a large amount of power shortage. Consequently, power grid enterprises must invest in the construction of Automatic Voltage Control (AVC) system and many reactive power compensation devices to meet the needs of reactive power balance.

8.3.2 Analysis of Key Factors Affecting Clean Energy Consumption

8.3.2.1 Mechanism Analysis

The generation, supply, and use of the power system are completed simultaneously. Power load presents obvious time-varying characteristics. At present, the peak-valley difference of China's regional power grid has reached about 30% and is gradually expanding. The principle of system balance is to adjust conventional power output to track load changes and maintain dynamic balance. A basic condition for the smooth operation of the power system is that the system regulation capacity is strong enough to adapt to the change of load (Jiang et al., 2017).

Due to the resource characteristics of wind and sunlight, the output of clean energy has randomness and fluctuation. Figure 8.8 shows wind power and load curve in a province on a typical day. As can be seen from Fig. 8.8, the maximum daily fluctuation of wind power can reach 80% of the installed capacity and presents certain anti-peak regulation characteristics. Photovoltaic power generation is also intermittent and volatile, subject to day and night, weather, and cloud changes.

When the output of clean energy exceeds the regulated range of the system, the output must be controlled to ensure the dynamic balance of the system, resulting in wind and solar power curtailment (Xue et al., 2014). From the perspective of system conditions, there are several key factors affecting clean energy consumption, including power regulation performance (maximum technical output and minimum technical output), power grid interconnection and delivery capacity, load scale, and peak-valley difference (Shu et al., 2017).

8.3.2.2 Qualitative Analysis

In China, wind and solar power curtailment are mainly concentrated in the Three Norths regions, that is, Northeast, North, and Northwest China. Table 8.3 shows the clean energy generation curtailment in representative provinces and regions in China in 2015. The wind power curtailment in North China (North Hebei, Shandong, Shanxi and West Inner Mongolia), Northeast China (Jilin, Heilongjiang, Liaoning, and East Inner Mongolia), and Northwest China (Xinjiang, Gansu,

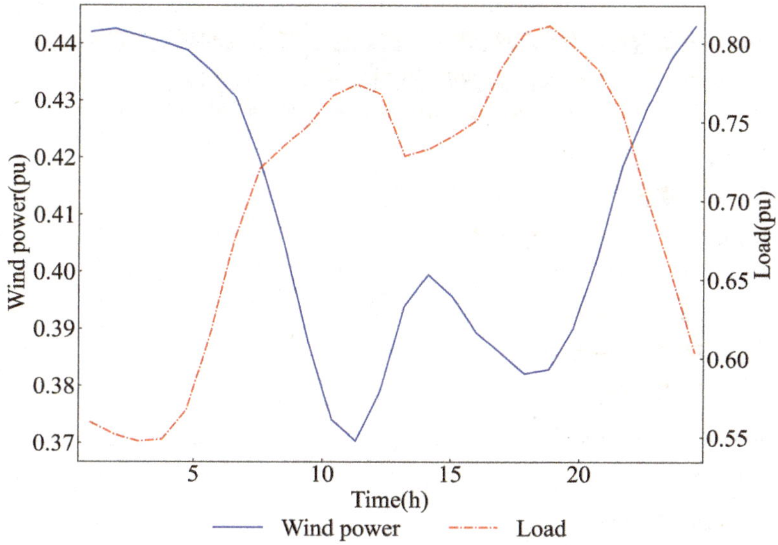

Fig. 8.8 Wind power and load curve in a province on a typical day

Table 8.3 Clean energy generation curtailment by region, 2015

Province/region	Wind power curtailment (TWh)	Solar power curtailment (TWh)	Curtailment rate (%)	The proportion of curtailment in China (%)
Gansu	8.19	2.62	36.8	27.5
Jilin	2.72	0.00	30.8	6.9
Xinjiang	7.11	1.51	30.7	21.9
West Inner Mongolia	7.25	–	21.6	18.4
Heilongjiang	1.86	0.00	20.5	4.7
Liaoning	1.91	0.00	14.5	4.9
Ningxia	1.25	0.28	11.0	3.9
East Inner Mongolia	1.85	0.00	10.9	4.7
North Hebei	1.89	0.00	9.9	4.8
Qinghai	0.00	0.25	2.9	0.6
Shanxi	0.26	0.00	2.4	0.7
Shandong	0.08	0.00	0.6	0.2

Source National Energy Administration

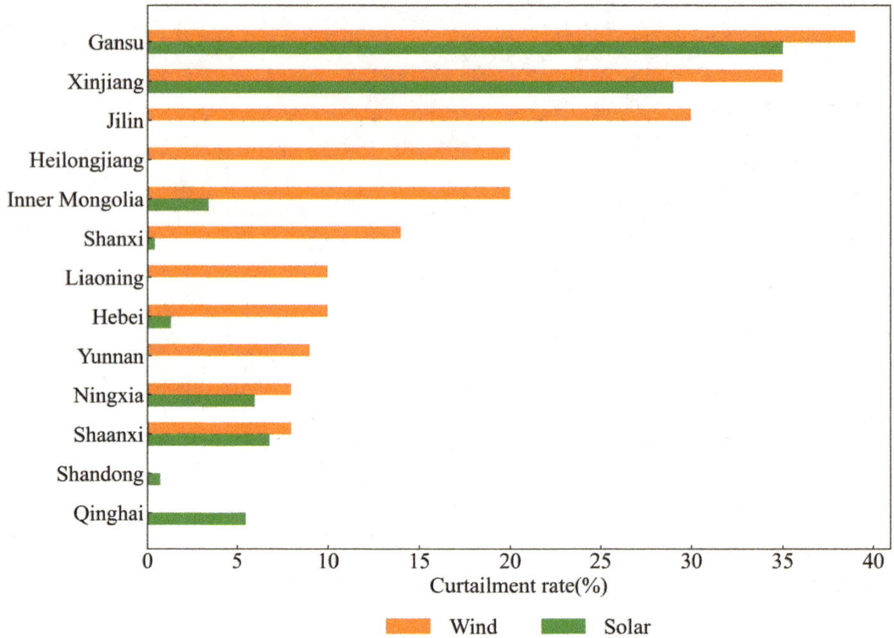

Fig. 8.9 Wind and solar power curtailment rate by region, 2015

and Ningxia) are 9.48TWh, 8.34TWh, and 16.55TWh respectively, accounting for 99.9% of the total wind power curtailment. The solar power curtailment in Northwest China is 4.66TWh, accounting for 95% of the total solar power curtailment.

The regional distribution of wind and solar power curtailment rate in China is shown in Fig. 8.9. There are five provinces and regions, i.e., Gansu, Xinjiang, Jilin, West Inner Mongolia, and Heilongjiang, whose power curtailment rates exceed 15%. While in most provinces and regions, there is not any power curtailment. Among them, the total installed capacity of clean energy in Shandong and Jiangsu is very large, reaching 8.54GM and 8.34GM respectively. The full consumption of clean energy has been achieved. In terms of time distribution, 67% of wind power curtailment in China and more than 90% of wind power curtailment in Northeast China occurs during the heating period. The wind power curtailment at low load accounts for 80% of the total. The wind and solar power curtailment in China reflect the characteristics of regional, seasonal, and period concentration.

The root causes of the wind and solar power curtailment in China can be analyzed from the following three perspectives:

From the perspective of the relationship between installed capacity and load, it can be found that the installed capacity of clean energy in China is unbalanced and inversely distributed with the load. The load in the Three Norths regions accounts for only 36% of the total load in China, but 75% of the country's installed clean

Table 8.4 Clean energy installed capacity and penetration rate by region, 2015

Province/region/country	Wind energy installed capacity (GW)	Solar energy installed capacity (GW)	Penetration rate (%)
Eastern Inner Mongolia	8.80	0.58	172
Gansu	12.52	6.10	143
Ningxia	8.22	3.09	102
Xinjiang	16.91	5.29	100
Western Inner Mongolia	15.45	4.31	97
Denmark	4.89	0.61	93
Spain	22.98	7.09	78
Qinghai	0.47	5.64	73
Portugal	4.86	0.42	63
North Hebei	9.93	1.11	53
Jilin	4.44	0.07	46
Heilongjiang	5.03	0.02	43
Shanxi	6.69	1.13	33
Liaoning	6.39	0.16	27
Shandong	7.21	1.33	14
Jiangsu	4.12	4.22	10

Source National Energy Administration

energy capacity is concentrated there. The penetration rates of clean energy in East Inner Mongolia, Gansu, Ningxia, and Xinjiang exceed 100%, which have exceeded that of developed countries such as Denmark, Spain, and Portugal, as shown in Table 8.4. Due to the limited scale of local load, clean energy consumption puts forward higher requirements for power regulation performance and power grid interconnection.

From the perspective of power regulation performance, the proportion of flexible regulation power supply such as pumping storage and gas turbine in China is low. The flexible power supply in the Three Norths regions is only 17.9% of the installed capacity of clean energy, including 7% in the Northeast and 2.4% in the Northwest. While the proportion of flexible power supply in Spain, Germany, and the United States in the total installed capacity is 31, 19, and 47% respectively. The installed capacity of thermal power in China exceeds 1TW, but the peak shaving capacity is generally only about 50%. In the Three Norths regions, the thermal power units have only 20% regulation capacity during the heating period. In contrast, thermal power units in Spain, Denmark, and other countries are capable of deep peak shaving, with adjustable output of up to 80%.

From the perspective of the power grid interconnection level, the needs of clean energy delivery in the Three Norths regions cannot be met. By the end of 2015, the installed capacity of the clean energy of the northwest power grid was 60GW, and the inter-regional power transmission was only 16.1GW. The installed capacity of clean energy in the northeast power grid had reached 25.5GW, but the inter-regional power transmission was only 3GW, with a power surplus of over 20GW. However, the transmission capacity between Denmark and Norway, Sweden, and other countries was 8GW, which was 1.6 times the installed capacity of wind power in China. Portugal has 5.3GW of clean energy installed capacity and 2.8GW of connected capacity with neighboring countries.

In general, clean energy consumption in China is strongly correlated with load scale, power regulation, grid interconnection, and other key factors. The existing clean energy consumption conditions in the Three Norths regions are insufficient, and the consumption space in the eastern and central regions is not fully utilized through grid interconnection, which are the main reasons for wind and solar power curtailment. In addition, China is building a national power market, and market mechanisms such as peak shaving compensation and price response have not been established. Plans are set by local governments, which give priority to local plants and are less willing to accept clean energy from outside. Provincial barriers are serious, which objectively aggravates the difficulty of clean energy consumption.

8.3.3 Reform Strategy of Clean Energy Consumption Patterns

The consumption and utilization of clean energy is a systematic problem involving power supply, power grid, and power load. Figure 8.10 gives the reform strategies of clean energy consumption patterns from the perspective of power, gird, and consumer.

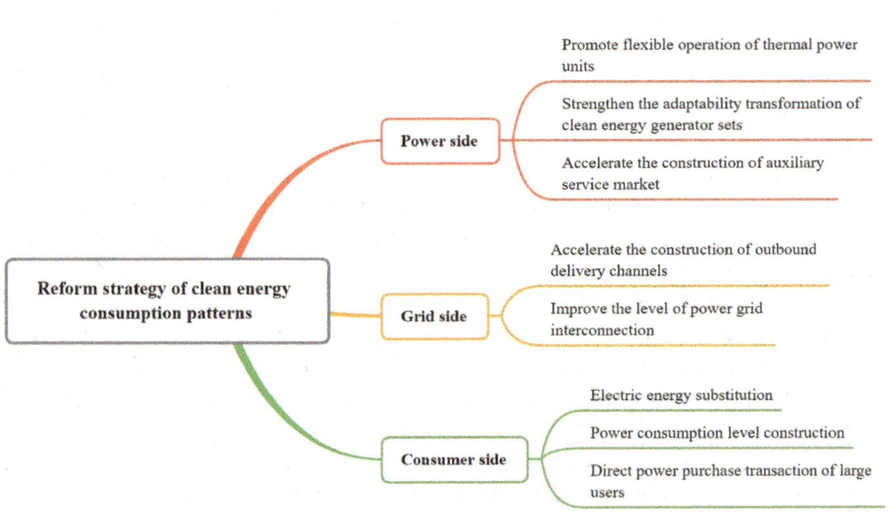

Fig. 8.10 Reform strategy of clean energy consumption patterns

8.3.3.1 Power Side

China will be in the era of mixed energy for a long time. The development and utilization of renewable and clean energy cannot be separated from the development and utilization of traditional fossil energy. To promote the consumption of clean energy, it must coordinate the relationship between clean energy and traditional energy. Traditional energy needs to gradually adapt to the development trend of energy structure. Only the two complement each other can achieve an energy environment of a lower carbon economy (Ramanathan, 2005). To fundamentally improve the consumption level of clean energy and make high-quality and efficient use of clean energy, it is necessary to tap the peak shaving potential and improve the power regulation capacity.

Promote Flexible Operation of Thermal Power Units

The thermal power units of northeast and northwest power grids are flexibly transformed to ensure that the regional power supply has practical peak shaving characteristics. At the same time, temperature monitoring also needs to be strengthened. The load rate of the thermal power unit is controlled in the whole process during the heating period. At the end of the heat supply, the thermal power unit shall be shut down in advance to release the load space to accommodate clean energy for power generation. The flexibility of thermal power unit dispatching is thus improved (Ren et al., 2009).

Strengthen the Adaptability Transformation of Clean Energy Generator Sets

The technical transformation of the wind farm is reflected in its adaptability to low voltage and high voltage, the frequency modulation performance of wind turbines, and the ability to improve power quality.

Accelerate the Construction of Auxiliary Service Market

The construction of peak shaving capacity of thermal power units is inseparable from technical transformation, which is inseparable from capital investment and technical reserve. Therefore, it is necessary to speed up the construction of the auxiliary service market. When thermal power plants can deeply adjust peak load to provide space for wind power accommodation, wind power enterprises need to make economic compensation to thermal power enterprises that make load concessions. When the peak shaving capacity of the thermal power plant is insufficient and it is difficult or unwilling to make an adaptive adjustment, some funds should be allocated to compensate those units that actively participate in the transformation (Huang et al., 2019).

8.3.3.2 Grid Side

Accelerate the Construction of Outbound Delivery Channels

Further speed up the construction of water and electricity delivery channels. Build an inter-provincial and inter-connected platform for optimal allocation of power resources. Establish a stronger power transmission terminal from the West to the East to solve the problems of clean energy collection and distribution. In the reform of power transmission and distribution price, include the "East–West Assistance" investment of power grid enterprises in the power transmission and distribution price. Adjust the proportion of capital in the central government budget based on actual conditions.

Improve the Level of Power Grid Interconnection

Strengthen the construction of the AC power grid at the transmitting and receiving ends to ensure that cross-regional DC can operate at full power. At the same time, pay attention to the application of computer network technology. Create a market environment for power information exchange. Research and develop a dispatching auxiliary decision-making system based on big data technology. Realize the automatic push of fault information, emergency plan, history record, and other information. Shorten the dispatcher's research and judgment time and disposal time for faults. Improve the integrity of power grid operation.

8.3.3.3 Consumer Side

Electric Energy Substitution

Terminal energy substitution refers to the mutual substitution of coal, oil, natural gas, electricity, and other energy used at the terminal. Replace scarce, inefficient, high-carbon, and high-risk energy with abundant, efficient, low-carbon, safe and convenient energy. Finally, realize the adjustment of energy consumption structure. In the future, with the continuous growth of China's total energy consumption, the electrification level in different fields of energy consumption such as industry, transportation (Dominković et al., 2018), commerce, and residents will be improved. This is of strategic significance for optimizing China's energy consumption structure and realizing sustainable energy development.

Whether from the perspective of development trends, practical requirements, or technical conditions, electric energy substitution is an important trend of terminal energy substitution. Adhere to reform and innovation. Improve market-oriented electricity trading. Accelerate the research and development of key equipment replaced by electric energy. For qualified power substitution projects, financial subsidies shall be given using rewards and subsidies. Guide enterprises to implement alternative transformation.

Through the special transmission and distribution price policy and the price guidance mechanism such as exemption of high-reliability cost, promote the substitution of electric energy. Strict access to environmental protection and energy efficiency standards. Strengthen the supervision and inspection of emissions from coal-fired boilers, kilns, port ship fuel oil, and so on. Accelerate the planning and construction of supporting facilities for electric vehicles. The power grid enterprises shall increase the investment and construction of supporting power grids and reduce the investment of enterprises in the implementation of electric energy substitution. Encourage financial institutions and social capital to participate in the operation of power substitution projects. Encourage power generation enterprises to sign strategic cooperation agreements with power substitution enterprises. Foster a market in which electricity can replace electricity.

Power Consumption Level Construction
Establish and improve load side response measures based on price incentives. Further, optimize and promote the peak valley electricity price mechanism on the generation side and user side. Establish a flexible electricity price mechanism on the selling side. Motivate users to participate. Formulate and refine subsidy standards for participants. Set up special funds. Explore the time-based dynamic price mechanism and try out electricity price strategies such as time of use electricity price, key peak load electricity price, and real-time electricity price. Establish an information system, integrate system operation, market transaction, and user power consumption data, and improve load side big data analysis ability. Improve the intelligence and automation level of equipment. Actively promote the demonstration, application, and promotion of high-capacity and distributed energy storage technology (Kittner et al., 2017).

Direct Power Purchase Transaction of Large Users
The direct purchase of electricity by large users of clean energy is to break the traditional mode of power grid companies buying electricity from power generation enterprises and then selling electricity to power customers. Instead, power generation enterprises and power customers conduct direct transactions using negotiation, listing, principal-agent, and so on. Establish a bidding mechanism for hydropower. By the principles of fairness, positivity, and prudence, gradually reduce the power access threshold. Expand the scope of direct trading partners and power scale. Attract more power buyers and sellers to participate in direct transactions and realize the optimal allocation of power resources.

Actively carry out large user transactions across provinces and regions. Improve the price formation mechanism of trans provincial transmission and distribution. Establish a trans provincial power market system. In this way, clean energy enterprises reduce electricity prices and further make small profits but quick turnover. Electricity customers also reduce electricity costs to promote production. In turn, it provides a wider range of load space for the local consumption of clean energy. Therefore, promoting the direct power purchase transaction of large users of clean energy is a more powerful load-side management measure (Huang & He, 2013).

8.4 Economic Analysis of Clean Energy Transformation

For the development of the low-carbon economy, clean energy occupies a core position. The effective development and utilization of clean energy can provide a fundamental guarantee for the development of the low-carbon economy. The existing energy structure is bound to change. By accelerating the research and development of clean energy, the modern energy industry can continue to develop in the direction of safety, low-carbon, economy, cleanliness, efficiency, and sustainability.

8.4.1 Mechanisms for Developing Clean Energy to Affect Economic Growth

Nowadays, the purpose of developing clean energy in various countries is generally clear. First, clean energy plays an irreplaceable role in controlling greenhouse gas emissions. Second, the whole world is facing the dilemma of energy shortage. Countries aim to ensure future energy security by developing clean and renewable energy. However, the third possible important role of clean energy, that is, its impact on economic growth, has not been paid enough attention (Bhattacharya et al., 2016). Although most policymakers believe that the development of clean energy and job creation can go hand in hand, some people believe that the possibility of large-scale job creation will be decentralized or even produce negative employment effects due to distribution effect and budget effect. Therefore, it is worth exploring whether promoting the investment and consumption of clean energy can stimulate economic growth.

8.4.1.1 Price and Cost

Output Effect

Generally, when discussing economic issues, the two main factors are labor and capital. For clean energy, its higher production cost relative to traditional energy has become the focus, causing various output effects. The most important one may be the reduction of output or the increase of demand for other alternative factors of production due to its high cost. This may lead to a reduction in the demand for productive labor factors overall. In addition, the increase of overall social production costs caused by the increase of energy costs will lead to the loss of a country's advantage in international competition, which will produce the pressure to reduce real wages in the future. If the market mechanism can reduce the real wage in the labor market, then full employment will reach a new equilibrium. However, if the real wage cannot be reduced, it may cause widespread unemployment.

Compensation Effect

Another issue that needs to be considered is what economic sector will bear the macroeconomic impact caused by changes in energy prices and costs, that is, the compensation effect. The impact on the economy if it is energy-intensive sectors, monopolized industries, private households that would primarily bear the high energy costs is respectively analyzed as follows:

For the energy-intensive sectors, the potential negative economic impact would be even greater. It is because these industries generally participate in international competition, the high cost makes them lose competitiveness in the world.

For the industries with monopoly profits, the negative economic impact may be smaller. It is because these industries are not effectively regulated and do not participate in market competition. They would be forced to adjust and optimize the industrial structure to improve efficiency.

For private households, the impact can be strong or weak, which depends on the price elasticity of household demand for energy consumer goods. If their energy expenditure increases, then households may change the energy consumption mode, or reduce the energy consumption level, or use other alternative products.

For the government, if its public budget is affected, then the government must reduce other expenditures or increase other tax revenue. This would further reduce the income of production departments or consumers. In this way, the impact of the public budget would spread to the private budget, which may lead to a reduction in private investment and consumption.

To sum up, the compensation effect generally occurs within the framework of structural adjustment. When the energy cost increases, the possible negative consumption impact must be considered when analyzing the change of structural demand.

The impact of changes in clean energy prices and costs on the economy depends on whether the government's support plan is effective and in place and the economy's adjustment mechanism. This mechanism itself depends on the relative supply and demand elasticity of different economic sectors. Therefore, it is difficult to conclude whether clean energy promotes or hinders economic growth only from the perspective of price and cost.

8.4.1.2 Structural Demand

In addition to the price and cost, the investment and consumption of clean energy also have an impact on the structural changes of the economy. The implementation of clean energy policies usually requires additional investment to assist and support. Take biomass energy and biofuels as an example. They not only require investment in technology and equipment but also increase the demand for forest resources and agricultural products, which is a direct and positive impact. However, at the same time, the supply of traditional energy and the investment demand for traditional energy will also be reduced, which is a direct negative impact (Sbia et al., 2014).

The investment cost of clean energy is much higher than that of traditional energy. This high cost is bound to eventually be passed on to consumers, affecting the private budget. Since the production and use of clean energy need the support of the output of many other industries, these direct positive or negative economic impacts mentioned above will form indirect economic impacts through the connection between industries, and ultimately affect the structure of the whole economy. Many of these impacts are related to technological progress or policy support. Therefore, it must distinguish the different manifestations of these economic impacts in different economic fields, such as investment, operation, and maintenance, household consumption, and trade.

Structural Investment

In some countries, the government's promotion of clean energy plans has affected the behavior of many industries directly related to clean energy production and consumption, including government planning, finance, construction, and manufacturing sectors. The birth of relevant production technologies or services based on clean energy is a direct investment promotion effect. However, some industries not directly related to clean energy will also be affected, such as clean energy suppliers, including steel manufacturing industry, transportation industry, service industry, and so on. The increased output of these suppliers increases income and provides more employment opportunities, thus forming an indirect investment promotion effect (Lee, 2013).

In addition, the impact of this investment promotion effect on the fossil energy production sector should also be considered. Since traditional energy will gradually be replaced by clean energy, investment in these sectors will be reduced. The income and employment rate of the traditional energy production sector, the technology sector, service sector, and supply sector will decline, forming a negative investment effect.

To sum up, the impact of clean energy investment and consumption on the overall economy depends partly on the competition between various direct and indirect positive and negative effects of structural investment.

Structural Operation and Maintenance

Investment activities are over after the installation of clean energy generation equipment. The rapid radiation effect driven by investment will slowly subside. However, the subsequent operation and maintenance of these clean energy power generation equipment need to be completed by certain employees. It will lead to the continuous increase of income in these industries, which is a positive economic driving force. There are indirect impacts, too. The income and employment rate of the supporting industries of upstream and downstream products or services related to operation and maintenance will increase. Similarly, there are negative direct and indirect impacts on the fossil energy industry. The operation and maintenance activities required by the fossil energy industry will be reduced due to the development of clean energy. The demand for upstream and downstream products and services related to these activities will also decrease, resulting in a continuous reduction in their income and employment opportunities.

Structural Trade

Trade-related to fossil energy will gradually decrease, while trade-related to clean energy will increase (Gómez et al., 2018). For most countries, fossil energy needs to be imported. Therefore, these countries will naturally tend to reduce the import of fossil energy. The import and export of technical products or intermediate products related to clean energy will increase. For energy importing countries, increasing the supply of domestic clean energy can reduce their demand for importing energy. But it will also lead to a reduction in the income of other energy-exporting countries. For clean energy-exporting countries, vigorously investing in clean energy, and developing clean energy technologies will increase their output and trade income.

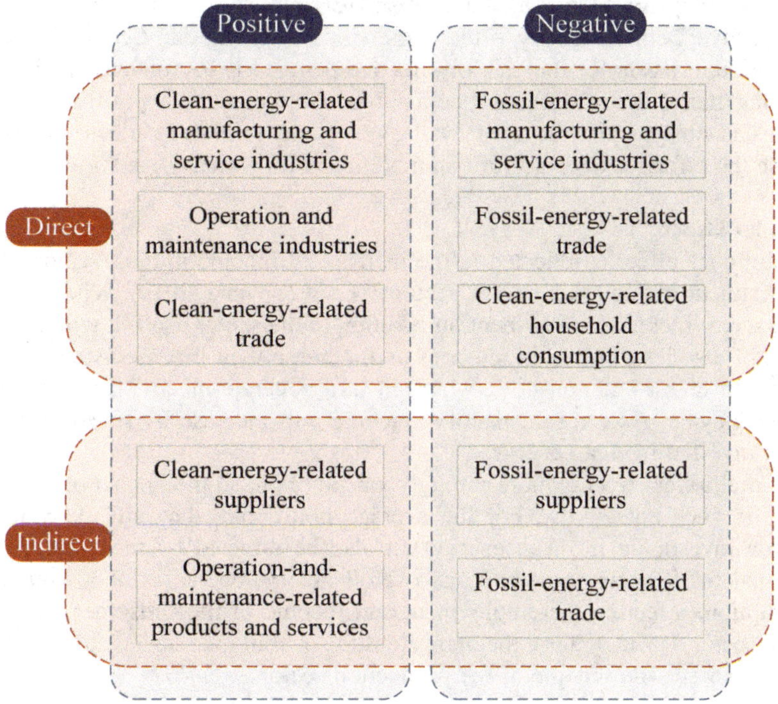

Fig. 8.11 Industries impacted by structural demand and forms of impacts

Structural Household Consumption

Raising the price of clean energy may lead to additional income growth in some industries. However, as a part of the final energy consumers, it is family consumers who bear the high price, which causes the economic burden of families and reduces the real income. Household consumers are spending more on clean energy and less on other aspects. This can also be seen as the impact of price changes on household consumption budget. Household demand for other products will decline, which may affect other industries, especially the consumer goods industry.

However, if the additional investment, operation, and maintenance activities of clean energy can promote employment, reduce unemployment, or increase the income of employees in relevant industries, it may also increase the overall household income. The negative impact of rising clean energy prices can be compensated, and household consumption in other aspects can remain unchanged. Nevertheless, this impact is more likely to be negative when the cost price of clean energy does not fall to a certain extent. Figure 8.11 shows some typical industries impacted by structural demand and how they are impacted.

8.4.1.3 Multiplier Effect and Acceleration Effect

The Effective Demand theory is the cornerstone of Keynesian Economics, which holds that the reason for the lack of effective demand is insufficient employment. Assuming that Keynes' effective demand is insufficient, the positive economic growth and employment brought by the policy of developing clean energy will increase the effective demand for commodities, with which come two effects.

Multiplier Effect

The multiplier effect is reflected in the increase of income in most sectors due to the investment in one clean energy sector (Cheng & Wang, 2019). When the clean energy sector increases investment, in addition to the sector itself, it will also drive the rise of a series of related industries and increase their income. After that, they continue to invest and consume, which in turn generate income in other sectors, creating a cycle. As a result, national income will increase by several times the amount invested in clean energy.

The multiplier effect is more through consumption, so the impact of household income is very noteworthy. For the average household, they are likely to benefit from investment in clean energy but also be affected by the decline of the traditional fossil energy sector. If the overall income decreases, they will reduce consumption, affecting the employment and income of the consumer goods production sector. It will achieve the opposite if the overall income increases, boosting the employment and income of the production sector.

Acceleration Effect

With the increase of national income, the demand for many products, including consumer goods and investment goods, will increase. The original social production capacity cannot keep up with the new demand, so various departments, including the clean energy department, will invest in production to meet these needs. This is how the acceleration effect occurs.

The acceleration effect is more reflected at the industrial level. When the demand and income of the clean energy industry increase, the product demand of its associated downstream industries will also increase, and then additional investment will be made for production. Through the vertical conduction of the industrial structure, an accelerating effect is formed.

Multiplier effect and acceleration effect interact to jointly promote sustained economic growth. Whether these two effects work depends on economic conditions and on whether the responses of various economic and social actors fit the basic assumptions of Keynesian Economics. An important assumption in applying this to clean energy is that policy-driven demand for clean energy does not crowd out aggregate demand in other sectors. So, some constraints must be considered. The influence of Keynesian demand policy on rational expectations, the internationalization of commodity and financial market is far more complex than the mechanism described above. Therefore, the successful application of Keynesian demand policy in clean energy investment and consumption needs the verification of the empirical model.

8.4.1.4 Technology Output Effect

In most cases, technological change is linked to prior investment. The new system integrates technological change and brings about the modernization of the capital stock. A country's national economy grows with the increase and renewal of capital stock. In this process, the spillover effect of technology must be considered. Here, the technology spillover effect of clean energy is discussed. Whether clean energy itself has an output effect, that is, whether it can increase the potential material output, is crucial. How the technology output effect works can be presented in Fig. 8.12.

In the case of fixed total investment in clean energy, two scenarios can be assumed:

(a) Assume that clean energy technologies show no output effects. Since investments have no output effect on their own, if the overall investment is fixed, then investment in clean energy producers will be used to improve technologies, and investment in production will be crowded out. In other words, due to the crowding-out effect of technology on investment, the output of the national economy will decline.

(b) Assume that clean energy technology has an output effect, which will increase the output of material products. The crowding-out effect of technology on investment will be alleviated.

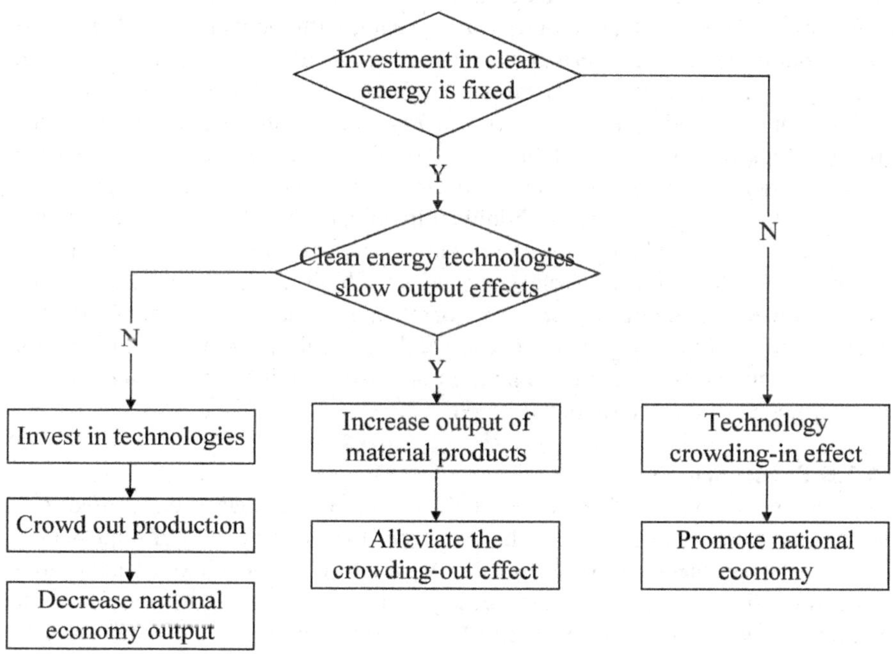

Fig. 8.12 Mechanism of clean energy technology output effect

In the case of increasing investment in clean energy, possibly due to the transfer from consumption tendency to investment tendency, or due to the increase of GDP, there is no investment crowding-out effect, but the technology crowding-in effect. A more favorable environment is provided for technology development, which in turn will have a positive effect on the national economy.

Whether the technological changes resulting from clean energy policies can work depends on the output effects of environmentally friendly technologies under the above two cases. The assumption that there is no output effect of investment is valid for end solutions, was also incorporated into production systems and dominated environmental protection research in the 1970s and 1980s. But, on the other hand, through the research in recent years, as the investment in environmental protection becomes more and more important, it also gradually shows a stronger output effect. It must be made clear, however, that the magnitude of this effect must depend on technological developments. The surge in output is due to a few green technologies, not all of them.

8.4.1.5 First Mover Advantage

The success of a country's foreign trade depends not only on price competitiveness but also on quality competitiveness. For knowledge-intensive products, including clean energy technologies, to obtain a high market share in the world, they must rely on the country's technological innovation ability, learning curve effect, and early market share of products (Pável & Rajagopal, 2017). If a country formulates the strategy of developing clean energy early on, it is bound to focus on the development of relevant technologies earlier. Later, with the increasing demand for these technologies in the international market, those that have mastered the core technologies and realized specialization earlier will gain a leading position.

Many countries today face increasing energy demand on the one hand and pressure to reduce the use of fossil fuels on the other. So, these countries began to follow the footsteps of the pioneers, and vigorously develop clean energy. Countries with a first-mover advantage should build competitive clusters of clean energy industries early and make it difficult for such clusters to move to low-cost production countries. These industrial clusters must have high technological development capacity, open innovation system, horizontal and vertical integrated production structure, etc. In this way, it can not only firmly control the domestic market without being occupied by foreign competitors but also benefit from the huge demand for its products in the international market.

8.4.1.6 Summary

From the above analysis, it can be concluded that clean energy does affect economic growth and the impact mechanisms are diverse. But the consequences of these effects are uncertain. It depends on a variety of socioeconomic conditions, on the strength of the positive and negative effects in various situations, and the comprehensive performance of the results of various influencing mechanisms.

8.4.2 Ways to Promote a Low-Carbon Economy Based on Clean Energy

8.4.2.1 Subsidy Policy

The effective use of financial and tax tools is conducive to reducing the cost of capital use of enterprises and forming a market-oriented incentive mechanism for enterprises and other subjects to participate in the clean energy industry. In terms of financial tools, financial funds should be used to support the Research and Development (RD) and innovation of the clean energy industry and the cultivation of the end market. For example, support for technological innovation and consumer markets can be strengthened in the form of discount interest on corporate loans and consumer subsidies. In terms of tax tools, on the one hand, China's clean energy-related tax structure should be further enriched, especially to reduce the tax burden on small and medium-sized clean energy enterprises for technological innovation. On the other hand, tax incentives for clean energy enterprises should be enhanced. The function of tax tools in cultivating consumer preferences should be given full play through tax incentives (Li et al., 2017).

8.4.2.2 Technological Innovation

On the one hand, through the establishment of RD centers and cooperation in RD projects, the degree of international participation can be enriched and the progress level of clean energy technology in China can be promoted. On the other hand, China should also establish a clean energy collaborative innovation system with enterprises as the core and the participation of universities and scientific research institutions (Lin & Zhu, 2019).

8.4.2.3 Market Cultivation

Nowadays, the markets of developed countries are relatively saturated, the Chinese market is backward and the development of the national market lags seriously. It is necessary to promote the development of the clean energy industry through market cultivation. In terms of Chinese market cultivation, the government can expand the consumption market of clean energy products by purchasing services and promoting power system reforms such as distributed energy application and free or subsidized access to the grid. In terms of international market development, Chinese enterprises should be encouraged to "go global". For example, provide customized products and services in the form of free assistance and application guidance for clean energy products to African, Latin American, and other countries.

8.4.2.4 Financial Support

The long investment cycle, high technical risk, and uncertain income of the clean energy industry determine the strategic value of financial support to promote the rational and healthy development of the clean energy industry. At present, China's financial support system for clean energy enterprises has not been effectively established. First, develop clean energy industry investment funds. Enrich angel investment, venture capital, and other investment forms. Subsequently, encourage the development of non-bank financial industries such as financial leasing and financial guarantee, forming a beneficial supplement to the financial system dominated by commercial banks. Finally, based on the rapid development of the internet financial industry, on the premise of strengthening supervision, use forms such as equity-based crowdfunding to support the clean energy industry.

8.4.2.5 Cross Border Mergers and Acquisitions

Cross-border Mergers and Acquisitions (MA) is an effective way for Chinese enterprises to obtain key technologies and enter the international market with the help of foreign enterprises. Take brands and technologies of high-tech clean energy enterprises such as the United States, Japan, and the European Union as the MA objects through diversified forms such as holding and equity participation, the technology generation gap between Chinese enterprises and developed countries can be narrowed.

8.5 Conclusions

Clean energy occupies the core position of low-carbon economic development. The development of the clean energy industry is the basis for China to take the road of sustainable energy development and achieve stable economic growth.

(a) Section 8.1 first introduces the status of the world energy balances from three aspects: production, supply, and power generation. This section concludes that clean energy such as nuclear energy and other renewables are developing strongly, and it is inevitable to shake and replace the status of traditional fossil fuels. The effective practices of developed countries in promoting clean energy transformation and development, such as establishing the concept, cultivating power market, and opening overseas markets, are summarized.

(b) Section 8.2 analyzes the factors that influence energy efficiency, such as technical and economic factors and the mechanisms by which they exert their influence. It is pointed out that technological progress is the decisive factor, and energy input, output level, and pollution emission are the direct factors. In addition, a comparative analysis of energy efficiency is carried out from two aspects of economy and technology. In terms of energy-economic efficiency, although China's energy intensity is approaching the world average, it is still much higher than that of developed countries. In terms of energy technical efficiency, the overall efficiency of the energy system is still low, and a large amount of energy is seriously wasted in the process of mining, conversion, storage, transportation, and terminal utilization.

(c) Section 8.3 analyzes the main difficulties faced by China's clean energy consumption. Mechanism analysis is done on time-varying power system balance regulation with clean energy participation. Qualitative analysis is done on wind and solar curtailment, clean energy capacity, and permeability in typical regions of China. From the perspectives of load scale, power regulation, and grid interconnection, the root causes of the difficulty of clean energy consumption are pointed out. Finally, the reform strategies of clean energy consumption modes for power supply, power grid, and consumers are put forward.

(d) Section 8.4 discusses mechanisms for developing clean energy to promote a low-carbon economy in the new era. Clean energy affects economic development in a variety of ways, depending on various socio-economic conditions and on the combined forces of positive and negative effects in various situations. It is considered that the development of China's clean energy industry should adhere to the strategies of fiscal and tax subsidies, technological innovation, market cultivation, financial support, cross-border MA.

References

Bhattacharya, M., Paramati, S. R., Ozturk, I., & Bhattacharya, S. (2016). The effect of renewable energy consumption on economic growth: Evidence from top 38 countries. *Applied Energy, 162*, 733–741.

Castagneto-Gissey, Giorgio. (2014). How competitive are EU electricity markets? An assessment of ETS Phase II. *Energy Policy 73*, 278–297.

Chang, T., Fang, W., & Wen, L.-F. (2001). Energy consumption, employment, output, and temporal causality: Evidence from Taiwan based on cointegration and error-correction modelling techniques. *Applied Economics, 33*(8), 1045–1056.

Cheng, H., & Wang, B. (2019). Multiplier effect of science and technology innovation in regional economic development: based on panel data of coastal cities. *Journal of Coastal Research, 94*, 883–890. https://doi.org/10.2112/si94-175.1

Cornillie, J., & Fankhauser, S. (2004). The energy intensity of transition countries. *Energy Economics, 26*(3), 283–295.

Diaconescu, I., Patrascu, R., Tutica, D., Ionescu, C., Minciuc, E. (2019). Influence of technical and economic factors in the assessment of energy efficiency projects in industry. In *9th International Conference on Energy and Environment (CIEM)* (pp. 359–363). Politehnica Univ Timisoara, Timisoara, Romania.

Dominković, D. F., Bačeković, I., Pedersen, A. S., & Krajačić, G. (2018). The future of transportation in sustainable energy systems: Opportunities and barriers in a clean energy transition. *Renewable Sustainable Energy Reviews, 82*, 1823–1838.

Fang, B., Wang, Q., Peec-Org, C. (2010). Low-carbon economy and development on biomass energy in China. In *Power and Energy Engineering Conference* (pp. 263–+). Wuhan, Peoples R China.

Fouquet, R. (2012). The demand for environmental quality in driving transitions to low-polluting energy sources. *Energy Policy, 50*, 138–149.

Gómez, M., Ciarreta, A., & Zarraga, A. (2018). Consumo de energía, crecimiento económico y comercio: Un análisis de causalidad para México (Energy consumption, economic growth and trade: A causality analysis for Mexico). *EconoQuantum, 15*(1), 53–72. https://doi.org/10.18381/eq.v15i1.7112

Han, S., Lin, C. Q., Zhang, D. S., & Farnoosh, A. (2019). Projections and recommendations for energy structure and industrial structure development in China through 2030: A system dynamics model. *Sustainability, 11*(18), 4901. https://doi.org/10.3390/su11184901

Hang, L., & Tu, M. (2007). The impacts of energy prices on energy intensity: Evidence from China. *Energy Policy, 35*(5), 2978–2988.

He, T. L. (2011). On the main task of china's industrial structure adjustment under the goal of low carbon economy. *In International Symposium on Technical Innovation of Industrial Transformation and Structural Adjustment* (pp. 89–93). Yantai City, Peoples R China.

Hewitt, R. J., Winder, N. P., Jiménez, V. H., Alonso, P. M., & Bermejo, L. R. (2017). Innovation, pathways and barriers in Spain and beyond: An integrative research approach to the clean energy transition in Europe. *Energy Research & Social Science, 34*, 260–271.

Hossain, M. K., Kamil, A. A., Baten, M. A., & Mustafa, A. (2012). Stochastic frontier approach and data envelopment analysis to total factor productivity and efficiency measurement of Bangladeshi rice. *PLoS ONE, 7*(10), e46081.

Huang, H. Y., Jiang, B. F., Xiong, H. Q., Xia, S. L., Cheng, Y. J., Wang, C. M. (2019). Research on trading mechanism and safety checking mechanism of trans-provincial peak-shaving auxiliary service. In *3rd IEEE International Electrical and Energy Conference (CIEEC)* (pp. 264–269). Beijing, Peoples R China. https://doi.org/10.1109/cieec47146.2019.cieec-2019134.

Huang, Y., & He, Z. (2013). Main experience of wind power consumption at abroad and its reference for China. *Shanxi Electric Power, 41*(10), 19–24.

Jacobsson, S., & Lauber, V. (2006). The politics and policy of energy system transformation—explaining the german diffusion of renewable energy technology. *Energy Policy, 34*(3), 256–276.

Jiang, J. H., Qiao, Y., Lu, Z. X., Wang, N. B., Li, J., Ding, K., China Electrotechn Soc. (2017). Research on peak load regulation capacity planning for long-term market. In *China International Electrical and Energy Conference (CIEEC)* (pp. 583–588). Beijing, Peoples R China.

Kittner, N., Lill, F., & Kammen, D. M. (2017). Energy storage deployment and innovation for the clean energy transition. *Nature Energy, 2*(9), 1–6.

Lanoie, P., Patry, M., & Lajeunesse, R. (2008). Environmental regulation and productivity: Testing the porter hypothesis. *Journal of Productivity Analysis, 30*(2), 121–128.

Lee, J. W. (2013). The contribution of foreign direct investment to clean energy use, carbon emissions and economic growth. *Energy Policy, 55*, 483–489. https://doi.org/10.1016/j.enpol.2012.12.039

Li, H., Bao, Q., Ren, X. S., Xie, Y. T., Ren, J. Z., & Yang, Y. K. (2017). Reducing rebound effect through fossil subsidies reform: A comprehensive evaluation in China. *Journal of Cleaner Production, 141*, 305–314. https://doi.org/10.1016/j.jclepro.2016.09.108

Li, K. M., Fang, L. T., & He, L. R. (2018). How urbanization affects China's energy efficiency: A spatial econometric analysis. *Journal of Cleaner Production, 200*, 1130–1141. https://doi.org/10.1016/j.jclepro.2018.07.234

Li, K., & Lin, B. (2018). How to promote energy efficiency through technological progress in China? *Energy, 143*, 812–821. https://doi.org/10.1016/j.energy.2017.11.047

Li, M., Jin, T., Liu, S., & Zhou, S. (2021). The cost of clean energy transition in rural China: Evidence based on marginal treatment effects. *Energy Economics, 1*, 105167.

Lin, B., & Zhu, J. (2019). The role of renewable energy technological innovation on climate change: Empirical evidence from China. *The Science of the Total Environment, 659*(1), 1505–1512.

Lu, X., Pan, J., & Ying, C. (2010). Sustaining economic growth in china under energy and climate security constraints. *China & World Economy, 14*(6), 85–97.

Luis, E. J., Rivera, M., Wheeler, P. (2019). Wind energy development and technology in the world: A brief overview. In *IEEE CHILEAN Conference on Electrical, Electronics Engineering, Information and Communication Technologies (CHILECON)*, Valparaiso, Chile.

Pável, R.-M., & Rajagopal. (2017). Adoption of renewable energy technologies in Mexico: The role of cognitive factors and innovation attributes. *International Journal of Energy Sector Management, 11*(4), 626–649.

Phylipsen, G., Blok, K., & Worrell, E. (1997). International comparisons of energy efficiency-Methodologies for the manufacturing industry. *Energy Policy, 25*(7), 715–725.

Quitzow, R., Thielges, S., Goldthau, A., Helgenberger, S., Mbungu, G. (2019). Advancing a global transition to clean energy—the role of international cooperation. *Economics-the Open Access Open-Assessment E-Journal 13*. https://doi.org/10.5018/economics-ejournal.ja.2019-48.

Ramanathan, R. (2005). An analysis of energy consumption and carbon dioxide emissions in countries of the Middle East and North Africa. *Energy, 30*(15), 2831–2842.

Ren, J. X., Li, F. Q., Shi, Q. G., Wu, J., Zhu, Q. Z., Pan, W. G., Li, Q. F., Jsme. (2009). Study on the thermal parameters' determination method of ultra supercritical thermal power generating units. In *9th International Conference on Power Engineering (ICOPE-09)* (pp. 455–457). Kobe, Japan.

Sbia, R., Shahbaz, M., & Hamdi, H. (2014). A contribution of foreign direct investment, clean energy, trade openness, carbon emissions and economic growth to energy demand in UAE. *Economic Modelling, 36*, 191–197. https://doi.org/10.1016/j.econmod.2013.09.047

Sebri, M., & Ben-Salha, O. (2014). On the causal dynamics between economic growth, renewable energy consumption, CO2 emissions and trade openness: Fresh evidence from BRICS countries. *Renewable and Sustainable Energy Reviews, 39*, 14–23.

Sharma, S. C., Sylwester, K., & Margono, H. (2007). Decomposition of total factor productivity growth in U.S. states. *Quarterly Review of Economics & Finance, 47*(2), 215–241.

Shen, X. B., & Lin, B. Q. (2017). Total factor energy efficiency of China's industrial sector: A stochastic frontier analysis. *Sustainability, 9*(4), 646. https://doi.org/10.3390/su9040646

Shu, Y., Zhang, Z., & Guo, J. (2017). Study on key factors and solution of renewable energy accommodation. *Zhongguo Dianji Gongcheng Xuebao/proceedings of the Chinese Society of Electrical Engineering, 37*(1), 1–8.

Song, M., Guo, X., Wu, K., & Wang, G. (2015). Driving effect analysis of energy-consumption carbon emissions in the Yangtze River Delta region. *Journal of Cleaner Production, 103*, 620–628. https://doi.org/10.1016/j.jclepro.2014.05.095

Tao, C. Q., & Li, C. (2018). Impact of environmental regulation on total-factor energy efficiency from the perspective of energy consumption structure. *International Energy Journal, 18*(1), 1–10.

Wu, L. F., Sun, L. W., Qi, P. X., Ren, X. W., & Sun, X. T. (2021). Energy endowment, industrial structure upgrading, and CO2 emissions in China: Revisiting resource curse in the context of carbon emissions. *Resources Policy, 74*, 102329. https://doi.org/10.1016/j.resourpol.2021.102329

Xiong, S. Q., Ma, X. M., & Ji, J. P. (2019). The impact of industrial structure efficiency on provincial industrial energy efficiency in China. *Journal of Cleaner Production, 215*, 952–962. https://doi.org/10.1016/j.jclepro.2019.01.095

Xue, Y., Lei, X., Xue, F., Chen, Y. U., Dong, Z., Wen, F., & Ping, J. U. (2014). A review on impacts of wind power uncertainties on power systems. *Proceedings of the Csee, 34*(29), 5029–5040.

Yan, J., Chou, S. K., Desideri, U., & Lee, D. J. (2015). Transition of clean energy systems and technologies towards a sustainable future (Part II). *Applied Energy, 160*(Dec.15), 619–622.

Yang, M., Yang, F.-X., & Chen, X.-P. (2011). Effects of substituting energy with capital on China's aggregated energy and environmental efficiency. *Energy Policy, 39*(10), 6065–6072.

Zhang, T. (2009). Frame work of data envelopment analysis—a model to evaluate the environmental efficiency of China's industrial sectors. *Biomedical and Environmental Sciences, 22*(1), 8–13.

Zheng, Y., Ligang, X. U. (2012). The process of developing green energyin developed countries and its lessons. *Resources Science*.

Zhi, C., Sai, D., Kun, Z., Junfeng, Y. F. (2016). Future power grid dispatch and control mode with large-scale clean energy integration in China. In *IEEE PES Asia-Pacific Power and Energy Engineering Conference (APPEEC)* (pp. 1749–1754). Xian, Peoples R China: Asia-Pacific Power and Energy Engineering Conference.

Global Energy Internet Green and Low-Carbon Energy Economic Innovation

9

9.1 Introduction

Fossil energy has dominated the global energy structure for centuries, which caused tremendous harm to the ecological environment. This kind of damage leads to a transformation that new energy owns more and more share of global energy structure. New energy is also called unconventional energy which includes solar energy, geothermal energy, wind energy, biomass energy, and nuclear fusion energy. Compared to conventional energy, the characteristics of new energy are followed. On the one side, new energies have rich resources in the earth, but their distributions are scattered. The most of new energies are renewable energies, so they are sustainable. On the other side, new energies contain low carbon or even no carbon, which means it has little negative effects on the ecological environment.

While the research of new energy technologies has made continuous progress in the twenty-first century, the process of energy structure transformation is accelerated. In the wake of rapid developments in modern industry, energy consumption also increases quickly. According to U.S. Energy Information Administration's International Energy Outlook 2020, world energy consumption will rise nearly 50% by 2050. The prediction of the world's primary energy consumption by energy sources and countries is shown in Fig. 9.1 (EIA, 2020). World energy consumption is growing quickly, and the new energy will replace petroleum and other liquids to become the most used energy in the future. The consumption of fuel energy will still increase, but its growth rate is low compared to the new energy. The consumption of nuclear energy is stable, it may be the potential radioactive contamination. The figure also reveals the energy consumption by region or country and focuses on the developing country. China has the highest energy consumption for a developing country, and the second-highest energy-consuming country is India. It is apparent that developing countries are gradually becoming the major energy consumers.

© Science Press 2022

H. Liu et al., *Big Data in Energy Economics*, Management for Professionals,
https://doi.org/10.1007/978-981-16-8965-9_9

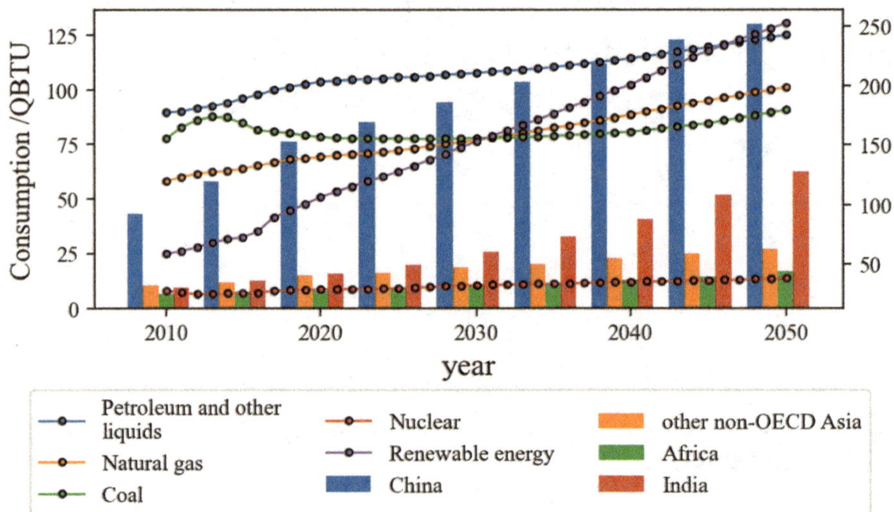

Fig. 9.1 Primary energy consumption

With the complication of energy structure and the increment of energy consumption, the traditional energy system architecture can no longer fit the present situation. Energy Internet (EI) is an advanced energy system architecture constructed by information concept which aims to promote the performance of energy system (Hussain et al., 2019). The information interaction is the key part of the EI. In accordance with function, the information interaction of the EI can be divided into three layers as follows: sensitive layer, network layer, and application layer. The sensitive layer consists of all kinds of sensors and drivers. The sensors are not limited to special sensors such as current sensors or voltage sensors, the vehicle navigation system which can be used for estimating traffic flow is also regarded as the sensor.

Those sensors are meant to supervise the system's status. Drivers take the role of driving devices and systems according to the order passed by the network layer, for example, from generators to any production units. The network layer composed of optical fiber communication net, wideband communication net, and mobile internet takes the responsibility to transmit huge amounts of data. With the development of edge computing (Khan et al., 2019b), the intelligent analysis based on small data set can be directly applied to the local system by the network layer, which alleviates the load of central computations and protects the sensitive data. The application layer is the pivot of analysis and decision. All kinds of computing resources are integrated to process large-scale data on various levels of quality. Meanwhile, with the cooperation of big data technologies and artificial intelligence technologies, high efficiency and smart decision services are provided for the application. The model of EI information interaction is shown in Fig. 9.2.

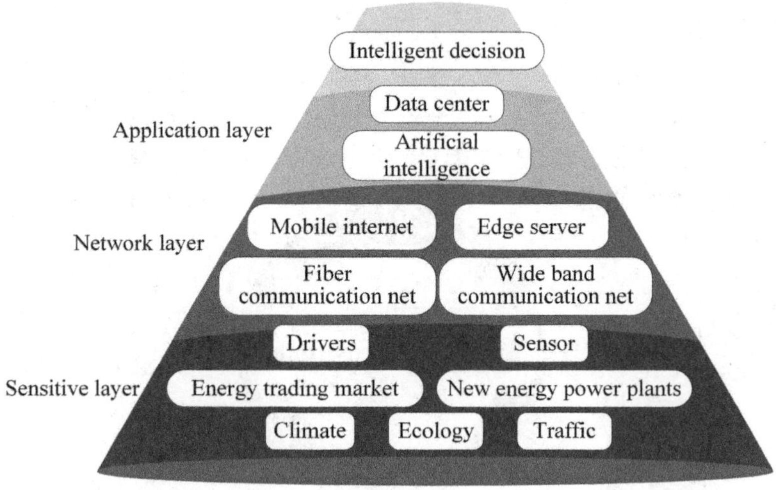

Fig. 9.2 The model of EI information interaction

The high-end communication and information technologies including the Internet of Things (Bedi et al., 2018), artificial intelligence, and big data are applied to realize the interaction between energy system, business system, traffic system, meteorology system, public sentiment system, etc. Besides the information interaction, by developing energy router technology (Kado et al., 2016), electricity to gas technology, other energy conversion technologies, and energy storage technologies, the interaction of different energy resources is also realized in the EI. The DC/AC converters are integrated with the energy router, energy flows coming from different energy sources or outputting for different loads can interexchange in the form of electricity. The model of energy flows interexchange is shown in Fig. 9.3.

According to the size of geographical area, the EI can be divided into City Energy Internet (CEI), National Energy Internet (NEI), and Global Energy Internet (GEI). The larger EI is composed of the smaller EI, their relationship is just like the internet and the local area internet.

Globalization enhances the interaction of technologies between different countries and districts. Those countries and districts where rich energy resources waiting for exploitation may lack corresponding technologies, or their energy supplies exceed energy demands. With the help of globalization, they can access relative technologies to develop their rich energy resources and sell their redundant energy supplies to the global market. More than that, some developing countries' energy structures are not sustainable, with the income of international capital, their energy structures can be modified. It is stated that a higher level of economic globalization can promote the development of renewable energy (Gozgor et al., 2020). Globalization also has influences on the world's carbon emissions, *Khan* et al. researched globalization's impacts on carbon emissions in Pakistan, and the results

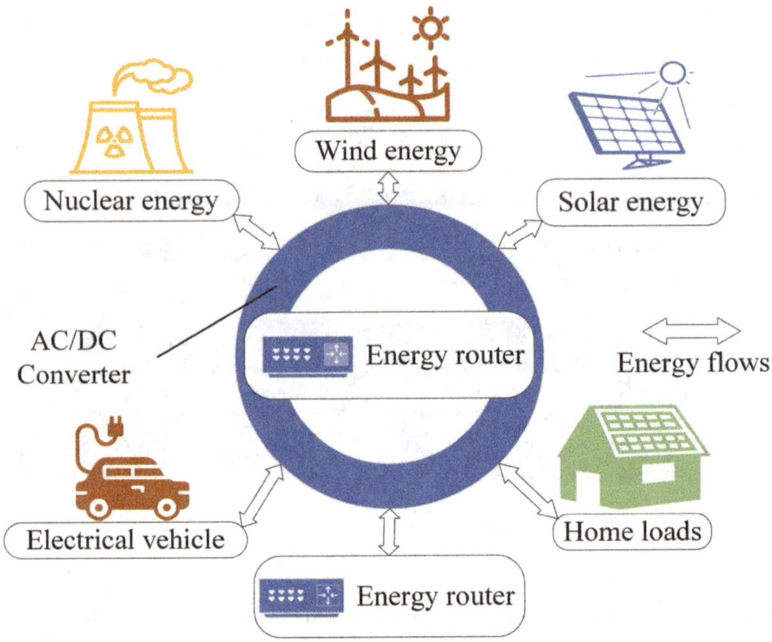

Fig. 9.3 Model of energy flows interexchange

indicated that foreign direct investment has a negative effect on carbon emissions (Khan et al., 2019a). Energy is the basic condition of modern industry and economy. The GEI can be regarded as the embodiment of globalization in the energy field.

9.2 Reform and Innovation of the New Energy System Under the Energy Internet

For the sake of discussion, the detailed definition of the energy system is given as follow: energy system, an integrated system of energy production, energy supply, and marketing, which is formed by coordination and optimization of energy generation, transmission, distribution, conversion, storage, consumption, and other segments. The energy system is the basis of the energy internet.

9.2.1 Comparison of Conventional Energy System and New Energy System

The energy system is decided by the energy structure. According to the BP Statistical Review of World Energy 2021, Fossil energy such as oil, natural gas, and coal still account for three fourth of the world's primary energy consumption (BP, 2021). But compared to the past share of fossil energy, it is obvious that conventional energy structure is transforming to new energy structure.

In the aspect of production in the conventional energy system, the distribution of fossil energy sources is relatively centralized, which means it is convenient to develop them in a centralized way, so large-scale energy exploitation factories are set in source areas. By processing the primary energy, secondary energy including coke, gasoline, diesel oil, electric energy, and other energy are gained. Taking electric energy as an example, most of the world's electric power systems use large units for power generation, and large-scale electric energy production methods with a single unit power exceeding 300 MW mainly include coal-fired thermal power generation, heavy-duty gas turbine, and nuclear power generation. The reason why used large-scale and centralized ways to produce energy instead of small-scale and decentralized way in the past is that small thermal generators have high energy consumption and heavy pollution. As for the new energy system, the production is small-scale and distributed around energy consumers. The production units in the new energy system are normally based on clear energy.

In the conventional energy system, there are few economical energy storage technologies that can contain such huge energy production. Corresponding to this kind of energy production mode, energy transmission is based on high power grids or pipelines. Only large state monopolies or other companies which have the same economic strength and organizational power can support large-scale energy production and transmission. Naturally, these organizations control the supply of energy, and the priority target of the conventional energy system that the balance of production and consumption should be achieved often leads to consumers passively participating in the market.

In the new energy system, small-scale energy production makes the qualification of the market access undemanding, even small private corporations can afford the cost of building a distributed energy factory, which can lead to healthy marketization. To gain competitive power, the energy consumers' demand will be fully considered by those private corporations, in this way, the consumers can participate the market actively.

9.2.2 Production in New Energy System

The new energy system is not completely based on new energy, where conventional energy production still occupies a significant position. It becomes more pluralistic. If conventional energy production serves the upstream industry like the large-scale power plants more, the new energy production serves the downstream industry more. The comparisons of the conventional energy system and new energy system are shown in Fig. 9.4.

A new characteristic of production in the new energy system is distribution. This concept first originated from the situation that emergency generators around energy consumers are started when large power gird fails. Conventional energy

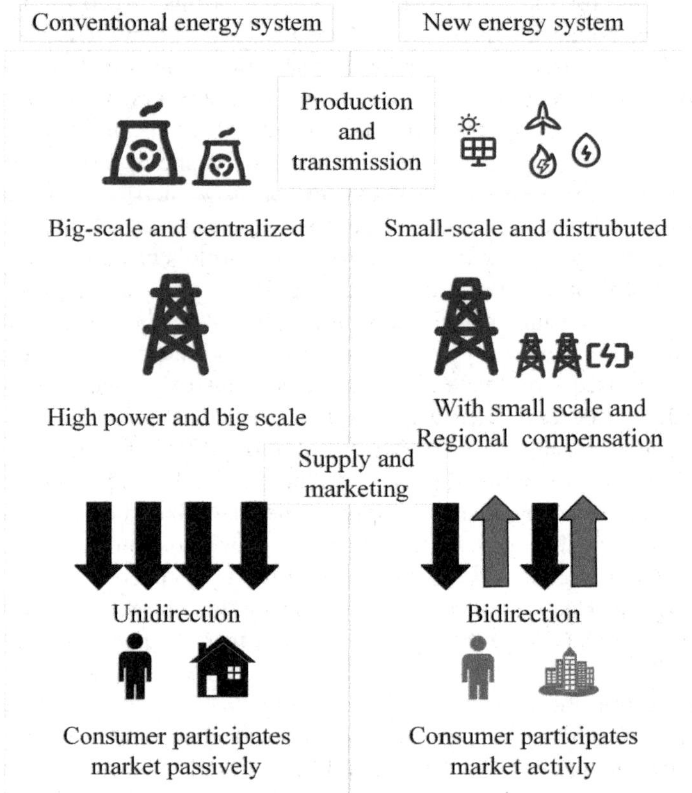

Fig. 9.4 The comparisons of the conventional energy system and new energy system

transmission is based on large-scale power grids, once a fault occurs in the power grid, it will affect the stability of the entire power system. In severe cases, the entire power grid may be paralyzed, leading to a large-scale blackout. Distributed energy is close to the user side, with few uncertainties and high stability, which also decreases the transmission loss (Alanne & Saari, 2006).

With the rapid development of new energy technologies, distributed energy production is not confined to only ensuring the safety of power grids, which also takes the responsibility of improving energy utilization efficiency and modifying the energy structure. The primary energies used in distributed energy production usually are renewable energy and clean energy such as wind energy, solar and gaseous fuel. Every single distributed energy production unit may use different energy sources. Combined Cooling Heating and Power (CCHP) system is a common distributed energy system, which based on the concept of cascade utilization of energy (Wu & Wang, 2006). The CCHP uses natural gas as the primary energy source to generate heat, electricity, and cold. The high-temperature flue gas obtained by burned natural gas is first used for power generation, and the rest heat is used to supply heating in winter. In summer, it drives absorption chillers for cooling. It also can provide domestic hot water, making full use of exhaust heat. The primary energy utilization rate can be increased to about 80%, saving a lot of primary energy, which brings significant economic benefits.

Compared to large-scale centralized energy production, distributed energy production is much more flexible. Not only does it mean that investment needed by distributed energy production is modest and relative projects can be started up immediately, but it also implies that it is convenient to modify the production plan and put it into practice quickly, which ensures the ability of distributed energy production Every single distributed energy production unit do not operate independently, it's production information such as the quantity of electricity and frequency of current and voltage is collected and uploaded to the EI, where the intelligent analysis and dispatch are driven by this information to offer reliable production. The sketch map of distributed energy production is shown in Fig. 9.5.

Intermittency is another characteristic of production in the new energy system, which is caused by the features of energy resources. Wind energy, solar energy, and tidal energy cannot achieve 24 h-production, which brings a challenge to the

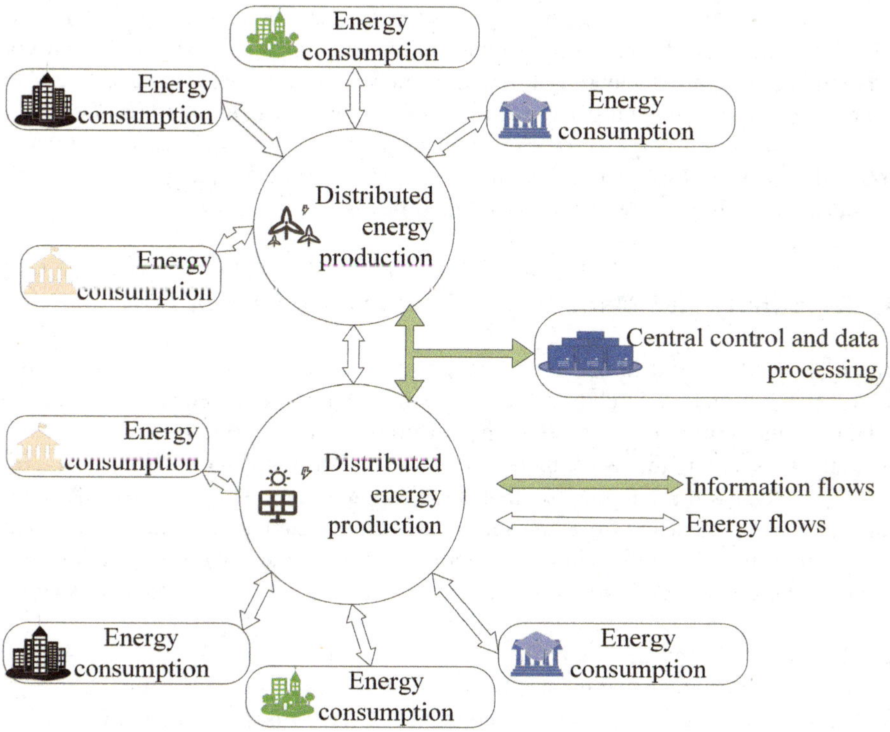

Fig. 9.5 The basic structure of distributed energy

stability of the whole power system. Wind energy is taken as an example. Asynchronous generators are used for wind power generation, which requires absorbing reactive power from the power grids. The parallel capacitor switching devices are needed to compensate for reactive power according to the voltage of access points which is mainly affected by variable wind speed. As the scale of wind power plants continues to expand, the proportion of wind power plants in the energy system continues to increase, and the impact of wind power output instability on the power grid is also increasing. In serious situations, the dynamic stability of the system will be broken, leading to the collapse of the entire energy system.

In the EI, these problems have been well solved. Firstly, the intermittency of wind energy and solar energy is caused by variable weather, and the weather forecasting system can transmit weather information such as wind speed and sunshine intensity to energy production units in real-time by the EI. In the weather forecasting process, big data and artificial intelligence are applied for analyzing the history of wind speed to predict future wind speed (Liu et al., 2018). All this information can be used to estimate the power and frequency information prepared for grid connection.

In addition, new energy storage technology also contributes to the stability of energy. The energy sources are uncertain and the transmission of energy is complex in the EI. Energy storage technology is an important technology that means to effectively alleviate the pressure of renewable energy grid connections. It can not only smooth the power generation output of intermittent renewable energy but also improve the frequency and peak regulation capabilities of generators (Evans et al., 2012). In addition, Energy storage undertakes the role of transition, matching, and optimization of multiple energy sources at various time and spatial scales, which provides more choices for energy reproduction.

9.2.3 Supply and Marketing in New Energy System

Based on the high-end information and communication technology in the EI, the direct and bidirectional relationship between suppliers and demanders is established. By integrating traffic system, meteorology system, public sentiment system, and energy system, the consumers' demand is thoroughly considered and predicted timely by the supplier so that suppliers and producers can modify their supply plans, at the same time, the supply also can guide the consumers' behavior. Energy trading is becoming more and more dynamics and complex, which brings a significant change in forms of supply and marketing in the new energy system.

9.2.3.1 Virtual Power Plant

To transfer distributed energy production into manageable and available energy supply, the concept Virtual Power Plant (VPP) is proposed. The VPP consists of distributed generators, energy storage system, information and communication system. The main idea of the VPP is that distributed energy production and controllable energy consumption are integrated as an aggregate that can participate

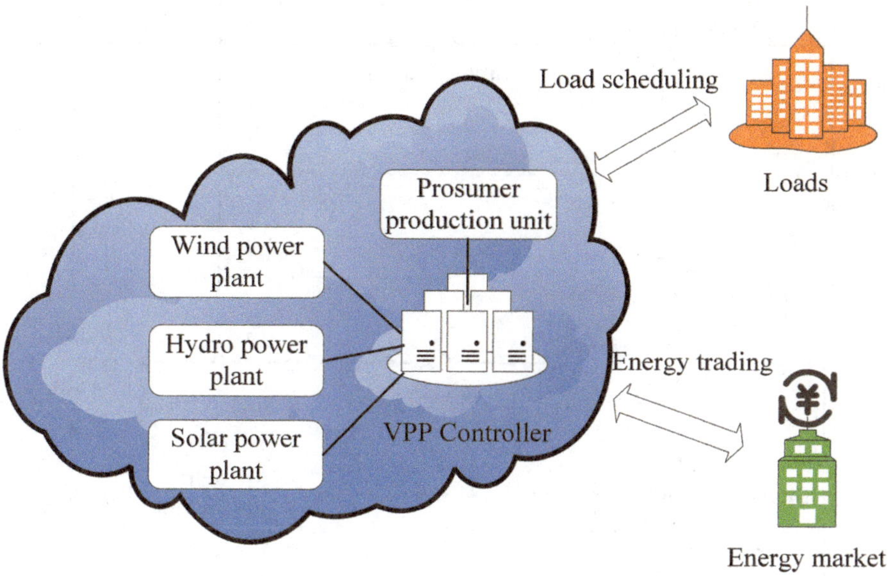

Fig. 9.6 The idea graph of the VPP

in the operation of the energy market, realize real-time energy trading, optimize resource utilization and improve power supply reliability (Zhang et al., 2018). The idea graph of the VPP is shown in Fig. 9.6.

The common trading form is called "VPP auctions", which means the tradeable goods are not the physical power plants but the energy outputs of power plants. In the EI, the VPP auctions are mainly held on the internet, because the sellers and buyers are all around the world, it's the inefficiency to gather them into one location, and the virtuality and openness of the internet can reduce the possibility that sellers or buyers collaborate with each other to monopolize the market (Ausubel & Cramton, 2010).

9.2.3.2 Microgrids

Besides the VPP, many activities of energy trading involve another concept called microgrids. As the name implies, microgrids are scaled-down grids that possess the basic architecture of traditional grids and gain the ability to integrate distributed energy resource systems by a decentralized approach as addition, which reduce the burden of central dispatch and permit every microgrid to operate in an efficient way. When there occurs some emergency, the microgrids can be divorced from the distributed energy system and operate individually. As for microgrids, the local demand and local energy supply are well combined.

The basic structure of microgrids is shown in Fig. 9.7. Feeders can connect to the main grid through Point of Common Coupling (PCC). Feeders A, B, and C are supplied by local distributed energy resources, they can separate from the

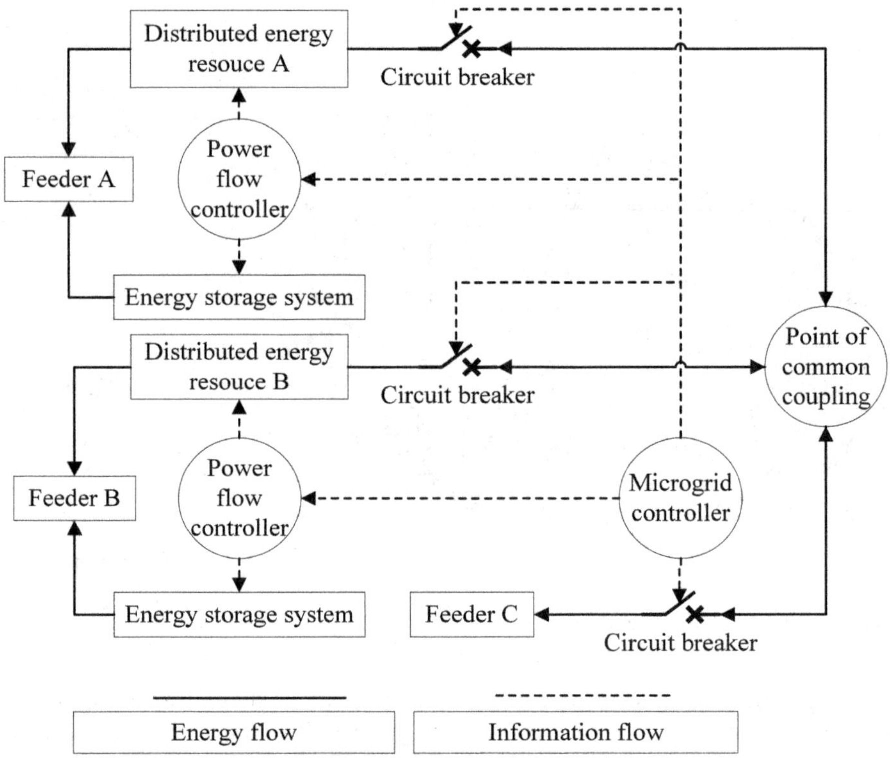

Fig. 9.7 The basic structure of microgrids

main grid and operate independently by enabling the circuit breaker. The feeder D represents interruptible loads, it does not have its own generation, but it can be supplied by other local grids through the PCC directly (Lasseter & Paigi, 2004).

The VPP and microgrids have some similar features, but there are still some differences between the VPP and microgrids (Nosratabadi et al., 2017). Firstly, the VPP usually operates in grid-connected form but microgrids can operate both in independent and grid-connected form. Secondly, microgrids need energy storage in case of emergency, while the VPP can operate without energy storage. Thirdly, the core technologies of the VPP are information and communication technologies, and the microgrids' core technologies are based on hardware. Fourthly, the VPP can integrate various energy resources in a large area, but the microgrids integrate modest energy resources in a relatively small area. Fifthly, microgrids offer an energy retail trading platform, and the VPPs can offer a platform for the wholesale market.

9.2.3.3 Prosumer

With the spread of distributed energy technology, energy consumers also have a chance to produce energy, which blurs the boundary between consumers and producers. A new type of economic individual called "Prosumer" appears in energy trading. The difference between prosumers and traditional consumers is that prosumers can not only carry out the consumption of energy, but they also can make contributions to energy production, while traditional energy consumers just consume energy in a passive way. The energy production of prosumers can be used to meet the need of themselves, the rest energy can be sold in the market. In this way, prosumers actively participate energy trading. The concept graph of the prosumer is shown in Fig. 9.8.

The trading modes of prosumers include prosumers to prosumers, prosumers to microgrids, or the VPP. It is researched that trading mode related to prosumers can improve the utilization of new energy and decrease the cost of energy storage (Luo et al., 2014). It is because that renewable energy can fit small-scale trading. The pricing of prosumers' trading is a very dynamic process, related works show that the strategy of bidding for prosumers is a Markov Pricing Equilibrium (Anand & Gupta, 2016). Considering the characteristics of prosumers' trading, A new local energy market mechanism is designed by *Morstyn* et al. which integrates peer-to-peer energy trading and probabilistic locational marginal pricing (Morstyn et al., 2020). The experiment shows that this mechanism can increase value for prosumers and the whole system by improving the share of renewable energy.

9.3 Energy Saving and Emission Reduction Under the Energy Internet

According to the current development intensity, the world's proven coal reserves can only be exploited for 100 years, and oil and natural gas can only be exploited for more than 50 years. The greenhouse gas produced by fossil energy accounts for over two-thirds of Global Greenhouse Gas (GHG) emissions, which is the main cause of climate change. In 2018, carbon dioxide emissions from the use of fossil energy were approximately 33.1 billion tons. If it is not effectively controlled,

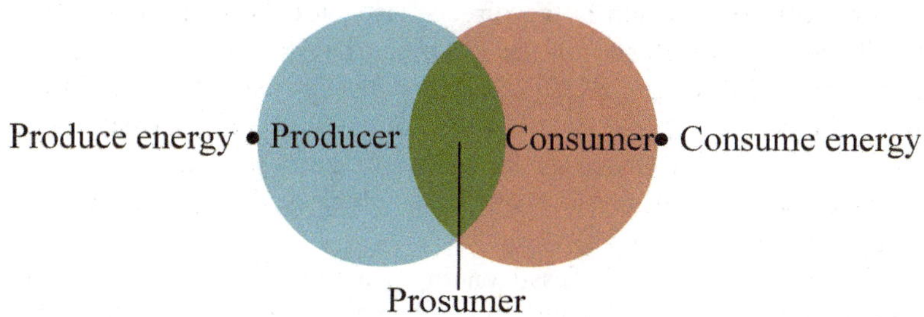

Fig. 9.8 The concept graph of the prosumer

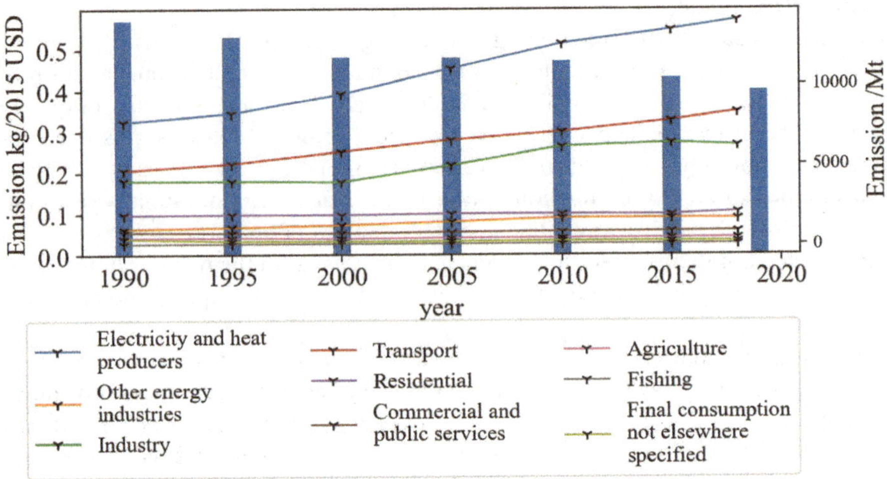

Fig. 9.9 World CO_2 emissions from 1990 to 2018

the global temperature rise will exceed 3°C by the end of this century, seriously threatening the survival of mankind.

The world CO_2 emissions from 1990 to 2018 are shown in Fig. 9.9 (IEA, 2021a). The top three high energy consumption sectors are the electricity and heating industry, other industry and transportation. The growth of these three parts has slowed down due to the realization of energy-saving and emission-reduction technologies. The carbon emissions per unit of Gross Domestic Product (GDP) steadily declines, revealing the development of the low carbon economy.

9.3.1 Energy Saving and Emission Reduction in Production Process

9.3.1.1 Energy Production

According to Fig. 9.9, about 40% of the world's CO_2 emissions is caused by electricity and heat producer. Among them, the electricity based on fossil energy accounts for over 60%, which is the direct reason for the high emission of CO_2 in electricity generation. Based on the log mean Divisia index, it is claimed that the major factor influencing the CO_2 emissions besides the economic output effect is energy structure (Xu et al., 2014). Reducing the share of fossil energy and increasing the share of new energy in the whole energy structure are the direct ways to reduce carbon emission in the source, which is also an important trend of the development in the EI.

Improving the efficiency of fossil energy generation is another way to save energy and reduce carbon emissions. The concept of the 4th generation of district heating and smart thermal grids is an example that focuses on the integration of distributed new energy resources (Lund et al., 2014). In accordance with the smart electrical grid, the smart thermal grids emphasis the mechanism that the interaction of energy flows and information flows will improve the efficiency of resource allocation. One of the typical applications is the CCHP system, where the resources are adequately utilized for electricity generation, heating, and cooling, transforming to the consumers in need flexibly.

9.3.1.2 Industry Production

The cement industry is a traditional energy-hungry industry. The electricity is mainly consumed in the grinding process and the cost of fuel is usually caused by heating. The burning of fuels also brings lots of GHG emissions. The steel industry is the fundamental industry of infrastructure which also has high energy consumption and high GHG emission industry. Similarly, the high energy consumption and high GHG emission of the steel industry are mainly caused by heating. The increment of world steel production is shown in Table 9.1 (WSA, 2021). According to the statistics of the World Steel Association, the total direct emissions caused by steel production in 2020 were of the order of 2.6 billion tons, representing between 7 and 9% of global anthropogenic CO_2 emissions, and the world steel production is poised to grow.

The common methods used for saving energy and reducing GHG emission in steel and cement industries are concluded, which include modifying raw material input flow, productive process flow, by-product output flow, and utilizing the waste heat (Madlool et al., 2013; Wang et al., 2020). In the EI, the factory can be regarded as an edge system, which has its own basic ability for modifying the parameters of production including the speed for kiln fan, temperature for heating, inputs quantity of raw materials, etc. The production parameters with data of energy consumption and CO_2 emission are collected and processed by big data technology so that some intelligent algorithms can be applied to calculate the optimized parameters. Besides, waste heat is an important energy resource in the cement and steel industry that cannot be neglected. But only little waste heat is recovered, which means that there are lots of development space.

In the EI, waste heat will transfer to another form of energy to enter the energy cycle again. According to temperature the recovery methods of waste heat can

Table 9.1 Total production of world crude steel Units/KT

Country	2015	2016	2017	2018	2019
World	1,625,141.5	1,632,780.4	1,735,874.6	1,825,485.7	1,875,155.4
China	803,825.0	807,609.4	870,854.5	928,260.0	996,341.7
Japan	105,134.4	104,775.3	104,661.1	104,318.8	99,284.1
India	89,026.4	95,476.6	101,454.7	109,272.0	111,350.7
United States	78,845.3	78,475.0	81,612.2	86,607.4	87,761.2
Russia	70,898.0	70,453.0	71,491.0	72,122.0	71,897.0
South Korea	69,670.5	68,575.8	71,030.0	72,464.0	71,411.9

be divided into three levels, they are low temperature (35–100), medium temperature (100–500), and high temperature (above 500). Lower temperature heat can be raised to higher temperature heat by heat pumps and the higher temperature heat also can be cooled down to lower temperature heat. Lower temperature heat is directly used for underground heat or district heat while higher temperature heat is usually used for power generation. The technology called Heat Recovery Power Generation (HRPG) is applied for energy saving and emission reduction. The factories change from pure consumers to prosumers. It is estimated that cement production with HRPG can save 1.13% of resources consumption and increase 2.84% of renewable energy. The cement production with HRPG can not only save energy but also modify the structure of energy structure slightly (Zhang et al., 2021).

9.3.2 Energy Saving and Emission Reduction in Supply and Marketing Process

9.3.2.1 Carbon Emissions Trading

With the deterioration of the ecological environment caused by excessive emission of the GHG, not only has it destroyed the heath of people, but also it has had a negative effect on the sustainable development of the world economy and human society. To address the environmental issues caused by excessive emission of the GHG, some Climate Change Conferences have been convened to deliberate over these global environmental problems so that practical policies can be formulated.

Carbon emissions trading has been regarded as a key measure in the reduction of GHG. It is a market mechanism, which based on the premise that organizations discharging carbon dioxide during their production process have to achieve their emission reduction targets otherwise they will get fined. The basic function of carbon emissions trading is that one organization pays the other to obtain a particular amount of greenhouse gas emissions. The buyer can use carbon emissions credits purchased for production. Briefly speaking, Carbon emissions will be regarded as a commodity for the transaction. Therefore, global unions, countries, corporations have to consider the carbon emissions to ensure their profits. On July 16, 2021, the Chinese national carbon emission trading market was set up, containing about 4.5 billion tons of carbon dioxide emissions, which is the biggest carbon emission trading market in the world. With the establishment of different regional carbon emission trading markets and GEI, the world carbon emission trading markets will gradually complete.

In the EI, big data can be used for analyzing the market conditions or verifying the market model, then a more effective market model can be proposed. Based on the big data of Community Independent Transaction Log (CITL), it is revealed that low emission companies have better trading performance compared to high emission companies in European Union Emission Trading Scheme (EU ETS) (Liu et al., 2017). Since carbon emissions are considered as a direct cost, it's necessary for corporations to restrategize about carbon emissions for ensuring max profits. A dynamic supplier selection model considering the carbon emission caused by ordering and transportation is proposed by Lamba, based on randomly generated big data (Lamba & Singh, 2019). This dynamic supplier selection model significantly improves the competitive power of companies in a green low-carbon economy.

9.3.2.2 Tradable Green Certificates

First of all, the electricity price subsidies are the main driving factor for the development of renewable energy, especially the development of wind and solar energy. The national renewable energy electricity price subsidies are paid by the renewable energy fund. The fund mainly comes from the renewable energy surcharge, which levies on the electricity consumption of the whole society (deducting residential electricity and agricultural electricity). However, due to the rapid growth of renewable energy power generation such as wind power, photovoltaics, and biomass, the renewable energy fund is still unable to make ends meet and cannot meet the subsidies demand brought by the massive grid connection.

To solve subsidies issues, The Tradable Green Certificates (TGC) are proposed. The TGC facilitates renewable resources into the liberalized market and reduces the burden of subsidies. Renewable energy generators apply for green certificates based on the amount of their on-grid power. Every green certificate has a unique identifier to qualified renewable energy companies based on their type of power generation, date of grid connection, place of production, and project name. A green certificate represents a certain quantity of renewable energy power. The green certificate supplier lists the green certificates on the green certificate trading platform, indicating the selling price and quantity, but the price shall not be higher than the corresponding subsidy price. During the transaction process, the EI takes the responsibility of information disclosure, which means it reduces the cost of each market entity to obtain market information and eliminates the asymmetry of information between buyers and sellers to improve the efficiency of resources allocation. Besides, it is necessary to monitor market transactions to maintain the fairness and safety of the green certificate market.

9.3.2.3 Demands-Side Structural Reform

New Energy Vehicle

According to Fig. 9.9, transportation has accounted for more than 20% of the whole. Faced with the environmental and energy issues brought by the development of the automobile industry, developing new energy vehicles has become a consensus in the international community. To actively occupy the dominant heights of this emerging market, all countries have coincidentally taken the development of new energy vehicles as the national strategy in the twenty-first century, and adopted subsidies, tax reductions, and other preferential measures to improve the country's technology of new energy vehicles. China's vehicle market accounts for a one-third share of the world market, the tendency that sales volume of new energy vehicles, especially electric vehicles, increases rapidly can be reflected in

Table 9.2 Sales of new energy vehicles in China, 2021. 1–7

	July (k)	January to July (k)	Month on month growth (%)	Year on year growth (%)
All	271	1478	5.8	164.4
EV	220	1227	5.6	233.3
PHEV	50.6	251	62.7	276.2

the Chinese new energy market. The sale data of the Chinese new energy market in 2021 is shown in Table 9.2 (CAAM, 2021).

The definition of new energy vehicles is that vehicles use new energy as power which includes fuel cell electric vehicles, hydrogen engine vehicles, battery electric vehicles, plug-in hybrid electric vehicles. Among these new energy vehicles, the electric vehicle has the closest connection to the EI. Electric vehicles need charging piles to replenish energy, which brings an extra burden to the existing power grid. On the other hand, the deployment of charging piles can influence the traffic flow. The irrational regulations of charging piles may cause an overload of power grids or traffic jams. The situation that the charging piles' influences loads of power grid will be considered in the EI and the Geographic Information System (GIS) can be applied for finding the intersection of traffic flow and energy flow, to modify the deployment of charging piles (Zhang et al., 2019).

Besides, users' demand, cost of investment, geographical positions, the charging mileage limit, the actual road condition, and service network reliability will also be taken into account, based on intelligent algorithms such as genetic algorithm (Gan et al., 2020; Liu et al., 2019). After the deployment of charging piles, it is important to build a charging system between drivers of EVs and new energy companies, the smart contract based on blockchain is proposed (Fu et al., 2020), and the multicenter and tamper-proof mechanisms of blockchain will ensure the security of trading between the EVs and charging piles (Huang et al., 2018).

Electrical Railway

Railway systems, especially electric railway systems, are a more efficient and less emissive transportation system compared to other modes of transportation, which is currently envisioned as the most promising transportation system for both people and freight to reduce atmospheric emission and combat climate change. The replacement of road transport with rail transport is commonly recognized as an efficient strategy to significantly reduce atmospheric emissions of transportation of both people and freight. A large-scale transition to the railway, however, requires the further improvement of its energy efficiency, to ensure long-term sustainability and the necessary reduction of carbon emissions.

The rail transportation system is a complex system, the methods of energy saving and emission reduction are varied, which mainly includes three aspects: train speed profile optimization, regenerative energy optimization, and train timetable optimization (Gao & Yang, 2019). The train speed is relative to the friction between track and wheel, it is feasible to predict the energy consumption through the rail friction modifier and wheel/rail conditions (Vander et al., 2014). The scale of train speed is usually affected by timetable, a model which integrates the train timetable and speed to decrease the net energy consumption is established by *Li and Lo* et al. The genetic algorithm is used to calculate the result which showed that this method can reduce the net energy consumption around 20% (Li & Lo, 2014). Regenerative braking energy is an indispensable part of energy saving. About half the energy consumption of rolling stock is caused by braking, and regenerative braking can bring one-thirds of energy back (González-Gil et al., 2014). The recovery energy is used to apply only for train operation due to the unidirectional flow of power (power to train).

In the EI, the reversible substations are established where bidirectional energy flows are permitted, the recovery energy from regenerative braking can be sold back to the power grid (González-Gil et al., 2013). In fact, the timetable of the train will affect the traction and braking for train, so the approach of energy saving should not be limited to one aspect but focus on the problem in a system sight, moreover, consider the externality of the system, which is also the core idea of the EI.

9.4 Healthy Construction of the Ecological Environment Under the Energy Internet

Ecological environment refers to the general term of the quantity and quality of water resources, land resources, biological resources, and climate resources that affect human survival and development. It is a composite ecosystem related to the sustainable development of society and the economy.

The development of society and the economy cannot go on with the absence of energy. The fact that rapid modernization demands plenty of energy conflicts with the situation that fossil energy will exhaust shortly and new energy just break the ice. The overexploitation of energy threatens the safety of the ecological environment and brings side effects to human being. Ironically, humans develop the economy at the cost of destroying the ecological environment, but they pay several times money to recover the destroyed ecological environment finally. Therefore, the coupling of the energy system, economy system, and ecological environment should be considered at the first of beginning so that the real sustainable development of these three parts can be realized.

9.4.1 Land Ecology and Photovoltaic Power

9.4.1.1 Desertification-Polar Energy

Desertification refers to land degradation in arid, semi-arid, and sub-humid arid areas caused by various factors including climate change and human activities. According to the main reason, it can be divided into wind erosion desertification, water erosion desertification, soil salinization, and freeze–thaw desertification. There are over two-thirds of countries and regions, one-fifth of the worldwide population, and a quarter of the entire land area is suffering from desertification, which is expanding at a rate of 50,000 to 70,000 km^2 per year. China is one of the countries most affected by desertification. Desertification brings lots of damage to the land, which results in the loss of soil productivity, degradation of vegetation cover, deterioration of hydrological conditions, climate disaster, etc.

In northwest China, the new energy technology coordinates well with the control of desertification. Similar to other desertified areas, land there is raged by wind and sand all the year round which degrades into the desert and semi-desert land on a large scale. But the rapid rise of photovoltaics has brought a huge investment stimulus effect and has also explored a new path for desertification sand control in terms of environmental protection and restoration.

Desertification plus extreme climate often result in sand storms which threaten the safety of agriculture and human life. The unbalance heat on the surface is an important reason for extreme climate. Unbalance heat in surface causes air pressure difference leading to gale. Photovoltaic power plants are conducive to absorbing heat from sunlight and reflecting rest heat to the air to balance the heat. It is researched that most of the gravels in the sand wind flow within a height of 30 cm near the surface. The desertified areas are covered with large-scale photovoltaic panels with the height above 1 m, which can against wind and fix the sand effectively. The function that photovoltaic panels can against extreme climate is verified by meteorological big data (Chang et al., 2020).

Besides, photovoltaic panels can collect rain, a new method of controlling desertification is proposed which is planting crops under the photovoltaic panels to make full use of their rain collecting ability. The electricity from photovoltaic power generation can be supplied to the agricultural facilities nearby, and the crops between photovoltaic panels can decline the impacts of sand wind on power generation. In this way, the control of desertification, energy production, and agricultural economy are combined organically.

9.4.1.2 Coal Mining Subsidence Area

Coal is an important primary energy source, but the over mining of coal has caused subsidence of surface, which leads to ground fissure, soil erosion, surface ponding, and building damage. These disasters threaten the life and property of the people severely. There is no doubt that the mining of coal will be suspended in the subsidence areas. The question is what to replace.

The subsidence areas without surface ponding have favorable sunlight and wind condition, where the wind power plants or photovoltaic power plants can be established. At Xintai, Shandong, China, the local government integrates the development of photovoltaic industry and the utilization of land in subsidence areas caused by coal mining and adopts a complementary model of agriculture and photovoltaics, which provides a new way for the transformation of cities' resources. The photovoltaic power generation basement in the coal mining subsidence area of Xintai plans to construct 2 million kilowatts of agricultural and photovoltaic complementary power stations, with a total investment of 20 billion yuan. The construction period is from 2016 to 2020. 30 photovoltaic power stations, 30,000 solar greenhouses, 20,000 arch sheds, and 100 multi-span greenhouses will be built. After the completion of the first phase of the agricultural facilities, 15,000 farmers can be employed in nearby places, and the annual comprehensive agricultural income can be 1 billion yuan (Xin, 2017).

For the surface ponding areas, float photovoltaic energy technology is applied. In Suzhou, Anhui, China, many spiral anchors of floating islands are embedded to match the underwater topography. According to the water regime, wind, and precipitation conditions of the water area, construction will be carried out after a strictly mechanical design to ensure the stability of each floating island. Over the past four years of operation, the power station has withstood the test of strong winds and other extreme weather. New energy technology empowers the transformation of coal mining subsidence area into a new energy basement, which not only revitalizes the abandoned land and realizes the reuse of resources, but also effectively promotes the adjustment of local industrial structure and brings tangible benefits to the local people.

9.4.2 Hydropower and Ecology

The construction of hydropower projects has achieved huge social and economic benefits, but at the same time, it has also permanently changed the water ecological environmental conditions of the river, affecting the river water ecosystem and the self-purification capacity of the water. Besides, the construction of hydropower requires land of the locals, the operation of hydropower causes noise and degradation of fishery resources, which brings side effects to the local people. The ecological impact of hydropower projects has gradually attracted the attention and attention of all sectors of society and becomes a major challenge facing the development of hydropower projects.

To guarantee the sustainable development of society and economy, the concept of ecological compensation is proposed (Yu & Xu, 2016), which is to coordinate the development of energy, the economy, and the environment by internalizing environmental externalities and adjusting for the relationships with stakeholders' benefits. The ecological compensation can be direct financial reimbursement or some actions to ameliorate the situation of ecological damage.

The direct impact of the construction of hydropower plants is the occupying of lands. The locals must leave their homes where they used to live and work, which is not only an economic issue but also a social issue, so it is important to establish a humanized financial compensation standard. A compensation accounting method is proposed by *Yu* et al. which is based on human welfare change for resettlement due to the construction of hydropower (Yu et al., 2016). The preference of the locals for culture, climates, and residential landform are thoroughly considered to build up various types of ecological compensation standards. The influence of power relations between the local governments, developers, and residents are studied by *Katus* et al. the research showed how these relations determine resettlement processes and outcomes (Katus et al., 2016). At present, the informatization of hydropower project resettlement is one of the important trends in the development of modern project management under the background of the information society. A scientific and comprehensive immigration informatization standard system can better promote hydropower immigration work, serve, and guide the construction of land acquisition and resettlement implementation work.

Another impact of hydropower is the decline of fish resources (Habit et al., 2007). To effectively evaluate the damage caused by hydropower, the Hydropower Biological Evaluation Toolset (HBET) is proposed, which is an integrated suite of science-based applications designed to evaluate whether the hydraulic conditions of hydropower structures are fish friendly by analyzing collected data and providing estimated injury and mortality rates. The data describes the process that fish go through hydropower facilities is collected by sensor fish, the combination of cloud computing and the HBET can offer a more scalable and flexible evaluation system (Hou et al., 2019). An evaluation standard is proposed by *Mueller* et al. which includes vitality, health grades, and severity of injuries (Mueller et al., 2017). There have been many studies on the water ecological restoration of huge dam reservoirs used for hydropower, especially the diversity of fish that has attracted wide attention. Measures related to the protection of fish diversity have been adopted in many dam reservoir areas, mainly including: setting up fish passing facilities, Man-made flood peaks to induce oviposition, establishing fish nature reserve, etc.

9.4.3 Biological Energy and Ecology

The negative impacts of the extraction and exploration of fossil fuel on the environment and its depletion that has led to environmental degradation, which has encouraged researchers, stakeholders, and the government to explore alternative and renewable energy sources such as biomass. The application of biological energy has been increasing rapidly in recent years, the detailed data of electricity generation which comes from biological energy is described in Table 9.3 (IEA, 2021b).

Biomass refers to various organisms formed through photosynthesis, including animals, plants, and microorganisms. In accordance with the sources of biomass, the biomass can be mainly divided into agricultural biomass, forest biomass, and contaminated biomass. Etc. the three kinds of biomass mentioned have close links with ecology. It has been proofed that phytoremediation has good performance towards soil treatment (Guo et al., 2014). The possibility that metal and energy can be recovered from contaminated biomass is demonstrated (Han et al., 2018). Hence, the cycle is formed. The importance of biomass in the circular economy and the significance of biomass in energy production are comprehensively studied and analyzed by *Sherwood*, he pointed that “A circular economy

Table 9.3 Electricity generation from biofuels and waste by source, World 1990–2018

Units/GWh	Industrial waste	Primary solid biofuels	Biogases	Municipal waste (renew)	Liquid biofuels
1990	7,665	101,783	3,652	8,313	–
1995	10,936	88,903	6,165	11,980	–
2000	15,259	100,656	13,125	17,374	–
2005	11,662	146,365	21,154	24,296	1,981
2010	26,607	225,946	46,801	33,008	4,994
2015	28,426	323,818	83,557	37,528	8,255
2018	42,301	421,131	88,986	38,648	8,349

is a vision of a sustainable society, demanding greater responsibility for sustainable biomass production and the recovery of material value from products at end-of-life." (Sherwood, 2020).

Through the thermochemical process such as pyrolysis, gasification, and liquefaction, the biomass can be transformed into fuel, biological materials, electricity, and heating. Plants absorb CO_2 from the air and transform solar energy into chemical energy to store, which is the source of biomass energy. During the process of thermochemical decomposition of biomass, the CO2 is produced back to air. Hence, there is no net increment of carbon emission, and the carbon–neutral is realized (Gyamfi et al., 2021). Due to the diversity of usages of biomass, it is necessary to evaluate the efficiency of every usage. It is verified by several exergy analyses of biomass that gasification is the most efficient method (Saidur et al., 2012). The process of utilization for biomass is shown in Fig. 9.10.

In the EI, all kinds of sensors are applied to collect useful data, which is used to evaluate hidden economic value so that subsequent development plans can be made. Airborne LiDAR and multispectral sensors are used for estimating the potential of forest biomass energy based on (Hernando et al., 2019). Big data technologies are applied to evaluate the economic utility of agricultural waste so that the sustainability of supply management can be modified (Inckot et al., 2011). The GIS can be used to optimize the location of the biomass conversion plants where contaminated willows are processed so that the transport cost can be reduced (Voets et al., 2013).

9.5 Conclusions

The summary for each section in this chapter is given as follows:

(a) Section 9.1 briefly introduces the concept of energy internet. The key idea, the interaction of energy flows and information flows, is discussed. The rest of this chapter is meant to introduce the coupling of the energy system and other systems.

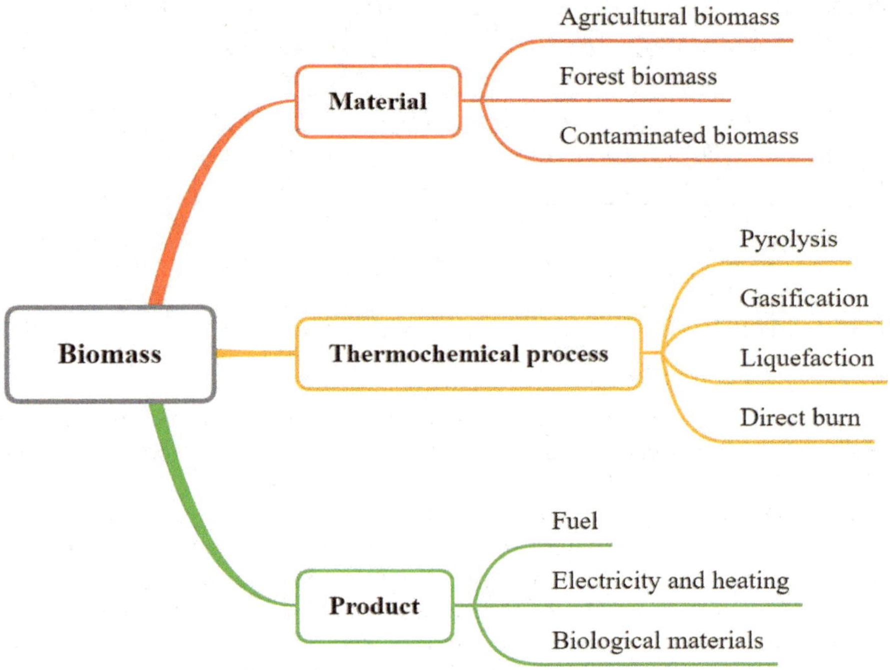

Fig. 9.10 The process of utilization for biomass

(b) Section 9.2 first defines the definition of the energy system. Then, differences between the conditional energy system and the new energy system are analyzed through several prospects including production, supply, and marketing. As for the new energy system, some characteristics of energy production, distribution, and intermittency are discussed, and issues brought by them and corresponding approaches in the EI are summarized. Several important concepts of energy supply and marketing, the VPP, microgrids, and Prosumer are introduced.

(c) Section 9.3 first mentions some innovations of energy saving and emission reduction in energy production industries and other high energy consumption industries. Some market mechanisms including carbon emission trading and tradable green certificates with relative researches based on big data are introduced. The new energy vehicles and electric railways are taken as examples to discuss the demand-side structural reform.

(d) Section 9.4 first analyzes the influences of the energy system on the ecological environment. The ecological compensation in hydropower is discussed. The combination of new energy and environmental remediation is introduced, which includes establishing photovoltaic power stations to control desertification and developing coal mining subsidence, planting botany to treat soil, and gaining energy or material from contaminated biomass.

References

Alanne, K., & Saari, A. (2006). Distributed energy generation and sustainable development. *Renewable and Sustainable Energy Reviews, 10*(6), 539–558. https://doi.org/10.1016/j.rser.2004.11.004

Anand, V., Gupta, V. (2016). Markov pricing equilibrium in a prosumer-aggregator dynamic game. In *2016 American Control Conference (ACC)* (pp. 4120–4125). https://doi.org/10.1109/ACC.2016.7525569.

Ausubel, L. M., & Cramton, P. (2010). Virtual power plant auctions. *Utilities Policy, 18*(4), 201–208. https://doi.org/10.1016/j.jup.2010.05.002

Bedi, G., Venayagamoorthy, G. K., Singh, R., Brooks, R. R., & Wang, K.-C. (2018). Review of Internet of Things (IoT) in electric power and energy systems. *IEEE Internet of Things Journal, 5*(2), 847–870. https://doi.org/10.1109/jiot.2018.2802704

BP British Petroleum Company (2021) Statistical Review of World Energy 2021 https://www.bp.com/content/dam/bp/business-sites/en/global/corporate/pdfs/energy-economics/statistical-review/bp-stats-review-2021-full-report.pdf. Accessed Aug 16, 2021.

CAAM China Association of Automobile Manufactures. (2021). Production of New Energy Vehicles in July 2021. http://en.caam.org.cn/Index/show/catid/53/id/1639.html. Accessed Sep 10, 2021.

Chang, Z., Liu, S., Fu, J., Wang, Q., Wang, F., Sun, T., Liu, S., Wang, F. (2020). Effect of desert photovoltaic on sand prevention and control—taking Gansu Gulang Zhenfa photovoltaic DC field as an example. In *IOP Conference Series: Earth and Environmental Science* (Vol. 601). https://doi.org/10.1088/1755-1315/601/1/012032

EIA Energy Information Administration. (2020). International Energy Outlook 2020. https://www.eia.gov/outlooks/ieo/. Accessed Aug 11, 2021.

Evans, A., Strezov, V., & Evans, T. J. (2012). Assessment of utility energy storage options for increased renewable energy penetration. *Renewable and Sustainable Energy Reviews, 16*(6), 4141–4147. https://doi.org/10.1016/j.rser.2012.03.048

Fu, Z., Dong, P., & Ju, Y. (2020). An intelligent electric vehicle charging system for new energy companies based on consortium blockchain. *Journal of Cleaner Production, 261*, 121219. https://doi.org/10.1016/j.jclepro.2020.121219

Gan, X. Y., Zhang, H. X., Hang, G., Qin, Z. D., & Jin, H. M. (2020). Fast-charging station deployment considering elastic demand. *IEEE Trans Transp Electrif, 6*(1), 158–169. https://doi.org/10.1109/tte.2020.2964141

Gao, Z. Y., & Yang, L. X. (2019). Energy-saving operation approaches for urban rail transit systems. *Frontiers of Engineering Management, 6*(2), 139–151. https://doi.org/10.1007/s42524-019-0030-7

González-Gil, A., Palacin, R., Batty, P., & Powell, J. P. (2014). A systems approach to reduce urban rail energy consumption. *Energy Conversion and Management, 80*, 509–524. https://doi.org/10.1016/j.enconman.2014.01.060

González-Gil, A., Palacin, R., & Batty, P. (2013). Sustainable urban rail systems: Strategies and technologies for optimal management of regenerative braking energy. *Energy Conversion and Management, 75*, 374–388. https://doi.org/10.1016/j.enconman.2013.06.039

Gozgor, G., Mahalik, M. K., Demir, E., & Padhan, H. (2020). The impact of economic globalization on renewable energy in the OECD countries. *Energy Policy, 139*, 13. https://doi.org/10.1016/j.enpol.2020.111365

Guo, Z. H., Liu, Y. N., Wang, F. Y., & Xiao, X. Y. (2014). Liquefaction of metal-contaminated giant reed biomass in acidified ethylene glycol system: Batch experiments. *J Cent South Univ, 21*(5), 1756–1762. https://doi.org/10.1007/s11771-014-2121-2

Gyamfi, B. A., Sarpong, S. Y., & Bein, M. A. (2021). The contribution of the anthropogenic impact of biomass utilization on ecological degradation: Revisiting the G7 economies. *Environmental*

Science and Pollution Research International, 28(9), 11016–11029. https://doi.org/10.1007/s11356-020-11073-8

Habit, E., Belk, M. C., & Parra, O. (2007). Response of the riverine fish community to the construction and operation of a diversion hydropower plant in central Chile. *Aquatic Conservation: Marine and Freshwater Ecosystems, 17*(1), 37–49. https://doi.org/10.1002/aqc.774

Han, Z. Y., Guo, Z. H., Zhang, Y., Xiao, X. Y., & Peng, C. (2018). Potential of pyrolysis for the recovery of heavy metals and bioenergy from contaminated broussonetia papyrifera biomass. *BioResources, 13*(2), 2932–2944. https://doi.org/10.15376/biores.13.2.2932-2944

Hernando, A., Puerto, L., Mola-Yudego, B., Manzanera, J. A., García, A., Maltamo, M., & Valbuena, R. (2019). Estimation of forest biomass components using airborne LiDAR and multispectral sensors. *iForest—Biogeosciences and Forestry, 12*(2), 207–213. https://doi.org/10.3832/ifor2735-012

Hou, H. F., Deng, Z. D., Martinez, J. J., Fu, T., Lu, J., Tan, L., Miller, J., Bakken, D. (2019). A cloud-based decision support system framework for hydropower biological evaluation. In *Future Technologies Conference (FTC)* (pp. 517–529). Vancouver, CANADA, Advances in Intelligent Systems and Computing. Springer International Publishing Ag, CHAM,. https://doi.org/10.1007/978-3-030-02686-8_39.

Huang, X., Xu, C., Wang, P., & Liu, H. (2018). LNSC: A security model for electric vehicle and charging pile management based on blockchain ecosystem. *IEEE Access, 6*, 13565–13574. https://doi.org/10.1109/access.2018.2812176

Hussain, S. M. S., Nadeem, F., Aftab, M. A., Ali, I., & Ustun, T. S. (2019). The emerging energy internet: architecture, benefits, challenges, and future prospects. *Electronics, 8*(9), 1037. https://doi.org/10.3390/electronics8091037

IEA International Energy Agency. (2021a). CO2 emissions by sector, World 1990–2018. https://www.iea.org/data-and-statistics/data-browser?country=WORLD&fuel=CO2%20emissions&indicator=CO2BySector. Accessed Sep 10, 2021.

IEA International Energy Agency. (2021b). Electricity generation from biofuels and waste by source, World 1990–2018. https://www.iea.org/data-and-statistics/data-browser?country=WORLD&fuel=Energy%20supply&indicator=WasteGenBySource. Accessed Sep 10, 2021.

Inckot, R. C., Santos, G. D., de Souza, L. A., & Bona, C. (2011). Germination and development of Mimosa pilulifera in petroleum-contaminated soil and bioremediated soil. *Flora, 206*(3), 261–266. https://doi.org/10.1016/j.flora.2010.09.005

Kado, Y., Shichijo, D., Wada, K., & Iwatsuki, K. (2016). Multiport power router and its impact on future smart grids. *Radio Science, 51*(7), 1234–1246. https://doi.org/10.1002/2016rs006041

Katus, S., Suhardiman, D., & Senaratna, S. S. (2016). When local power meets hydropower: Reconceptualizing resettlement along the Nam Gnouang River in Laos. *Geoforum, 72*, 6–15. https://doi.org/10.1016/j.geoforum.2016.03.007

Khan, M. K., Teng, J. Z., Khan, M. I., & Khan, M. O. (2019). Impact of globalization, economic factors and energy consumption on CO2 emissions in Pakistan. *Science of the Total Environment, 688*, 424–436. https://doi.org/10.1016/j.scitotenv.2019.06.065

Khan, W. Z., Ahmed, E., Hakak, S., Yaqoob, I., & Ahmed, A. (2019). Edge computing: A survey. *Future Generation Computer Systems, 97*, 219–235. https://doi.org/10.1016/j.future.2019.02.050

Lamba, K., & Singh, S. P. (2019). Dynamic supplier selection and lot-sizing problem considering carbon emissions in a big data environment. *Technological Forecasting and Social Change, 144*, 573–584. https://doi.org/10.1016/j.techfore.2018.03.020

Lasseter, R. H., Paigi, P. (2004). Microgrid: A conceptual solution. In *35th Annual IEEE Power Electronics Specialists Conference (PESC 04), Aachen, GERMANY, Jun 20–25 2004. IEEE Power Electronics Specialists Conference Records* (pp. 4285–4290). IEEE, New York

Li, X., & Lo, H. K. (2014). An energy-efficient scheduling and speed control approach for metro rail operations. *Transportation Research Part B: Methodological, 64*, 73–89. https://doi.org/10.1016/j.trb.2014.03.006

Liu, H., Mi, X.-W., & Li, Y.-F. (2018). Wind speed forecasting method based on deep learning strategy using empirical wavelet transform, long short term memory neural network and Elman neural network. *Energy Conversion and Management, 156*, 498–514. https://doi.org/10.1016/j.enconman.2017.11.053

Liu, W., Tang, Y., Yang, F., Dou, Y., & Wang, J. (2019). A multi-objective decision-making approach for the optimal location of electric vehicle charging facilities. *Computers, Materials & Continua, 60*(2), 813–834. https://doi.org/10.32604/cmc.2019.06754

Liu, Y.-P., Guo, J.-F., & Fan, Y. (2017). A big data study on emitting companies' performance in the first two phases of the European Union Emission Trading Scheme. *Journal of Cleaner Production, 142*, 1028–1043. https://doi.org/10.1016/j.jclepro.2016.05.121

Lund, H., Werner, S., Wiltshire, R., Svendsen, S., Thorsen, J. E., Hvelplund, F., & Mathiesen, B. V. (2014). 4th generation district heating (4GDH). *Energy, 68*, 1–11. https://doi.org/10.1016/j.energy.2014.02.089

Luo, Y., Itaya, S., Nakamura, S., Davis, P. (2014). Autonomous cooperative energy trading between prosumers for microgrid systems. In *Paper presented at the 2014 IEEE 39th Conference on Local Computer Networks Workshops.*

Madlool, N. A., Saidur, R., Rahim, N. A., & Kamalisarvestani, M. (2013). An overview of energy savings measures for cement industries. *Renewable and Sustainable Energy Reviews, 19*, 18–29. https://doi.org/10.1016/j.rser.2012.10.046

Morstyn, T., Teytelboym, A., Hepburn, C., & McCulloch, M. D. (2020). Integrating P2P energy trading with probabilistic distribution locational marginal pricing. *IEEE Transactions on Smart Grid, 11*(4), 3095–3106. https://doi.org/10.1109/TSG.2019.2963238

Mueller, M., Pander, J., & Geist, J. (2017). Evaluation of external fish injury caused by hydropower plants based on a novel field-based protocol. *Fisheries Management and Ecology, 24*(3), 240–255. https://doi.org/10.1111/fme.12229

Nosratabadi, S. M., Hooshmand, R.-A., & Gholipour, E. (2017). A comprehensive review on microgrid and virtual power plant concepts employed for distributed energy resources scheduling in power systems. *Renewable and Sustainable Energy Reviews, 67*, 341–363. https://doi.org/10.1016/j.rser.2016.09.025

Saidur, R., BoroumandJazi, G., Mekhilef, S., & Mohammed, H. A. (2012). A review on exergy analysis of biomass based fuels. *Renewable and Sustainable Energy Reviews, 16*(2), 1217–1222. https://doi.org/10.1016/j.rser.2011.07.076

Sherwood, J. (2020). The significance of biomass in a circular economy. *Bioresource Technology, 300*, 122755. https://doi.org/10.1016/j.biortech.2020.122755

Vander, M. J., Eadie, D. T., Oldknow, K. D., & Iwnicki, S. (2014). A predictive model of energy savings from top of rail friction control. *Wear, 314*(1–2), 155–161. https://doi.org/10.1016/j.wear.2013.11.037

Voets, T., Neven, A., Thewys, T., & Kuppens, T. (2013). GIS-BASED location optimization of a biomass conversion plant on contaminated willow in the Campine region (Belgium). *Biomass and Bioenergy, 55*, 339–349. https://doi.org/10.1016/j.biombioe.2013.02.037

Wang, R. Q., Jiang, L., Wang, Y. D., & Roskilly, A. P. (2020). Energy saving technologies and mass-thermal network optimization for decarbonized iron and steel industry: A review. *Journal of Cleaner Production, 274*, 122997. https://doi.org/10.1016/j.jclepro.2020.122997

WSA World Steel Association. (2021).s Total production of crude steel. https://www.worldsteel.org/steel-by-topic/statistics/annual-production-steel-data/P1_crude_steel_total_pub/CHN/IND. Accessed Sep 13, 2021.

Wu, D. W., & Wang, R. Z. (2006). Combined cooling, heating and power: A review. *Progress in Energy and Combustion Science, 32*(5–6), 459–495. https://doi.org/10.1016/j.pecs.2006.02.001

Xin RDZ (2017) The first photovoltaic power station built in coal mining subsidence area in China is connected to Xintai grid for power generation. http://news.bandao.cn/news_html/201709/20170929/news_20170929_2768810.shtml. Accessed Sep 8, 2021.

Xu, S.-C., He, Z.-X., & Long, R.-Y. (2014). Factors that influence carbon emissions due to energy consumption in China: Decomposition analysis using LMDI. *Applied Energy, 127*, 182–193. https://doi.org/10.1016/j.apenergy.2014.03.093

Yu, B., & Xu, L. (2016). Review of ecological compensation in hydropower development. *Renewable and Sustainable Energy Reviews, 55*, 729–738. https://doi.org/10.1016/j.rser.2015.10.038

Yu, B., Xu, L., & Wang, X. (2016). Ecological compensation for hydropower resettlement in a reservoir wetland based on welfare change in Tibet, China. *Ecological Engineering, 96*, 128–136. https://doi.org/10.1016/j.ecoleng.2016.03.047

Zhang, G., Jiang, C., & Wang, X. (2018). Comprehensive review on structure and operation of virtual power plant in electrical system. *IET Generation, Transmission & Distribution, 13*(2), 145–156. https://doi.org/10.1049/iet-gtd.2018.5880

Zhang, X., Xiang, N., Pan, H., Yang, X., Wu, J., Zhang, Y., Luo, H., & Xu, C. (2021). Performance comparison of cement production before and after implementing heat recovery power generation based on emergy analysis and economic evaluation: A case from China. *Journal of Cleaner Production, 290*, 125901. https://doi.org/10.1016/j.jclepro.2021.125901

Zhang, Y., Zhang, Q., Farnoosh, A., Chen, S., & Li, Y. (2019). GIS-based multi-objective particle swarm optimization of charging stations for electric vehicles. *Energy, 169*, 844–853. https://doi.org/10.1016/j.energy.2018.12.062